MATRIX ALGEBRA
FOR BUSINESS AND ECONOMICS

MATRIX ALGEBRA FOR BUSINESS AND ECONOMICS

S. R. SEARLE

Associate Professor, Cornell University

and

W. H. HAUSMAN

Associate Professor, Cornell University

with assistance from

H. BIERMAN, Jr., J. E. HASS AND L. J. THOMAS

also of

Cornell University, Ithaca, N.Y.

Wiley-Interscience / A Division of John Wiley & Sons

New York · London · Sydney · Toronto

10 9 8 7 6 5 4 3 2 1

Library of Congress Catalogue Card Number: 79-93298

SBN 471 76941 X

Printed in the United States of America

PREFACE

Business administration and economics have in recent years witnessed many attempts by theorists and practitioners to conduct empirical research and to build models designed to assist in decision-making. Matrix algebra is a basic tool in these activities.

Managers and economists who can intelligently use matrix algebra have a distinct advantage in applying the tools reported in current literature in such diverse areas as econometrics, linear programming, business decision-making, and in the development of economic theory. Matrices are also used in finance and investment analysis where one must consider elements of uncertainty, for example, in portfolio problems. Similarly, matrix algebra is a part of the application of Markov processes in sequential decision situations. Matrix algebra is also a prime tool in the use of computers.

While matrices have a long history, their application in quantitative problems of economics and business administration is relatively recent. Today's uses of matrix algebra in these disciplines are many, but few compared with what they are likely to be a decade hence. Thus learning matrix algebra brings significant rewards now and increasing rewards for the future.

This book opens with a discussion of introductory concepts including subscripts and summation notation in Chapter 1, followed by five chapters dealing with elementary matrix operations, determinants, inverse matrices, and rank and linear independence. The seventh chapter considers the problem of solving simultaneous linear equations (especially those not of full rank) in which the concept of a generalized inverse matrix is introduced and its value demonstrated. Chapter 12 is concerned with characteristic roots and vectors. Throughout these eight chapters basic concepts are illustrated by numerical examples and applications to business and economics. In addition, four chapters (8–11) deal specifically with major applications of matrix algebra to business and economic problems. Chapter 8 uses matrix algebra

[v]

to solve problems involving Markov chains; Chapter 9 covers linear programming and the revised simplex method of solving linear programming
problems; and Chapters 10 and 11 treat uses of matrix algebra in analyzing
business and economic data, namely statistical analyses relating to regression
and linear models. Only a brief outline of statistical theory, based on examples,
is given in these two chapters. The book closes with a discussion of special
topics (Chapter 13).

We have attempted throughout to keep the mathematical discussion
relatively informal; high-school algebra is the basic prerequisite for using
the book, and differential calculus is used on only a few pages, which can be
omitted if desired. The more advanced mathematical material has been
placed in chapter appendices.

A typical course using this book might cover Chapters 1–7 and 12, involving
the basic concepts of matrix algebra and, depending on the wishes of the
instructor, selected chapters or topics from those chapters representing
applications of matrix algebra in Markov chains, linear programming,
regression analysis, and linear models. The chapter on Markov chains could
be covered after Chapter 3, and the chapters on linear programming and
regression analysis could follow Chapter 6; in this manner earlier coverage
of the application chapters could be attained if desired. Chapter 12 also
contains significant applications of characteristic roots and vectors to the
solution and stability analysis of systems of difference equations in business
and economic models.

Numerous exercises have been provided at the end of chapters. Some
exercises extend the applications of matrix algebra covered in the respective
chapters; for example, exercises 14 through 19 in Chapter 3 present the
application of matrix algebra to the problem of calculating the mean and the
variance of returns from a financial portfolio. Exercises are presented in
the order of increasing difficulty, with the first five to ten exercises being
most suitable for students with limited backgrounds in mathematics.

Chapters, and sections within chapters, are numbered with Arabic
numerals $1, 2, 3, \ldots$. References to sections are made by means of the
decimal system, with an appropriate Section number shown in the heading
of each page; e.g. Section 1.3 is Section 3 of Chapter 1, with 1.3 being part
of the heading on page 3. Numbered equations have the numbers $(1), (2), \ldots$
within each chapter. Those of one chapter are seldom referred to from
another, but when they are the chapter reference is explicit; but a reference
such as "equation (3)," or more simply "(3)," relates to the equation
numbered (3) in the chapter containing the reference.

Readers familiar with *Matrix Algebra for the Biological Sciences* by S. R.
Searle (John Wiley & Sons, Inc., 1966) will find liberal adaptations from it
here.

We gratefully acknowledge the helpful assistance of Mrs. Jean M. Tubbs and Mrs. Elva R. Lovell in secretarial work, with particular thanks to Mrs. Betty J. Holman for typing most of the initial manuscript.

Ithaca, N.Y.
December, 1969

S. R. SEARLE
W. H. HAUSMAN

H. BIERMAN, JR.
J. E. HASS
L. J. THOMAS

CONTENTS

MATRIX ALGEBRA
FOR BUSINESS AND ECONOMICS

CHAPTER 1

INTRODUCTION

---

1. THE SCOPE OF MATRIX ALGEBRA

Economics and the art of management are in the process of becoming increasingly quantitative. As a result, economists and managers find themselves confronted more and more with large amounts of numerical data, measurements of one form or another gathered from such sources as accounting, production, market research and government bureaus. But the mere collecting and recording of data achieve nothing; having been collected, data must be investigated to see what information may be obtained concerning the problem at hand. Any investigation so made, being quantitative in nature, will involve mathematics in some way, even if only the calculating of percentages or averages. Frequently, however, one has to subject data to more complex calculations, requiring procedures that involve a higher level of mathematical skill. In order to carry out the mathematics the economist and manager must either learn the procedures or at least learn something of the language of mathematics, so that they may communicate satisfactorily with the mathematician whose aid is enlisted. In either case there will be numerous mathematical tools that it will be valuable for them to understand. Matrices and their algebra provide one of these tools.

This book deals with the application of matrix algebra to problems of economics and administration. Specific applications range from the solution of problems concerning the allocation of scarce resources such as equipment (assuming conditions of certainty) to the investigation of portfolio decisions under conditions of uncertainty. While we have attempted to include examples of current practice, the reader should understand that the development of new applications and theories making use of the mathematical techniques introduced in this book is growing so rapidly that we can only sample from the rich population of potential applications.

The manipulative skill required for using matrices is neither great nor excessively demanding, whereas in terms of brevity, simplicity and clarity the value of matrix algebra is often appreciable. Furthermore, the almost universal nature of matrix expressions has great appeal, for often the same results can be applied, with only minor changes, to situations involving both small amounts of data and extremely large amounts. Matrix algebra is therefore a vehicle by which mathematical procedures for many problems, both large and small, can be described independently of the size of the problem. Size does not affect the understanding of the procedures, only the amount of calculating involved, which in turn determines time and cost, factors whose importance in today's world of high-speed computers is rapidly diminishing. The presence of such computers is additional reason for enhancing the value of matrix algebra. Expressing a set of calculations in terms of matrices simplifies the problem of making easy and efficient use of a high-speed computer, and therefore matrix algebra and high-speed computers go hand in hand.

2. GENERAL DESCRIPTION OF A MATRIX

A matrix is a rectangular array of numbers arranged in rows and columns. As such, it is a device frequently used in organizing the presentation of numerical data so that they may be handled with ease mathematically. One value of the matrix is that it provides a means of condensing into a small set of symbols a wealth of mathematical manipulations. As a result, matrices greatly assist in problems of data analysis arising from the increasing growth of quantitative methods in economics and business. When mathematical analysis is needed, matrices are often found useful both in organizing the calculations involved and in clarifying an understanding of them.

The aspect of a matrix which makes it an aid in organizing data provides a convenient starting point for its description. We begin with an illustration.

Illustration. Chow (1960) reports the average retail prices of automobiles classified by age. In doing so he finds it convenient to summarize the results in tabular form, as shown in Table 1.

TABLE 1. AVERAGE RETAIL PRICES OF AUTOMOBILES
CLASSIFIED BY AGE

Age (in years)	Year		
	1950	1951	1952
1	$1,881	$2,120	$2,445
2	1,512	1,676	1,825
3	1,261	1,397	1,484
4	1,054	1,144	1,218

The array of numbers in the table may be extracted and written simply as:

$$\begin{bmatrix} 1,881 & 2,120 & 2,445 \\ 1,512 & 1,676 & 1,825 \\ 1,261 & 1,397 & 1,484 \\ 1,054 & 1,144 & 1,218 \end{bmatrix}$$

where the position of an entry in this array determines its meaning. For example, the entry 1,825 in the second row and third column represents the price of two-year-old cars in 1952. Similarly, a row represents the prices of cars of a given age for different years, and a column represents the prices of cars of different ages for a given year. Thus the position of a number in the array identifies the age of the car and the year in which the price applies.

An array of numbers of this sort is called a *matrix*. It is identified as such by enclosing it in square brackets [ ].

The first thing to notice about any matrix is that it is a rectangular (or square) array, with the entries therein set out in rows and columns. The elements of each row usually have something in common, as do those of each column. The individual entries in the array are called the *elements* or *terms* of the matrix; in general they can be numbers of any sort or even functions of one or more variables. For our purposes we will usually think of them as being real numbers, positive, negative or zero.

Matrix algebra, being the algebra of matrices, is the algebra of arrays such as we have just described, each treated as an entity and denoted by a single symbol. The algebra involves the elements within the arrays, but it is the handling of the arrays themselves, as separate entities, that constitutes matrix algebra.

Before proceeding to a formal definition of a matrix we present some mathematical notation that is widely used in matrix algebra. (Readers familiar with subscript and summation notation may omit the next three sections.)

3. SUBSCRIPT NOTATION

Algebra is arithmetic with letters of the alphabet representing numbers. Thus the first two rows of the matrix given previously, namely

$$\begin{bmatrix} 1,881 & 2,120 & 2,445 \\ 1,512 & 1,676 & 1,825 \end{bmatrix}$$

might be written as

$$\begin{bmatrix} a & b & c \\ d & e & f \end{bmatrix}$$

where a, b and c represent the numbers 1,881, 2,120 and 2,445 respectively. This matrix, the array thought of as a self-contained entity, might be given the name A. We could then write

$$A = \begin{bmatrix} a & b & c \\ d & e & f \end{bmatrix}.$$

Since a matrix described in this manner using the letters a through z could have only twenty-six elements, a less restrictive notation is needed. Instead of specifying each number by a letter we use one letter for a whole series of numbers, modified by attaching to it a series of integers to provide the necessary identification. Thus the matrix A could be written as

$$A = \begin{bmatrix} a_1 & a_2 & a_3 \\ b_1 & b_2 & b_3 \end{bmatrix}.$$

The integers 1, 2 and 3 are called *subscripts*, and the elements of the first row, for example, are read as "a, one," "a, two" and "a, three" or as "a, sub one," "a, sub two" and "a, sub three." In this instance the subscripts describe the column in which each element belongs and all elements in the same column have a common subscript; for example, the elements in the second column have the subscript 2.

In practice a single letter of the alphabet is commonly used for all elements. Two subscripts are used for each element and written alongside one another, the first specifying the row the element is in and the second the column. Thus A could be written as

$$A = \begin{bmatrix} a_{11} & a_{12} & a_{13} \\ a_{21} & a_{22} & a_{23} \end{bmatrix}.$$

The elements are read as "a, one, one," "a, one, two" or as "a, sub one, one," "a, sub one, two," and so on. In this way an element's subscripts uniquely locate its position in the matrix.

Just as we have used the letter a to denote the value of an element of a matrix, so we can use other letters to denote other things; for example, we might refer to a row as "row i," and if i were equal to 2 "row i" would be the second row of the matrix. In the same way we can use j for referring to columns, and "column j" or the "jth column" is, for $j = 3$, the third column. By this means we refer to the element in the ith row and jth column of a matrix as a_{ij}, using i and j as subscripts to the letter a to denote the row and column the element is in; a_{12}, for example, is in the first row and second column. In general, then, a_{ij} is in the ith row and jth column, e.g. a_{ij} for $i = 2$ and $j = 3$ is a_{23}, the element in the second row and third column. No comma is placed between the two subscripts unless it is needed to avoid confusion. For example, were there twelve columns in a matrix, the element in row 1 and column 12 could be written as $a_{1,12}$ to distinguish it from $a_{11,2}$, the element in the eleventh row and second column.

This notation provides ready opportunity to detail the elements of a row, a column or even a complete matrix, in a very compact manner. Thus the first row of A,

$$a_{11} \quad a_{12} \quad a_{13}$$

can be described as $\qquad a_{1j} \qquad$ for $\quad j = 1, 2, 3.$

This representation means that the elements of the row are a_{1j} with j taking in turn each one of the integer values 1, 2 and 3. Columns can be described in like fashion. The whole matrix is accounted for by enclosing a_{ij} in curly brackets and writing

$$A = \{a_{ij}\} \qquad \text{for} \quad i = 1, 2 \qquad \text{and} \quad j = 1, 2, 3.$$

This notation completely specifies the elements of the matrix, its name and its size. More generally,

$$A = \{a_{ij}\} \qquad \text{for} \quad i = 1, 2, \ldots, r, \qquad \text{and} \quad j = 1, 2, \ldots, c$$

is the general matrix of r rows and c columns, the dots indicating that i takes all integer values from 1 through r and j takes all integer values from 1 through c.

4. SUMMATION NOTATION

The most frequently used arithmetic operation is that of adding numbers together. Subscript notation provides a means of describing this operation in a very succinct manner. Suppose we wish to add five numbers represented as a_1, a_2, a_3, a_4 and a_5. Their sum is

$$a_1 + a_2 + a_3 + a_4 + a_5.$$

These symbols can be put in words as

"the sum of all values of a_i for $i = 1, 2, \ldots, 5$."

The phrase "the sum of all values of" is customarily represented by Σ, the capital form of the Greek letter sigma. Accordingly the sum of the a's is written as

$$\Sigma \, a_i \qquad \text{for} \quad i = 1, 2, \ldots, 5.$$

A further abbreviation is $\sum_{i=1}^{i=5} a_i$, where the phrase "for $i = 1, 2, \ldots, 5$" has been replaced by the "$i = 1$" and "$i = 5$" below and above the Σ sign.

This indicates that the summation is for all integer values of i from 1 through 5. Thus

$$\sum_{i=1}^{i=5} a_i = a_1 + a_2 + a_3 + a_4 + a_5.$$

Many variations on this summation notation are to be found. Frequently the "$i = 5$" above the Σ is reduced to just 5, so that a commonly seen form is

$$\sum_{i=1}^{5} a_i = a_1 + a_2 + a_3 + a_4 + a_5.$$

Similarly

$$\sum_{i=1}^{3} x_i = x_1 + x_2 + x_3$$

and

$$\sum_{i=1}^{n} x_i = x_1 + x_2 + x_3 + \cdots + x_{n-2} + x_{n-1} + x_n,$$

the sum of n x's. Sometimes the terms above and below the Σ sign are omitted when they are clearly obvious from the context, and sometimes they are used as subscripts and superscripts to the Σ:

$$\sum_1^4 y_i = y_1 + y_2 + y_3 + y_4.$$

The lower limit of i need not necessarily be the value 1: with $i = 3$ written below the Σ sign we have

$$\sum_{i=3}^{7} y_i = y_3 + y_4 + y_5 + y_6 + y_7.$$

Furthermore, although subscripts in a summation are usually consecutive integers, omissions can be indicated as follows:

$$\sum_{\substack{i=3 \\ i \neq 4}}^{7} y_i = y_3 + y_5 + y_6 + y_7.$$

Summation notation has so far been described in terms of simple sums, but it also encompasses sums of squares and sums of products, and indeed sums of any series of expressions that can be identified by subscript notation. Thus the sum of squares $c_1^2 + c_2^2 + c_3^2 + c_4^2$ can be written as

$$\sum_{i=1}^{4} c_i^2 = c_1^2 + c_2^2 + c_3^2 + c_4^2.$$

Similarly, if we have two series of numbers p_1, p_2, p_3 and q_1, q_2, q_3 the sum of their products term by term, $p_1 q_1 + p_2 q_2 + p_3 q_3$, is written

$$\sum_{i=1}^{3} p_i q_i = p_1 q_1 + p_2 q_2 + p_3 q_3.$$

The notation we have described applies equally well to cases involving two subscripts: in summing over either subscript the other remains unchanged.

For examples, consider summing a_{ij} over j for $i = 1$:

$$\sum_{j=1}^{3} a_{1j} = a_{11} + a_{12} + a_{13};$$

and summing a_{ij} over i for $j = 2$:

$$\sum_{i=1}^{2} a_{i2} = a_{12} + a_{22}.$$

The two general results corresponding to these are

$$\sum_{j=1}^{3} a_{ij} = a_{i1} + a_{i2} + a_{i3} \quad \text{and} \quad \sum_{i=1}^{2} a_{ij} = a_{1j} + a_{2j}.$$

Summing the first of these with respect to i leads to the operation of double summation:

$$\sum_{i=1}^{2} \left(\sum_{j=1}^{3} a_{ij} \right) = \sum_{i=1}^{2} (a_{i1} + a_{i2} + a_{i3})$$
$$= (a_{11} + a_{12} + a_{13}) + (a_{21} + a_{22} + a_{23}).$$

Similarly,

$$\sum_{j=1}^{3} \left(\sum_{i=1}^{2} a_{ij} \right) = \sum_{j=1}^{3} (a_{1j} + a_{2j})$$
$$= (a_{11} + a_{21}) + (a_{12} + a_{22}) + (a_{13} + a_{23}),$$

which is seen to be the same as $\sum_{i=1}^{2} \left(\sum_{j=1}^{3} a_{ji} \right)$. Removing the brackets gives the important result

$$\sum_{i=1}^{2} \sum_{j=1}^{3} a_{ij} = \sum_{j=1}^{3} \sum_{i=1}^{2} a_{ij}.$$

The order of summation in double summation is of no consequence. In general

$$\sum_{i=1}^{m} \sum_{j=1}^{n} a_{ij} = \sum_{j=1}^{n} \sum_{i=1}^{m} a_{ij}.$$

In terms of a matrix of m rows and n columns the left-hand side of the above expression is the sum of row totals and the right-hand side is the sum of column totals, both sums equaling the total of all elements.

Sums of squares and products can be written in a similar fashion:

$$\sum_{j=1}^{4} a_{ij}^2 = a_{i1}^2 + a_{i2}^2 + a_{i3}^2 + a_{i4}^2,$$

$$\sum_{i=1}^{3} a_{ij} b_{ij} = a_{1j} b_{1j} + a_{2j} b_{2j} + a_{3j} b_{3j},$$

and

$$\sum_{j=1}^{3} a_{1j} b_{j1} = a_{11} b_{11} + a_{12} b_{21} + a_{13} b_{31}.$$

Some readers may question the use of j in b_{j1} in this last expression since until now we have used i as the first subscript. There is, however, nothing authoritarian about i and j as subscripts; any letters of the alphabet may be so used. While i and j are commonly found in this role, both in this book and elsewhere, they are by no means the only letters used as subscripts; for example,

$$\sum_{i=1}^{2}\sum_{j=1}^{3} a_{ij} = \sum_{p=1}^{2}\sum_{q=1}^{3} a_{pq} = \sum_{k=1}^{2}\sum_{t=1}^{3} a_{kt}$$

$$= a_{11} + a_{12} + a_{13} + a_{21} + a_{22} + a_{23}.$$

The expression $\sum_{j=1}^{3} a_{1j}b_{j1}$ used above is an example of the more general form

$$\sum_{j=1}^{n} a_{ij}b_{jk} = a_{i1}b_{1k} + a_{i2}b_{2k} + a_{i3}b_{3k} + \cdots + a_{in}b_{nk},$$

an expression used extensively in multiplying two matrices together. It is discussed in detail in the next chapter (Section 2.5), but to anticipate for a moment we may note that if A is the matrix

$$A = \begin{bmatrix} a_{11} & a_{12} & a_{13} \\ a_{21} & a_{22} & a_{23} \end{bmatrix} \quad \text{and } B \text{ is} \quad B = \begin{bmatrix} b_{11} & b_{12} & b_{13} \\ b_{21} & b_{22} & b_{23} \\ b_{31} & b_{32} & b_{33} \end{bmatrix},$$

then

$$\sum_{j=1}^{3} a_{1j}b_{j1} = a_{11}b_{11} + a_{12}b_{21} + a_{13}b_{31}$$

is the sum of the term by term products of the elements of the first row of A and the first column of B. Similarly

$$\sum_{j=1}^{3} a_{ij}b_{jk} = a_{i1}b_{1k} + a_{i2}b_{2k} + a_{i3}b_{3k}$$

is the sum of the term by term products of the elements of the ith row of A and the kth column of B.

Two other attributes of summation operations may be noted. One involves summation of terms that do not have subscripts; for example,

$$\sum_{i=1}^{4} x = x + x + x + x = 4x.$$

The other involves summation of several terms each of which is multiplied by a common constant; for example,

$$\sum_{i=1}^{3} ky_i = ky_1 + ky_2 + ky_3 = k\left(\sum_{i=1}^{3} y_i\right).$$

The generalizations of these results are

$$\sum_{i=1}^{n} x = nx \quad \text{and} \quad \sum_{i=1}^{n} ky_i = k\left(\sum_{i=1}^{n} y_i\right).$$

Thus, summation from 1 to n of a constant is n times the constant; and a constant multiplier may be factored outside a summation sign.

There is also a multiplication procedure analogous to Σ. It is denoted by the symbol capital pi, Π, and involves multiplying together all the terms to which it applies. Thus, whereas

$$\sum_{i=1}^{5} b_i = b_1 + b_2 + b_3 + b_4 + b_5,$$

$$\prod_{i=1}^{5} b_i = b_1 b_2 b_3 b_4 b_5,$$

the product of all the b's. Operationally, Π is equivalent to Σ except that it denotes multiplication instead of addition.

5. DOT NOTATION

A further abbreviation often used is

$$\sum_{i=1}^{m} a_{ij} = a_{.j}$$

The dot subscript in place of i denotes that summation has taken place over the i subscript. Since the notation $a_{.j}$ shows no indication of the limits of i over which summation has occurred, it is used only when these limits are clear from the context of its use. In line with $a_{.j}$ we also have

$$a_{i.} = \sum_{j=1}^{n} a_{ij}$$

and

$$a_{..} = \sum_{i=1}^{m} a_{i.} = \sum_{j=1}^{n} a_{.j} = \sum_{i=1}^{m} \sum_{j=1}^{n} a_{ij}$$

6. DEFINITION OF A MATRIX

A matrix is a rectangular (or square) array of numbers arranged in rows and columns. All rows are of equal length, as are all columns. In terms of the notation of Section 3 we will let a_{ij} denote the element in the ith row and jth column of a matrix A, and if A has r rows and c columns it is written

$$
\text{as} \qquad A =
\begin{bmatrix}
a_{11} & a_{12} & a_{13} & \cdots & a_{1j} & \cdots & a_{1c} \\
a_{21} & a_{22} & a_{23} & \cdots & a_{2j} & \cdots & a_{2c} \\
\cdot & & & \cdot & & \cdot & \\
\cdot & & & \cdot & & \cdot & \\
\cdot & & & \cdot & & \cdot & \\
a_{i1} & a_{i2} & a_{i3} & \cdots & a_{ij} & \cdots & a_{ic} \\
\cdot & & & \cdot & & \cdot & \\
\cdot & & & \cdot & & \cdot & \\
\cdot & & & \cdot & & \cdot & \\
a_{r1} & a_{r2} & a_{r3} & \cdots & a_{rj} & \cdots & a_{rc}
\end{bmatrix}.
$$

The three dots indicate, in the first row for example, that the elements a_{11}, a_{12} and a_{13} continue in sequence up to a_{1j} and on up to a_{1c}; likewise in the first column the elements a_{11}, a_{21}, continue in sequence up to a_{r1}. The use of three dots to represent omitted values of a long sequence in this manner is standard and will be used extensively. This form of writing a matrix clearly specifies its terms, and also its size, namely the number of rows and columns. An alternative and briefer form is

$$A = \{a_{ij}\}, \qquad \text{for} \quad i = 1, 2, \ldots, r, \qquad \text{and} \quad j = 1, 2, \ldots, c,$$

the curly brackets indicating that a_{ij} is a typical element, the limits of i and j being r and c respectively.

We will call a_{ij} the ijth *element*, the first subscript referring to the row the element is in and the second to the column. Thus a_{23} is the element in the second row and third column. The size of the matrix, i.e. the number of rows and columns, is referred to as its *order* (or sometimes as its *dimensions*). Thus A with r rows and c columns has order $r \times c$ (read as "r by c") and, to emphasize its dimensions, the matrix can be written $A_{r \times c}$. The first term in the first row of a matrix, a_{11} in this case, is called the *leading term* of the matrix. An example of a 2×3 matrix (a matrix of order 2×3) is

$$
A_{2 \times 3} =
\begin{bmatrix}
4 & 0 & -3 \\
-7 & 2.73 & 1
\end{bmatrix}.
$$

Notice that zero is legitimate as an element, that the elements need not all have the same sign, and that integers and decimal numbers can both be elements of the same matrix.

When $r = c$, i.e. when the number of rows equals the number of columns, A is square and is referred to as a *square matrix* of order r. In this case the elements $a_{11}, a_{22}, a_{33}, \ldots, a_{rr}$ are referred to as the *diagonal elements* of the matrix, and the sum of them is the *trace* of the matrix; that is, when A is square the trace of A equals $\sum_{i=1}^{r} a_{ii}$. The trace of a rectangular matrix is not defined.

When all the non-diagonal elements of a square matrix are zero, the matrix is described as a *diagonal matrix;* for example,

$$A = \begin{bmatrix} 3 & 0 & 0 \\ 0 & -17 & 0 \\ 0 & 0 & 99 \end{bmatrix}$$

is a diagonal matrix. Another variant of a square matrix is that in which all elements above (or below) the diagonal are zero; for example,

$$B = \begin{bmatrix} 1 & 5 & 13 \\ 0 & -2 & 9 \\ 0 & 0 & 7 \end{bmatrix} \quad \text{and} \quad C = \begin{bmatrix} 2 & 0 & 0 \\ 8 & 3 & 0 \\ 1 & -1 & 2 \end{bmatrix}.$$

Such forms are usually given the name *triangular matrix*, and in particular B is called an *upper triangular matrix* and C is a *lower triangular matrix*.

Illustration. Matrices often arise in business decision-making. One use of them is illustrated, for example, by the work of Derman (1963), who has designed a model for making maintenance or replacement decisions. He defines each stage in the deterioration of the system as being a state of the system. Suppose the states are listed both across the top and down the left side of a matrix. The *ij*th element of the matrix represents the probability that the system, being in state i (row i) at some instant of time, will be in state j (column j) one unit of time later. The matrix of such elements is called the *transition probability matrix* of the system. Its elements, the probabilities, are affected by the various policy decisions that may be employed regarding the operation of the system.

For a specific illustration, assume a piece of equipment can either be working properly or be in need of adjustment. If it is working properly suppose there is .90 probability that it will continue to do so and .10 probability that it will need adjustment after a unit of time. If it needs adjustment suppose there is .99 probability that it will continue to need adjustment and .01 probability that it will adjust itself after a unit of time. Then the tabular array of probabilities is

STATE OF EQUIPMENT

At Beginning of Time Period	At End of Time Period	
	Operating Properly	Needing Adjustment
	Probabilities	
Operating Properly	.90	.10
Needing Adjustment	.01	.99

and the transition probability matrix is

$$\begin{bmatrix} .90 & .10 \\ .01 & .99 \end{bmatrix}.$$

Note that the sum of the probabilities in any row is unity. This is true for all transition probability matrices, since the system must be in one of the allowable states after a transition.

The general form of a transition probability matrix is

$$P = \begin{bmatrix} p_{11} & p_{12} & p_{13} & \cdots & p_{1m} \\ p_{21} & p_{22} & p_{23} & \cdots & p_{2m} \\ \cdot & \cdot & \cdot & \cdot & \cdot \\ \cdot & \cdot & \cdot & \cdot & \cdot \\ \cdot & \cdot & \cdot & \cdot & \cdot \\ p_{m1} & p_{m2} & p_{m3} & \cdots & p_{mm} \end{bmatrix}$$

$$= \{p_{ij}\} \quad \text{for} \quad i, j = 1, 2, \ldots, m,$$

where p_{ij} is the probability that a system in state i will, within some given unit of time, transfer to state j. Assumptions about the values of the p's, together with the application of statistical theory and matrix algebra, lead to a variety of conclusions about the relative desirability of different decision rules. The use of a matrix as an entity for handling an array of transition probabilities is one we shall return to on numerous occasions, for it is a useful tool for a wide range of business decisions.

7. VECTORS AND SCALARS

A matrix consisting of only a single column is called a *column vector;* for example,

$$x = \begin{bmatrix} 3 \\ -2 \\ 0 \\ 1 \end{bmatrix}$$

is a column vector of order 4. Likewise a matrix that is just a single row is a *row vector.* We use a prime to indicate a row vector:

$$x' = [3 \quad -2 \quad 0 \quad 1].$$

We could also describe x as a 4×1 matrix and x' as a 1×4 matrix.

A single number such as 2, 6.4, or -4 is called a *scalar.* The elements of a matrix are usually scalars although a matrix is sometimes expressed as a matrix of smaller matrices. In some situations it is convenient to think of a scalar as a matrix of order 1×1.

8. GENERAL NOTATION

In this book matrices will be denoted by upper case letters and their elements by the lower case counterparts with appropriate subscripts. Vectors are denoted by lower case letters, usually from the end of the alphabet, using the prime superscript to distinguish a row vector from a column vector. Thus x is a column vector and x' is a row vector.

Throughout this book square brackets are used for displaying a matrix:

$$A = \begin{bmatrix} 1 & 4 & 6 \\ 0 & 2 & 3 \end{bmatrix}.$$

A variety of other forms can be found in the literature, some of which are

$$\begin{pmatrix} 1 & 4 & 6 \\ 0 & 2 & 3 \end{pmatrix}, \quad \begin{Bmatrix} 1 & 4 & 6 \\ 0 & 2 & 3 \end{Bmatrix} \quad \text{and} \quad \begin{Vmatrix} 1 & 4 & 6 \\ 0 & 2 & 3 \end{Vmatrix}.$$

Single vertical lines are seldom used, since they are usually reserved for indicating a determinant (see Chapter 4).

Another valuable notation that has already been used is

$$A = \{a_{ij}\} \quad \text{for} \quad i = 1, 2, \ldots, r \quad \text{and} \quad j = 1, 2, \ldots, c.$$

The curly brackets indicate that a_{ij} is a typical term of the matrix A for all pairs of values of i and j from unity up to the limits shown, in this case r and c; i.e. A is a matrix of r rows and c columns. This is by no means a universal notation and several variants of it can be found in the literature. There is nothing sacrosanct about the letter A for denoting a matrix; any letter may be used. Lower case letters are widely used to denote vectors, but not universally so. Nor is the procedure of using a prime superscript for distinguishing a row vector from a column vector to be considered absolute. In some texts matrices and vectors are found printed in bold face type to distinguish them from scalars denoted by the same letters, but this is not done here, since most of the book deals with matrices and vectors and there is little likelihood of confusion. Whenever a letter represents a scalar this fact is stated, unless it is clear from the context.

9. ILLUSTRATIVE EXAMPLES

A text on matrix algebra designed for pure mathematicians would deal with many topics that do not appear in this book because they have little connection with matrix applications in business and economics. The text for mathematicians might also have few numerical examples and illustrations. This will not be the case here, however, and indeed we have already used some illustrations and examples in this chapter. In doing so we have capitalized on the great advantage of numerical examples, namely, that to many people arithmetic involving actual numbers is easier to follow than long expositions in algebra. In addition, the use of numerical illustrations describing situations in business and economics also has the advantage of demonstrating the applicability of matrix algebra to business and economic problems. Since this book stresses the use of matrix algebra in business and economics, it would be most desirable to demonstrate each new topic by a numerical problem drawn from some business or economic situation (what we call an "illustration"), thereby obtaining both of the advantages mentioned above. However, in the development of the material there are many intermediate mathematical steps for which few, if any, real-world business or economic problems can be used as illustrations. For this part of the work we rely heavily on numerical "examples," namely, numerical exercises which have the sole purpose of demonstrating mathematical procedures. These in turn lead up to the more advanced techniques that will later be found useful in solving more complex real-life problems in business and economics.

10. EXERCISES

1. For $a_{11} = 17 \quad a_{12} = 31 \quad a_{13} = 26 \quad a_{14} = 11$
   $ a_{21} = 19 \quad a_{22} = 27 \quad a_{23} = 16 \quad a_{24} = 14$
   $ a_{31} = 21 \quad a_{32} = 23 \quad a_{33} = 15 \quad a_{34} = 16$

   show that

   (a) $a_{1.} = 85$, $a_{2.} = 76$ and $a_{3.} = 75$,
   (b) $a_{.1} = 57$, $a_{.2} = 81$, $a_{.3} = 57$ and $a_{.4} = 41$,
   (c) $a_{..} = 236$.

2. For the above series of values show that

   (a) $\displaystyle\sum_{i=1}^{3} a_{ii} = 59$,

   (b) $\displaystyle\sum_{\substack{j=1 \\ j \neq 2}}^{4} a_{ij} = 54$, 49 and 52 for $i = 1$, 2 and 3 respectively,

   (c) $\displaystyle\sum_{\substack{i=1 \\ i \neq 2}}^{3} \sum_{\substack{j=1 \\ j \neq 3}}^{4} a_{ij} = 119$.

3. (a) Show that, for the values of a_{ij} given in Exercise 1, trace(A) = 59 where
   $A = \{a_{ij}\}$ for $i, j = 1, 2, 3$.
   (b) What is the value of trace(B) for $B = \{a_{i,j+1}\}$ for $i, j = 1, 2, 3$?
   (c) Is there a value for trace(M) when

   $$M = \{a_{ij}\} \quad \text{for} \quad i = 1, 2, 3 \quad \text{and} \quad j = 1, 2, \ldots, 4?$$

4. Which of the following matrices are diagonal matrices, upper triangular matrices,
   and lower triangular matrices?

   (a) $\begin{bmatrix} 10 & 0 & 4 \\ 0 & -7 & 0 \\ 0 & 0 & 6 \end{bmatrix}$
   (b) $\begin{bmatrix} 4 & 7 & 6 & 1 \\ 0 & 3 & 0 & -8 \\ 0 & 0 & 1 & 2.7 \end{bmatrix}$

   (c) $\begin{bmatrix} 7 & 0 & 0 \\ 0 & .38 & 0 \\ 0 & 0 & \sqrt{2} \end{bmatrix}$
   (d) $\begin{bmatrix} 6 & 0 & 0 \\ 0 & \sqrt{3} & 0 \\ 0 & 0 & -\sqrt{2} \end{bmatrix}$

   (e) $\begin{bmatrix} 1 & 0 & 0 \\ 4 & 2 & 0 \\ 6 & 5 & 3 \end{bmatrix}$
   (f) $\begin{bmatrix} 7 & 0 & -74 \\ 2.7 & 0 & 17 \\ 0 & 0 & 4 \end{bmatrix}$

5. Show that

   (a) $\displaystyle\sum_{i=3}^{5} 3^i = 351$,
   (b) $\displaystyle\sum_{k=2}^{7} 2^k = 252$,

   (c) $\displaystyle\sum_{r=1}^{5} r = 15$,
   (d) $\displaystyle\sum_{\substack{s=1 \\ s \neq 2}}^{6} s(s+1) = 106$,

   (e) $\displaystyle\prod_{i=1}^{4} 2^i = 1024$.
   (f) $\displaystyle\prod_{i=1}^{3} \sum_{j=1}^{4} (i+j) = 5544$.

6. Consider a purchasing situation where there are only 2 possible products.
   Initially, someone who uses product 1 will buy it again in the next period with
   probability .7. Someone who uses product 2 stays with product 2 with probability
   .6. However, if $1,200,000 is spent on advertising product 1, those probabilities
   change to .8 and .4, respectively. Write down the transition probability matrices
   for the 2 situations, before and after the expenditure on advertising. Does the
   advertising alter the market in favor of brand 1?

7. A ticket seller for an out-of-town show wants to decide how many tickets he
   should buy. He can only place one order. Each ticket costs $5 and can be sold for
   $8; leftover tickets have no value. The number of tickets he is able to sell is
   known to be between one and four. Prepare a matrix of profits associated with
   his different decisions and the possible outcomes, letting rows represent decisions
   and columns represent outcomes.

8. Show that $\displaystyle\sum_{i=1}^{2} a_i b_i \neq \left(\sum_{i=1}^{2} a_i\right)\left(\sum_{i=1}^{2} b_i\right)$.

9. Show that the following identities hold by computing both sides of each expression for the numerical example in Exercise 1. $\left[\text{The square of } a_{i.} \text{ is } a_{i.}^2 = \left(\displaystyle\sum_{j=1}^{n} a_{ij}\right)^2.\right]$

(a) $\displaystyle\sum_{i=1}^{m}\sum_{j=1}^{n} a_{ij}^2 = \sum_{j=1}^{n}\sum_{i=1}^{m} a_{ij}^2$

(b) $\displaystyle\sum_{i=1}^{m}\left(\sum_{j=1}^{n} a_{ij}\right)^2 = \sum_{i=1}^{m} a_{i.}^2$

(c) $\displaystyle\sum_{j=1}^{n}\sum_{k=1}^{n} a_{ij}a_{hk} = a_{i.}a_{h.}, \text{ for } i \neq h$

(d) $\displaystyle a_{..}^2 = \sum_{i} a_{i.}^2 + 2\sum_{i=1}^{n-1}\sum_{h=i+1}^{n} a_{i.}a_{h.}$

     $\displaystyle = \sum_{i} a_{i.}^2 + 2\sum_{i=1}^{n-1}\sum_{h>i}^{n} a_{i.}a_{h.}$

     $\displaystyle = \sum_{i} a_{i.}^2 + \sum_{i=1}^{n}\sum_{\substack{h=1 \\ h\neq i}}^{n} a_{i.}a_{h.}$

(e) $\displaystyle\sum_{\substack{i=1 \\ i\neq p}}^{m}\sum_{\substack{j=1 \\ j\neq q}}^{n} a_{ij} = a_{..} - a_{p.} - a_{.q} + a_{pq}$

(f) $\displaystyle\sum_{i=1}^{m}\sum_{j=1}^{n} (a_{ij} - 1) = a_{..} - mn$ (g) $\displaystyle\sum_{j=1}^{n}\sum_{\substack{k=1 \\ k\neq j}}^{n} a_{ij}a_{ik} = a_{i.}^2 - \sum_{j=1}^{n} a_{ij}^2$

(h) $\displaystyle\left(\sum_{j=1}^{n} a_{ij}\right)^2 = \sum_{j=1}^{n} a_{ij}^2 + 2\sum_{j=1}^{n-1}\sum_{p=j+1}^{n} a_{ij}a_{ip}$

     $\displaystyle = \sum_{j=1}^{n} a_{ij}^2 + 2\sum_{j=1}^{n-1}\sum_{p>j}^{n} a_{ij}a_{ip}$

     $\displaystyle = \sum_{j=1}^{n} a_{ij}^2 + \sum_{j=1}^{n}\sum_{\substack{p=1 \\ p\neq j}}^{n} a_{ij}a_{ip}$

(i) $\displaystyle\sum_{i=1}^{m}\sum_{j=1}^{n} 4a_{ij} = 4a_{..}$

REFERENCES

Chow, G. C. (1960). Statistical demand functions for automobiles and their use for forecasting. In *The Demand for Durable Goods*, A. C. Harberger (ed). University of Chicago Press.

Derman, C. (1963). Optimal replacement and maintenance under Markovian deterioration with probability bounds on failure. *Management Science*, **9**, 478–481.

CHAPTER 2

BASIC ARITHMETIC

This chapter describes the arithmetic operations of addition, subtraction and multiplication as they apply to matrices. The counterparts of zero and unity are also described.

1. ADDITION

We begin with the operation of addition, introducing it by means of an illustration.

Illustration. A manufacturing company classified its total annual sales by product and sales region. The distribution of sales for the company's three products in its three sales regions is contained in Table 1.

TABLE 1. ANNUAL UNIT SALES

Product	Sales Region		
	1	2	3
I	98	24	42
II	39	15	22
III	22	15	17

Let us write the body of the table as a 3×3 matrix

$$A = \begin{bmatrix} 98 & 24 & 42 \\ 39 & 15 & 22 \\ 22 & 15 & 17 \end{bmatrix},$$

[*17*]

so that with the same frame of reference the data for a subsequent year can also be written as a matrix:

$$B = \begin{bmatrix} 55 & 19 & 44 \\ 43 & 53 & 38 \\ 11 & 40 & 20 \end{bmatrix}.$$

Then, over the two years, the total number of units of product I sold in region 1 is the sum of the elements in the first row and first column of each matrix, namely $98 + 55 = 153$, and the total number of units of product III sold in region 2 is $15 + 40 = 55$. In this way the matrix of all such sums

$$\begin{bmatrix} 98 + 55 & 24 + 19 & 42 + 44 \\ 39 + 43 & 15 + 53 & 22 + 38 \\ 22 + 11 & 15 + 40 & 17 + 20 \end{bmatrix} = \begin{bmatrix} 153 & 43 & 86 \\ 82 & 68 & 60 \\ 33 & 55 & 37 \end{bmatrix}$$

represents the sales made over the two years by product and sales region. This is the matrix sum A plus B; it is the matrix formed by adding the matrices A and B element by element. Hence if we write $A = \{a_{ij}\}$ and $B = \{b_{ij}\}$ for $i = 1, 2, \ldots, r$ and $j = 1, 2, \ldots, c$ the matrix representing the sum of A and B is

$$A + B = \{a_{ij} + b_{ij}\} \qquad \text{for} \quad i = 1, 2, \ldots, r \qquad \text{and} \quad j = 1, 2, \ldots, c;$$

i.e. the sum of two matrices is the matrix of sums, element by element.

It should be evident that matrix addition can take place only when the matrices involved are of the same order; i.e. two matrices can be added only if they have the same number of rows and the same number of columns. Thus it is meaningless to attempt the addition of

$$\begin{bmatrix} 1 & 2 \\ 6 & -4 \end{bmatrix} \quad \text{and} \quad \begin{bmatrix} 3 & 9 \\ -6 & 1 \\ 4 & 3 \end{bmatrix}.$$

Matrices that have the same order and can therefore be added together are said to be *conformable for addition*. For example,

$$\begin{bmatrix} 1 & 2 & 3 \\ 6 & -4 & 5 \end{bmatrix} \quad \text{and} \quad \begin{bmatrix} -3 & 1 & 9 \\ 2 & 4 & 6 \end{bmatrix}$$

are conformable for addition and their sum is $\begin{bmatrix} -2 & 3 & 12 \\ 8 & 0 & 11 \end{bmatrix}.$

2. SCALAR MULTIPLICATION

We have just described matrix addition. A simple use of it shows that

$$A + A = \{a_{ij}\} + \{a_{ij}\}$$
$$= \{2a_{ij}\}$$
$$= 2A.$$

Extending this to the case where λ is a positive integer, we have

$$\lambda A = A + A + A + \cdots + A,$$

there being λ A's in the sum on the right. Carrying out these matrix additions gives

$$\lambda A = \{\lambda a_{ij}\} \quad \text{for} \quad i = 1, 2, \ldots, r \quad \text{and} \quad j = 1, 2, \ldots, c.$$

This result, extended for any scalar, is the definition of *scalar multiplication*. Thus the matrix A multiplied by the scalar λ is the matrix A with every element multiplied by λ. For example,

$$3\begin{bmatrix} 1 & -7 \\ 3 & 5 \end{bmatrix} = \begin{bmatrix} 3 & -21 \\ 9 & 15 \end{bmatrix}.$$

3. SUBTRACTION

Illustration. The cumulative sales for a company's products from January 1 to March 31 in a given year are described in Table 2.

TABLE 2. CUMULATIVE SALES AS OF MARCH 31

Product	Sales Region			
	1	2	3	4
I	910	1,275	1,210	1,304
II	860	967	667	1,048

If the entries in the table are represented by the matrix

$$A = \begin{bmatrix} 910 & 1,275 & 1,210 & 1,304 \\ 860 & 967 & 667 & 1,048 \end{bmatrix},$$

suppose the cumulative sales of the same products as of June 30 are

$$B = \begin{bmatrix} 2,050 & 1,340 & 1,344 & 1,384 \\ 1,380 & 1,058 & 1,011 & 1,189 \end{bmatrix}.$$

Actual sales between March 31 and June 30 of product I in sales region 1 are represented by the difference between cumulative sales at these dates, namely $2,050 - 910 = 1,140$, and corresponding sales of product II in region 4 are $1,189 - 1,048 = 141$. Hence the matrix

$$\begin{bmatrix} 2,050 - 910 & 1,340 - 1,275 & 1,344 - 1,210 & 1,384 - 1,304 \\ 1,380 - 860 & 1,058 - 967 & 1,011 - 667 & 1,189 - 1,048 \end{bmatrix}$$

$$= \begin{bmatrix} 1,140 & 65 & 134 & 80 \\ 520 & 91 & 344 & 141 \end{bmatrix}$$

represents the sales from March 31 to June 30 of both products for all four sales regions. This illustrates the matrix operation of subtraction. Thus if $A = \{a_{ij}\}$ and $B = \{b_{ij}\}$ for $i = 1, 2, \ldots, r$ and $j = 1, 2, \ldots, c$, subtraction in matrix algebra is defined as

$$A - B = \{a_{ij} - b_{ij}\} \quad \text{for} \quad i = 1, 2, \ldots, r \quad \text{and} \quad j = 1, 2, \ldots, c;$$

i.e. the difference between two matrices is the matrix of differences, element by element.[1]

Example.

$$\begin{bmatrix} 3 & 6 \\ 8 & 2 \\ 4 & 1 \end{bmatrix} - \begin{bmatrix} 1 & 1 \\ 0 & -3 \\ 2 & -5 \end{bmatrix} = \begin{bmatrix} 2 & 5 \\ 8 & 5 \\ 2 & 6 \end{bmatrix}.$$

As with addition, only matrices that are of the same order can be subtracted from one another. Thus matrices that are conformable for addition are also conformable for subtraction, and vice versa.

4.　EQUALITY AND THE NULL MATRIX

Two matrices are said to be equal when they are identical element by element. Thus $A = B$ when $\{a_{ij}\} = \{b_{ij}\}$, meaning that $a_{ij} = b_{ij}$ for $i = 1$, $2, \ldots, r$ and $j = 1, 2, \ldots, c$. If

$$A = \begin{bmatrix} 2 & 6 & -4 \\ 3 & 0 & 1 \end{bmatrix}, \quad B = \begin{bmatrix} 2 & 6 & -4 \\ 3 & 0 & 1 \end{bmatrix} \quad \text{and} \quad C = \begin{bmatrix} 2 & 6 & -4 \\ 2 & 0 & 1 \end{bmatrix},$$

---

[1] We can prove this more formally as a direct outcome of the rules of addition and scalar multiplication. Since $-B = (-1)B$,

$$A - B = A + (-1)B = \{a_{ij}\} + (-1)\{b_{ij}\}$$
$$= \{a_{ij}\} + \{-b_{ij}\} = \{a_{ij} - b_{ij}\}.$$

A is equal to B but not equal to C. The equality of two matrices has no meaning unless they are of the same order.

Combining the ideas of subtraction and equality leads to the definition of zero in matrix algebra. For, when

$$A = B,$$
$$a_{ij} = b_{ij}$$

for $i = 1, 2, \ldots, r$ and $j = 1, 2, \ldots, c,$ and so

$$A - B = \{a_{ij} - b_{ij}\} = 0.$$

The matrix on the right is a matrix of zeros; i.e. every element is zero. Such a matrix is called a *null matrix;* it is the zero of matrix algebra and as such is referred to by some writers as a *zero matrix.* It is not a unique zero because for a matrix of any order there is a null matrix of the same order. For example, null matrices of order 2×4 and 3×3 are respectively

$$\begin{bmatrix} 0 & 0 & 0 & 0 \\ 0 & 0 & 0 & 0 \end{bmatrix} \quad \text{and} \quad \begin{bmatrix} 0 & 0 & 0 \\ 0 & 0 & 0 \\ 0 & 0 & 0 \end{bmatrix}.$$

5. MULTIPLICATION

Before describing the multiplication of matrices a brief discussion is given of two simpler multiplication operations involving vectors. Once again, numerical illustrations are used to introduce general methods.

a. A product of vectors

Illustration. Suppose the retail unit sales of products I, II and III in a given year were 58, 26 and 8, respectively, and that the selling prices of the products were \$1, \$2 and \$3, respectively. The total sales revenue for all three products for the year is therefore $58(1) + 26(2) + 8(3) = 134$ dollars. Suppose that the respective unit sales for the products are written as a row vector

$$a' = [58 \quad 26 \quad 8],$$

and the respective prices for the products are written as a column vector

$$x = \begin{bmatrix} 1 \\ 2 \\ 3 \end{bmatrix}.$$

Then the total sales revenue obtained from the three products, 134 dollars, is the sum of products of the elements of a' (quantity sold) each multiplied by the corresponding element of x (price). This is the definition of the product $a'x$. It is written as

$$a'x = [58 \quad 26 \quad 8] \begin{bmatrix} 1 \\ 2 \\ 3 \end{bmatrix}$$

without any multiplication symbol between the vectors, and is calculated as

$$a'x = 58(1) + 26(2) + 8(3) = 134.$$

This example illustrates the general procedure for obtaining $a'x$: multiply each element of the row vector a' by the corresponding element of the column vector x and add the resultant products. The sum is $a'x$. Thus if

$$a' = [a_1 \quad a_2 \quad \cdots \quad a_n]$$

and

$$x = \begin{bmatrix} x_1 \\ x_2 \\ \cdot \\ \cdot \\ \cdot \\ x_n \end{bmatrix}$$

their product $a'x$ is defined as

$$a'x = a_1 x_1 + a_2 x_2 + \cdots + a_n x_n$$

$$= \sum_{i=1}^{n} a_i x_i.$$

The definition applies only when a' and x have the same number of elements; if they do not, the product $a'x$ is undefined and does not exist.

b. A matrix-vector product

Illustration. The preceding illustration described the retail domestic sales of a company. The company also has an institutional division and an international division, and total sales of all three divisions are contained in Table 3. The sales revenue of the domestic division was previously calculated

TABLE 3. SALES BY DIVISION AND PRODUCT

Division	Product and Selling Price		
	I($1)	II($2)	(III)($3)
	Unit Sales		
Domestic	58	26	8
Institutional	52	58	12
International	1	3	9

as $58(1) + 26(2) + 8(3) = 134$ dollars: similar calculations can be made for the other two divisions, as summarized in Table 4.

TABLE 4. SALES REVENUE BY DIVISION

Division	Sales Revenue
Domestic	$58(1) + 26(2) + 8(3) = 134$
Institutional	$52(1) + 58(2) + 12(3) = 204$
International	$1(1) + 3(2) + 9(3) = 34$

Let us write the entries of Table 3 as a matrix

$$A = \begin{bmatrix} 58 & 26 & 8 \\ 52 & 58 & 12 \\ 1 & 3 & 9 \end{bmatrix}$$

and the results of Table 4 as a vector $\begin{bmatrix} 134 \\ 204 \\ 34 \end{bmatrix}$. As can be seen from Table 4, the elements of this vector are derived in exactly the same way as the product $a'x$ developed earlier, using each successive row of A as the vector a'. The result is the product Ax; that is, Ax is obtained by repetitions of the product $a'x$ using the rows of A successively for a' and writing the results as a column vector. Thus

$$Ax = \begin{bmatrix} 58 & 26 & 8 \\ 52 & 58 & 12 \\ 1 & 3 & 9 \end{bmatrix} \begin{bmatrix} 1 \\ 2 \\ 3 \end{bmatrix}$$

$$= \begin{bmatrix} 58(1) + 26(2) + 8(3) \\ 52(1) + 58(2) + 12(3) \\ 1(1) + 3(2) + 9(3) \end{bmatrix} = \begin{bmatrix} 134 \\ 204 \\ 34 \end{bmatrix}.$$

General subscript notation for this example is as follows:

$$A = \begin{bmatrix} a_{11} & a_{12} & a_{13} \\ a_{21} & a_{22} & a_{23} \\ a_{31} & a_{32} & a_{33} \end{bmatrix}, \qquad x = \begin{bmatrix} x_1 \\ x_2 \\ x_3 \end{bmatrix},$$

$$\text{and} \qquad Ax = \begin{bmatrix} a_{11}x_1 + a_{12}x_2 + a_{13}x_3 \\ a_{21}x_1 + a_{22}x_2 + a_{23}x_3 \\ a_{31}x_1 + a_{32}x_2 + a_{33}x_3 \end{bmatrix} = \begin{bmatrix} \sum_{k=1}^{3} a_{1k}x_k \\ \sum_{k=1}^{3} a_{2k}x_k \\ \sum_{k=1}^{3} a_{3k}x_k \end{bmatrix}.$$

We see that the first element of Ax is the sum of products of the a_{ij}'s of the first row of A with the elements of x, and likewise for the other elements of Ax. Generalizing, the product Ax of a matrix A and a column vector x is a column vector whose ith term is the sum of products of the elements of the ith row of A each multiplied by the corresponding element of x.

Example.

$$\begin{bmatrix} 4 & 2 & 1 & 3 \\ 2 & 0 & -4 & 7 \end{bmatrix} \begin{bmatrix} 1 \\ 0 \\ -1 \\ 3 \end{bmatrix} = \begin{bmatrix} 4(1) + 2(0) + 1(-1) + 3(3) \\ 2(1) + 0(0) + -4(-1) + 7(3) \end{bmatrix} = \begin{bmatrix} 12 \\ 27 \end{bmatrix}.$$

Both from this example and from the definition we see that the product Ax is defined only when the number of elements in the rows of A (i.e., the number of columns) is the same as the number of elements in the column vector x. When this occurs, Ax is a column vector having the same number of elements as there are rows in A. Therefore, when A has r rows and c columns and the column vector x is of order c, Ax is a column vector of order r; its ith element is $\sum_{k=1}^{c} a_{ik}x_k$ for $i = 1, 2, \ldots, r$.

Similarly, the product $x'P$ is defined only if the number of elements in the row vector x' equals the number of elements in the columns of P (i.e., the number of rows), and the result is a row vector having as many elements as there are columns in P. The product $x'P$ is not equal to Px; in fact, the product Px may not exist when $x'P$ does, and vice versa.

Illustration. In a study of the life of newspaper subscriptions, Deming and Glasser (1968) describe the probabilities of moving from one subscription category to another by a transition probability matrix. A simplified version of their matrix is:

$$P = \begin{bmatrix} 0 & .7 & 0 & .3 \\ 0 & 0 & .8 & .2 \\ 0 & 0 & .9 & .1 \\ 0 & 0 & 0 & 1.0 \end{bmatrix}$$

where the rows and columns refer to four categories of the duration of a subscription: less than one year, between one and two years, more than two years, or cancelled.

Suppose that of 1,000 current subscribers it is known that 500 are currently in category 1, 200 are in category 2 and 300 are in category 3. The group of 1,000 subscribers may be described by the row vector

$$x' = [500 \quad 200 \quad 300 \quad 0].$$

In order to calculate the number of these subscribers in the various categories one year from now we multiply the row vector x' by the transition probability matrix P:

$$x'P = [500 \quad 200 \quad 300 \quad 0] \begin{bmatrix} 0 & .7 & 0 & .3 \\ 0 & 0 & .8 & .2 \\ 0 & 0 & .9 & .1 \\ 0 & 0 & 0 & 1.0 \end{bmatrix} = [0 \quad 350 \quad 430 \quad 220].$$

The resulting vector indicates that one year from now 350 of the initial 1,000 subscribers are expected to be in category 2, 430 in category 3 and 220 in category 4.

c. A product of two matrices

Multiplying two matrices can be explained as a simple repetitive extension of multiplying a matrix by a vector. To obtain the product of two matrices A and B think of the matrix B as being a series of column vectors. Then the product AB is the matrix obtained by setting alongside one another each of the product vectors of A with the columns of B.

Example.

If

$$A = \begin{bmatrix} 1 & 0 & 2 \\ 3 & 1 & 1 \\ 1 & 2 & 1 \\ -1 & 3 & 2 \end{bmatrix}, \quad \text{and} \quad B = \begin{bmatrix} 1 & 2 \\ 0 & 1 \\ 0 & -1 \end{bmatrix},$$

we think of B as being composed of the two column vectors

$$x = \begin{bmatrix} 1 \\ 0 \\ 0 \end{bmatrix} \quad \text{and} \quad w = \begin{bmatrix} 2 \\ 1 \\ -1 \end{bmatrix}.$$

The products of A with each of the columns of B are

$$Ax = \begin{bmatrix} 1(1) + 0(0) + 2(0) \\ 3(1) + 1(0) + 1(0) \\ 1(1) + 2(0) + 1(0) \\ -1(1) + 3(0) + 2(0) \end{bmatrix} = \begin{bmatrix} 1 \\ 3 \\ 1 \\ -1 \end{bmatrix}$$

and

$$Aw = \begin{bmatrix} 1(2) + 0(1) + 2(-1) \\ 3(2) + 1(1) + 1(-1) \\ 1(2) + 2(1) + 1(-1) \\ -1(2) + 3(1) + 2(-1) \end{bmatrix} = \begin{bmatrix} 0 \\ 6 \\ 3 \\ -1 \end{bmatrix}.$$

Setting these vectors alongside each other yields the matrix product AB:

$$AB = \begin{bmatrix} 1 & 0 \\ 3 & 6 \\ 1 & 3 \\ -1 & -1 \end{bmatrix},$$

and the complete operation is written as

$$AB = \begin{bmatrix} 1 & 0 & 2 \\ 3 & 1 & 1 \\ 1 & 2 & 1 \\ -1 & 3 & 2 \end{bmatrix} \begin{bmatrix} 1 & 2 \\ 0 & 1 \\ 0 & -1 \end{bmatrix} = \begin{bmatrix} 1 & 0 \\ 3 & 6 \\ 1 & 3 \\ -1 & -1 \end{bmatrix}.$$

To obtain this product think of moving from element to element across the ith row of A and simultaneously down the jth column of B, summing the products of corresponding elements. The resulting sum is the ijth element of AB. For example, as we move across the second row of A and simultaneously down the second column of B we get the sum:

$$3(2) + 1(1) + 1(-1) = 6 + 1 - 1 = 6.$$

Thus the element in the second row and second column of the product AB is 6. Hence it can be seen that in the product AB the ith element of the first column is the sum of products of the elements of the ith row of A with those of the first column of B, and the ith element in the second column of AB is the sum of products of the elements of the ith row of A with those of the second column of B.

The general multiplication procedure can now be stated precisely: the ijth element (the element in the ith row and jth column) of the product AB of two matrices A and B is the sum of products (element by element) of the elements of the ith row of A with the corresponding elements of the jth column of B. Thus if the ith row of A is $[a_{i1} \quad a_{i2} \quad \cdots \quad a_{ic}]$ and the jth

column of B is $\begin{bmatrix} b_{1j} \\ b_{2j} \\ \cdot \\ \cdot \\ \cdot \\ b_{cj} \end{bmatrix}$, the ijth element of AB is

$$a_{i1}b_{1j} + a_{i2}b_{2j} + \cdots + a_{ic}b_{cj} = \sum_{k=1}^{c} a_{ik}b_{kj}.$$

Example.

For $\qquad A = \begin{bmatrix} 1 & 0 & 2 \\ -1 & 4 & 3 \end{bmatrix} \qquad$ and $\quad B = \begin{bmatrix} 0 & 6 & 1 & 5 \\ 1 & 1 & 2 & 7 \\ 2 & 4 & 4 & 3 \end{bmatrix}$

the element in the first row and first column of the product AB is based on the first row of A and the first column of B and is

$$1(0) + 0(1) + 2(2) = 4;$$

the element in the first row and second column is

$$1(6) + 0(1) + 2(4) = 14;$$

and the element of AB in the second row and third column is

$$-1(1) + 4(2) + 3(4) = 19.$$

In this way AB is obtained as

$$AB = \begin{bmatrix} 4 & 14 & 9 & 11 \\ 10 & 10 & 19 & 32 \end{bmatrix}.$$

The reader should satisfy himself of the validity of this result.

The definition of the matrix product AB holds only if a certain condition is met, namely if the jth column of B (and hence all columns) has the same number of elements as does the ith row of A (and hence all rows). Since the number of elements in a column of a matrix is the number of rows in the matrix (and the number of elements in a row is the number of columns), this means that there must be exactly as many rows in B as there are columns in A. Thus the matrix product AB is defined only if the number of columns of A equals the number of rows of B. It might also be noted, as in the numerical example given previously, that the product AB has the same number of rows as A and the same number of columns as B.

When the number of columns in A is equal to the number of rows in B the matrices are said to be *conformable for multiplication for the product AB*, and AB has the same number of rows as A and the same number of columns as B. Hence if A is $r \times c$ and B is $c \times s$, i.e.,

$$A = \{a_{ij}\} \quad \text{for} \quad i = 1, 2, \ldots, r, \quad \text{and} \quad j = 1, 2, \ldots, c,$$
$$\text{and} \quad B = \{b_{ij}\} \quad \text{for} \quad i = 1, 2, \ldots, c, \quad \text{and} \quad j = 1, 2, \ldots, s,$$

the product AB is $r \times s$, and its ijth element is $\sum_{k=1}^{c} a_{ik}b_{kj}$. Hence we write

$$AB = \left\{ \sum_{k=1}^{c} a_{ik}b_{kj} \right\} \quad \text{for} \quad i = 1, 2, \ldots, r, \quad \text{and} \quad j = 1, 2, \ldots, s.$$

The expression for the ijth element is the sum of products of the elements of the ith row of A with those of the jth column of B, and as well as being the ijth element of AB it is sometimes called the *inner product* of the ith row of A and the jth column of B. The matrix product AB is therefore obtained by calculating the inner product of each row of A with each column of B, the inner product of the ith row of A with the jth column of B being the ijth element of AB.

Example.

For
$$A = \begin{bmatrix} 1 & 5 \\ 3 & 0 \end{bmatrix} \quad \text{and} \quad B = \begin{bmatrix} 3 & 6 & 1 \\ 2 & 2 & 4 \end{bmatrix}$$

the product AB exists, since A has two columns and B has two rows:

$$AB = \begin{bmatrix} 1 & 5 \\ 3 & 0 \end{bmatrix}\begin{bmatrix} 3 & 6 & 1 \\ 2 & 2 & 4 \end{bmatrix} = \begin{bmatrix} 13 & 16 & 21 \\ 9 & 18 & 3 \end{bmatrix}.$$

The order of the product AB is 2×3. Note, however, that the product BA does not exist, since B has three columns while A has only two rows.

d. Existence of matrix products

As noted in Section 1.6, subscript notation can be used to denote the order of a matrix; viz. $A_{r \times c}$ is a matrix of order r by c (r rows and c columns). The product AB can be written in this notation as

$$A_{r \times c} B_{c \times s} = P_{r \times s},$$

a form which provides the opportunity both for checking the conformability of A and B for multiplication and for ascertaining the order of their product.

Products of more than two matrices are derived by direct analogy from the product of two. Thus the product ABC exists if AB is conformable for multiplication with C. This is so if AB has the same number of columns as C has rows. But AB has the same number of columns as does B. Hence B and C must be conformable for multiplication in order that the product ABC exist. And, in general, a product $ABCD$ . . . exists provided each pair of adjacent matrices in the product is conformable for multiplication. Repeated use of subscript notation for the order of a matrix, as indicated above, simplifies determining the order of such a product. The adjacent subscripts (which must be equal for conformability) "cancel out," leaving the first and last subscripts as the order of the product. For example, the product $A_{2\times3}B_{3\times5}C_{5\times10}D_{10\times4}$ is a matrix of order 2×4. In general, $A_{r\times c}B_{c\times s}C_{s\times t}D_{t\times k}$ is a matrix or order $r \times k$, with direct extension to the case of more than four matrices in the product.

As illustrated above, the product BA does not necessarily exist, even if AB does. For BA can be written as $B_{c\times s}A_{r\times c}$, which is a legitimate product only if $s = r$. Otherwise BA is not defined. There are therefore three situations regarding the product of two matrices A and B. If A is of order $r \times c$,

(i) AB exists only if B has c rows,

(ii) BA exists only if B has r columns,

and   (iii) AB and BA both exist only if B is $c \times r$.

A corollary to (iii) is that $AA = A^2$ exists only when A is square. Another corollary is that both AB and BA always exist when A and B are square and of the same order. But, as shall be shown subsequently, the two products are not necessarily equal. As a means of distinguishing between them, AB is described as A *post-multiplied* by B (or *multiplied on the right* by B), and BA is described as A *pre-multiplied* by B (or *multiplied on the left* by B).

Illustration. Consider the piece of equipment described in Section 1.6. The equipment is either working properly or in need of adjustment, with the following transition probability matrix for periodic changes:

$$P = \begin{bmatrix} .90 & .10 \\ .01 & .99 \end{bmatrix}.$$

Suppose the equipment starts out in adjustment in period one. We can let the row vector x_1' represent the state of the equipment in period one as follows:

$$x_1' = [1 \quad 0].$$

Here the subscript "1" on x' indicates the time period referred to and is not a subscript referring to an element of any matrix.

With this notation the state of the equipment in period two, x_2', may be calculated by multiplying the state vector of period one by the transition probability matrix P:

$$x_2' = x_1'P = [1 \quad 0]\begin{bmatrix} .90 & .10 \\ .01 & .99 \end{bmatrix} = [.90 \quad .10].$$

The state of the equipment in period three may be calculated as follows:

$$x_3' = x_2'P = x_1'P^2.$$

Similarly,

$$x_4' = x_1'P^3 \quad \text{and} \quad x_n' = x_1'P^{n-1}.$$

That is, the state of the equipment in period n is equal to the state vector in period 1 multiplied by the $(n-1)$th power of the transition probability matrix, P^{n-1}.

The power of a matrix exists only if the matrix is square, and its power may be computed by repetitive use of the rules for matrix multiplication. For example,

$$P^2 = PP = \begin{bmatrix} .90 & .10 \\ .01 & .99 \end{bmatrix}\begin{bmatrix} .90 & .10 \\ .01 & .99 \end{bmatrix} = \begin{bmatrix} .8110 & .1890 \\ .0189 & .9811 \end{bmatrix}.$$

Higher powers may be computed similarly.

e. Vector products

The topics discussed in parts **a** and **b** of this section are special cases of the general matrix product $A_{r \times c} B_{c \times s} = P_{r \times s}$. Thus in **a** we considered the case for $r = 1$ and $s = 1$, which means that $A_{r \times c}$ becomes $A_{1 \times c}$, a row vector, and $B_{c \times s}$ becomes $B_{c \times 1}$, a column vector. Retaining the convention that a' is a row vector, further emphasized by writing its order as subscripts $a_{1 \times c}'$, we then have

$$a_{1 \times c}' b_{c \times 1} = p_{1 \times 1},$$

a scalar or equivalently a 1×1 matrix. However,

$$b_{c \times 1} a_{1 \times c}' = P_{c \times c},$$

a square matrix. In **b** we considered a column vector pre-multiplied by a matrix; that is,

$$A_{r \times c} b_{c \times 1} = p_{r \times 1},$$

the product being a column vector; similarly a row vector post-multiplied by a matrix is

$$a_{1 \times c}' B_{c \times r} = p_{1 \times r}',$$

a row vector.

In words these four results are as follows:

(i) A row vector post-multiplied by a column vector is a scalar.
(ii) A column vector post-multiplied by a row vector is a matrix.
(iii) A matrix post-multiplied by a column vector is a column vector.
(iv) A row vector post-multiplied by a matrix is a row vector.

If at all times we retain the idea of moving across a row and simultaneously down a column to obtain an element of a product, then the above results follow directly.

Examples. Given

$$A = \begin{bmatrix} 1 & 2 \\ 3 & 4 \end{bmatrix}, \qquad B = \begin{bmatrix} 6 & 3 & 1 \\ -1 & 2 & 5 \end{bmatrix}, \qquad x' = [1 \quad 5], \qquad y = \begin{bmatrix} 3 \\ 1 \end{bmatrix},$$

the following results are true:

$$AB = \begin{bmatrix} 1(6) + 2(-1) & 1(3) + 2(2) & 1(1) + 2(5) \\ 3(6) + 4(-1) & 3(3) + 4(2) & 3(1) + 4(5) \end{bmatrix} = \begin{bmatrix} 4 & 7 & 11 \\ 14 & 17 & 23 \end{bmatrix},$$

$$x'B = [1 \quad 13 \quad 26], \qquad Ay = \begin{bmatrix} 5 \\ 13 \end{bmatrix},$$

$$x'Ay = 70, \qquad x'y = 8 \qquad \text{and} \qquad yx' = \begin{bmatrix} 3 & 15 \\ 1 & 5 \end{bmatrix}.$$

f. Products with diagonal matrices

A diagonal matrix was defined in Section 1.6 as a square matrix which has all non-diagonal elements zero. (Some but not all of the diagonal elements may also be zero.)

The procedure for multiplying matrices results in multiplication by diagonal matrices being particularly easy: pre-multiplication of a matrix A by a diagonal matrix D gives a matrix whose rows are those of A multiplied by the diagonal element of D in the same row. Thus with

$$D = \begin{bmatrix} 1.3 & 0 \\ 0 & 2.1 \end{bmatrix} \qquad \text{and} \qquad A = \begin{bmatrix} 2 & -1 & 7 \\ -1 & 0 & 1 \end{bmatrix},$$

$$DA = \begin{bmatrix} 2.6 & -1.3 & 9.1 \\ -2.1 & 0 & 2.1 \end{bmatrix}.$$

Post-multiplication leads in a similar way to the columns being multiplied by the respective diagonal elements of D, for example, with

$$A = \begin{bmatrix} 2 & 1 \\ 0 & -5 \\ 12 & 7 \end{bmatrix} \qquad \text{and} \qquad D = \begin{bmatrix} -7 & 0 \\ 0 & 4 \end{bmatrix}, \qquad AD = \begin{bmatrix} -14 & 4 \\ 0 & -20 \\ -84 & 28 \end{bmatrix}.$$

g. Some special products

In ordinary algebra, where a, b, c and d are scalars, we know that if $a^2 = 0$ then $a = 0$; or if $cd = 2c$ then $d = 2$. But matrix analogies of these common results are not necessarily true. Thus for

$$A = \begin{bmatrix} 1 & 2 & 5 \\ 2 & 4 & 10 \\ -1 & -2 & -5 \end{bmatrix}, \quad \text{we have } A^2 = \begin{bmatrix} 0 & 0 & 0 \\ 0 & 0 & 0 \\ 0 & 0 & 0 \end{bmatrix},$$

i.e. $A^2 = 0$ although $A \neq 0$. Similarly, $CD = 2C$ does not imply $D = 2$, nor does it imply anything about the product DC. For example, if

$$C = \begin{bmatrix} 1 & 1 \\ -1 & -1 \end{bmatrix} \quad \text{and} \quad D = \begin{bmatrix} 1 & 1 \\ 1 & 1 \end{bmatrix}$$

then $CD = 2C$ and $DC = 0$, as may be verified. Reasons why familiar results of scalar algebra do not always hold true with matrices will become clearer after the analogue of division, the inverse of a matrix, has been discussed in Chapter 5. In the meantime the reader should be warned that consequences of multiplication in scalar algebra do not always apply in matrix algebra.

6. THE LAWS OF ALGEBRA

We now give formal consideration to the associative, commutative and distributive laws of algebra as they relate to the addition and multiplication of matrices.

(a) The addition of matrices is associative provided the matrices are conformable for addition. For, if A, B and C have the same order,

$$(A + B) + C = \{a_{ij} + b_{ij}\} + \{c_{ij}\}$$
$$= \{a_{ij} + b_{ij} + c_{ij}\}$$
$$= A + B + C.$$

Also, $\quad \{a_{ij} + b_{ij} + c_{ij}\} = \{a_{ij}\} + \{b_{ij} + c_{ij}\}$
$$= A + (B + C),$$

so proving the associative law of addition, namely,

$$(A + B) + C = A + (B + C) = A + B + C.$$

(b) The associative law is also true for multiplication provided the matrices are conformable for multiplication. For, if A is $p \times q$, B is $q \times r$ and C is $r \times s$, then

$$(AB)C = \left\{ \sum_{k=1}^{r} \left(\sum_{j=1}^{q} a_{ij} b_{jk} \right) c_{kh} \right\}$$

$$= \left\{ \sum_{k=1}^{r} \sum_{j=1}^{q} a_{ij} b_{jk} c_{kh} \right\}$$

$$= \left\{ \sum_{j=1}^{q} a_{ij} \left(\sum_{k=1}^{r} b_{jk} c_{kh} \right) \right\}$$

$$= A(BC),$$

thus proving the associative law of multiplication,

$$(AB)C = A(BC) = ABC.$$

(c) The distributive law holds true also, i.e.,

$$A(B + C) = AB + AC$$

provided both B and C are conformable for addition (necessarily of the same order) and A and B are conformable for multiplication (and hence A and C also). If A is $p \times q$ and B and C are both $q \times r$

$$A(B + C) = \left\{ \sum_{j=1}^{q} a_{ij}(b_{jk} + c_{jk}) \right\}$$

$$= \left\{ \sum_{j=1}^{q} a_{ij} b_{jk} + \sum_{j=1}^{q} a_{ij} c_{jk} \right\} = AB + AC.$$

(d) Addition of matrices is commutative provided the matrices are conformable for addition. If A and B are of the same order

$$A + B = \{a_{ij} + b_{ij}\} = \{b_{ij} + a_{ij}\} = B + A.$$

(e) Multiplication of matrices is not in general commutative; i.e. AB does not always equal BA. As seen earlier, there are two possible products that can be derived from matrices A and B, AB and BA, and if A is of order $r \times c$ both products exist only if B is of order $c \times r$. AB is then square, of order r, and BA is also square, of order c. Possible equality of AB and BA can therefore be considered only where $r = c$, in which case A and B are both square and have the same order r. The products are then

$$AB = \left\{ \sum_{k=1}^{r} a_{ik} b_{kj} \right\} \quad \text{and} \quad BA = \left\{ \sum_{k=1}^{r} b_{ik} a_{kj} \right\}$$

$$\text{for} \quad i \text{ and } j = 1, 2, \ldots, r.$$

The ijth elements of these products do not necessarily have even a single term in common in their sums of products. Therefore, even when AB and BA both exist and are of the same order, they are not in general equal.

This does not mean that AB and BA are never equal, for they can be in particular cases. For example,

$$\begin{bmatrix} 3 & 2 \\ 2 & 3 \end{bmatrix}\begin{bmatrix} -1 & 2 \\ 2 & -1 \end{bmatrix} = \begin{bmatrix} 1 & 4 \\ 4 & 1 \end{bmatrix} = \begin{bmatrix} -1 & 2 \\ 2 & -1 \end{bmatrix}\begin{bmatrix} 3 & 2 \\ 2 & 3 \end{bmatrix},$$

whereas the general non-commutative property of matrix multiplication is illustrated by the example

$$\begin{bmatrix} 1 & 2 \\ 3 & 4 \end{bmatrix}\begin{bmatrix} 0 & -1 \\ 1 & -1 \end{bmatrix} = \begin{bmatrix} 2 & -3 \\ 4 & -7 \end{bmatrix} \neq \begin{bmatrix} 0 & -1 \\ 1 & -1 \end{bmatrix}\begin{bmatrix} 1 & 2 \\ 3 & 4 \end{bmatrix} = \begin{bmatrix} -3 & -4 \\ -2 & -2 \end{bmatrix}.$$

Multiplication of matrices is commutative in two instances of special importance: (i) multiplication involving a null matrix, that is, if A_p is a square matrix of order p, and if 0_p is likewise with all elements zero, then

$$0_p A_p = A_p 0_p = 0_p;$$

(ii) multiplication involving a diagonal matrix having all diagonal elements equal to unity. Such a matrix is called an *identity matrix*, or occasionally a *unit matrix*, and is usually denoted by the letter I, sometimes with a subscript for the order. Thus

$$I_3 = \begin{bmatrix} 1 & 0 & 0 \\ 0 & 1 & 0 \\ 0 & 0 & 1 \end{bmatrix}.$$

In general, if A_p is of order p, then

$$I_p A_p = A_p I_p = A_p.$$

Just as a null matrix is a zero of matrix algebra, so is an identity matrix a "one" or unit of matrix algebra. For example,

$$\begin{bmatrix} 1 & 0 \\ 0 & 1 \end{bmatrix}\begin{bmatrix} 1 & 2 & -4 \\ 9 & 7 & 2 \end{bmatrix} = \begin{bmatrix} 1 & 2 & -4 \\ 9 & 7 & 2 \end{bmatrix}.$$

A matrix of the form λI where λ is a scalar is sometimes called a *scalar matrix*. Thus $4I = \begin{bmatrix} 4 & 0 \\ 0 & 4 \end{bmatrix}$ is a scalar matrix.

For any matrix $A_{r \times s}$, pre- or post-multiplication by a null matrix of appropriate order results in a null matrix. Thus, if $0_{c \times r}$ is a null matrix of order $c \times r$, $0_{c \times r} A_{r \times s} = 0_{c \times s}$. Likewise

$$A_{r \times s} 0_{s \times p} = 0_{r \times p}.$$

For example, $[0 \ \ 0] \begin{bmatrix} 1 & 2 & -4 \\ 9 & 7 & 2 \end{bmatrix} = [0 \ \ 0 \ \ 0].$

Another situation in which the commutative law of multiplication does hold is when a matrix is multiplied by a scalar, λ say, for

$$\lambda A = \{\lambda a_{ij}\} = \{a_{ij}\lambda\} = A\lambda.$$

7. CONCLUSIONS

This discussion of the laws of matrix algebra emphasizes the need for considering the conformability of matrices before attempting the operations of addition, subtraction and multiplication. In practice this must always be so—conformability must always be considered, even if not mentioned explicitly. When the product AB is used we seldom write "AB, where A and B are conformable for multiplication," but conformability is assumed and should be demonstrable at all times.

8. EXERCISES

1. Using rows for seasons of the year (spring, summer, fall, winter) and columns for type of product (dresses, suits and shoes) the sales of a multi-store firm were:

$$\begin{bmatrix} 17 & 4 & 12 \\ 6 & 4 & 13 \\ 11 & 4 & 8 \\ 7 & 4 & 6 \end{bmatrix} \text{ at store 1,} \qquad \begin{bmatrix} 20 & 5 & 10 \\ 10 & 5 & 15 \\ 20 & 5 & 8 \\ 10 & 5 & 10 \end{bmatrix} \text{ at store 2}$$

and

$$\begin{bmatrix} 12 & 3 & 4 \\ 8 & 3 & 4 \\ 10 & 3 & 6 \\ 4 & 3 & 7 \end{bmatrix} \text{ at store 3.}$$

(a) Show that stores 1 and 3 together sell more of every item in every season than does store 2.

(b) Show that the total sales in all three stores is:

$$\begin{bmatrix} 49 & 12 & 26 \\ 24 & 12 & 32 \\ 41 & 12 & 22 \\ 21 & 12 & 23 \end{bmatrix}.$$

2. Mine production of selected minerals is usually classified by principal producing countries. Letting rows denote country 1, 2 and 3, and columns denote mineral 1, 2 and 3, production in thousands of tons for 1960 and 1965 respectively can be represented by

$$A = \begin{bmatrix} 450 & 780 & 210 \\ 1{,}050 & 240 & 90 \\ 1{,}500 & 120 & 590 \end{bmatrix} \quad \text{and} \quad B = \begin{bmatrix} 520 & 910 & 220 \\ 1{,}080 & 580 & 290 \\ 1{,}460 & 830 & 600 \end{bmatrix}.$$

(a) Determine the matrix of increases in production between 1960 and 1965.

(b) Using matrix arithmetic show that the matrix of average production is:

$$\begin{bmatrix} 485 & 845 & 215 \\ 1{,}065 & 410 & 190 \\ 1{,}480 & 475 & 595 \end{bmatrix}$$

3. Show that

(a)
$$\begin{bmatrix} 6 & 3 \\ 0 & 7 \\ -5 & 1 \end{bmatrix} + \begin{bmatrix} 3 & 8 \\ 2 & -1 \\ 6 & -4 \end{bmatrix} = \begin{bmatrix} 12 & 17 \\ -11 & 2 \\ 3 & 9 \end{bmatrix} - \begin{bmatrix} 3 & 6 \\ -13 & -4 \\ 2 & 12 \end{bmatrix}$$

(b)
$$4\begin{bmatrix} 3 & 8 \\ 1 & 9 \end{bmatrix} - 3\begin{bmatrix} 1 & -2 \\ 0 & 1 \end{bmatrix} = 2\begin{bmatrix} 2 & -1 \\ 7 & 4 \end{bmatrix} + 5\begin{bmatrix} 1 & 8 \\ -2 & 5 \end{bmatrix}.$$

4. One part of game theory deals with one-person games against nature, using a matrix called a payoff matrix, whose rows represent possible actions, and whose columns represent possible outcomes (degrees of success or failure), with the corresponding matrix elements representing the payoffs. A criterion used in studying such games is the minimax regret criterion for which a new matrix is formed from the payoff matrix. Every element in column j of the new matrix equals the largest element in column j of the payoff matrix. From this new matrix the payoff matrix is then subtracted, the difference being called the regret matrix. The elements of the regret matrix represent the maximum losses incurred by not choosing the best decision for a particular outcome.

One possible strategy for a player in this type of game is to choose the action corresponding to the row of the regret matrix which contains the minimum of the maximum entries of all the rows. Such a strategy is called a "minimax" strategy, one which minimizes the maximum regret which may occur.

(a) Determine the regret matrix corresponding to the payoff matrix below.

(b) Ascertain the minimax strategy to employ.

$$P = \begin{bmatrix} 0 & 2 & 100 & 0 \\ 7 & 8 & 100 & 2 \\ 10 & 40 & 1 & 90 \\ 1 & 3 & 1 & 1 \end{bmatrix}.$$

5. Another part of game theory deals with two-person games where there are two payoff matrices, one for each person. In both matrices, rows represent player 1's actions and columns represent player 2's actions. For example the two payoff matrices may be

$$P = \begin{bmatrix} 10 & 1 & 8 \\ 6 & 7 & 17 \\ 22 & 4 & 10 \end{bmatrix} \quad \text{and} \quad Q = \begin{bmatrix} 4 & 18 & 12 \\ 8 & 8 & 14 \\ 1 & 10 & 13 \end{bmatrix}.$$

  (a) What combination of actions by the two players would maximize player 1's payoff?
  (b) What combination would maximize player 2's payoff?
  (c) If the players combined as a team, what combination of actions would maximize the sum of their payoffs? What is the payoff matrix in this case?
  (d) What combination of actions would player 1 prefer if he wanted to maximize the difference between his payoff and player 2's payoff?

6. Consider the costs of shipping quantities of a product from a group of plants to a group of destinations. Consider a 3×4 example of a matrix of shipping costs (in dollars), where rows represent plants and columns represent destinations:

$$S = \begin{bmatrix} 10 & 15 & 9 & 7 \\ 14 & 8 & 12 & 8 \\ 6 & 14 & 22 & 17 \end{bmatrix}.$$

Production costs in dollars are 40 at plant 1, 38 at plant 2 and 41 at plant 3.
  (a) Write the matrix of production costs P as a 3×4 matrix with rows and columns defined the same as in S.
  (b) Determine the matrix of total costs (shipping and production combined).

7. For

$$A = \begin{bmatrix} 3 & 6 \\ 2 & 1 \end{bmatrix}, \quad B = \begin{bmatrix} 1 & 0 & 3 & 2 \\ 0 & -1 & -1 & 1 \end{bmatrix}, \quad x = \begin{bmatrix} 1 \\ 1 \\ 0 \\ -1 \end{bmatrix} \quad \text{and} \quad y = \begin{bmatrix} 1 \\ -1 \end{bmatrix}$$

show that

  (a) $AB = \begin{bmatrix} 3 & -6 & 3 & 12 \\ 2 & -1 & 5 & 5 \end{bmatrix},$

  (b) $Bx \quad \begin{bmatrix} -1 \\ -2 \end{bmatrix}, \quad B'Bx = \begin{bmatrix} -1 \\ 2 \\ -1 \\ -4 \end{bmatrix}, \quad \text{if } B' = \begin{bmatrix} 1 & 0 \\ 0 & -1 \\ 3 & -1 \\ 2 & 1 \end{bmatrix}$

(c) $Ay = \begin{bmatrix} -3 \\ 1 \end{bmatrix}$, $A'Ay = \begin{bmatrix} -7 \\ -17 \end{bmatrix}$,   if $A' = \begin{bmatrix} 3 & 2 \\ 6 & 1 \end{bmatrix}$

(d) $A^2 - 4A - 9I = 0$,

(e) $\frac{1}{9}A \begin{bmatrix} -1 & 6 \\ 2 & -3 \end{bmatrix} = \frac{1}{9} \begin{bmatrix} -1 & 6 \\ 2 & -3 \end{bmatrix} A = \begin{bmatrix} 1 & 0 \\ 0 & 1 \end{bmatrix}$.

8. Confirm:

(a) $\begin{bmatrix} 1 & 1 \\ 1 & 1 \end{bmatrix} \begin{bmatrix} 2 & 3 \\ 3 & 2 \end{bmatrix} = \begin{bmatrix} 2 & 3 \\ 3 & 2 \end{bmatrix} \begin{bmatrix} 1 & 1 \\ 1 & 1 \end{bmatrix}$,

(b) that if $A = \frac{1}{3} \begin{bmatrix} 1 & 1 & 1 \\ 1 & 1 & 1 \\ 1 & 1 & 1 \end{bmatrix}$, $A^2 = A$,

(c) that if $C = \begin{bmatrix} 6 & -4 \\ 9 & -6 \end{bmatrix}$, $C^2 = 0$,

(d) $\frac{1}{9} \begin{bmatrix} 4 & -5 & -1 \\ 1 & 1 & 2 \\ 4 & 4 & -1 \end{bmatrix} \begin{bmatrix} 1 & 1 & 1 \\ -1 & 0 & 1 \\ 0 & 4 & -1 \end{bmatrix} = I_3$.

9. For $X = \begin{bmatrix} 1 & 2 & 3 \\ 0 & -1 & -2 \\ -1 & 0 & 7 \end{bmatrix}$ and $Y = \begin{bmatrix} 6 & 0 & 0 \\ -3 & 4 & 0 \\ 0 & -5 & 2 \end{bmatrix}$ find X^2, Y^2, XY

and YX, and show that

$$(X + Y)^2 = X^2 + XY + YX + Y^2 = \begin{bmatrix} 40 & 5 & 44 \\ -28 & 13 & -33 \\ -1 & -62 & 88 \end{bmatrix}.$$

10. Given $A = \begin{bmatrix} 1 & 0 & 2 \\ 0 & 1 & 1 \\ 2 & 0 & 2 \end{bmatrix}$, $B = \begin{bmatrix} 1 & 3 & 0 \\ 0 & 4 & -1 \\ 2 & 3 & 0 \end{bmatrix}$, $X = \begin{bmatrix} 6 & 5 & 7 \\ 2 & 2 & 4 \\ 3 & 3 & 6 \end{bmatrix}$

show that $AX = BX$ even though $A \neq B$.

11. For $A = \{a_{ij}\}$ for $i, j = 1, 2, \ldots, r$,

and $D = \{d_{ij}\}$, $d_{ij} = 0$ for $i \neq j$,   for $i, j = 1, 2, \ldots, r$,

show that $AD = \{d_{jj}a_{ij}\}$ and $DA = \{d_{ii}a_{ij}\}$ for $i, j = 1, 2, \ldots, r$.

12. Bierman (1966) discusses bond refunding under the assumption that changes in interest rates are given by a transition probability matrix such as

$$P = \begin{bmatrix} .6 & .2 & .2 \\ .3 & .5 & .2 \\ .1 & .3 & .6 \end{bmatrix},$$

where the rows and columns both refer to interest rates of 5%, 6% and 7% respectively.

(a) The vector of current probabilities of each interest rate in period n, x_n', is determined by post-multiplying the probability vector in period 1, x_1', by P^{n-1}. Suppose the initial interest rate is 6%; x_1' then equals [0 1 0]. What are the probability vectors for the next two succeeding periods?

(b) Verify that

$$x_3' = x_2'P = x_1'P^2.$$

(c) The particular transition probability matrix P given above has a special form, in which columns as well as rows sum to unity. Such a matrix is said to be *doubly stochastic*. For all such matrices $x_1'P^{n-1}$ tends to a vector of a special form as n tends to infinity. Calculate x_4', x_5' and x_6' and suggest what this special form might be.

(d) What do you think this special form would be if P were of order 10?

13. Consider the following two transition probability matrices in which $p + q = 1$:

$$P = \begin{bmatrix} p^2 & 2pq & q^2 \\ p^2 & 2pq & q^2 \\ p^2 & 2pq & q^2 \end{bmatrix} \quad \text{and} \quad T = \begin{bmatrix} p & q & 0 \\ \frac{1}{2}p & \frac{1}{2} & \frac{1}{2}q \\ 0 & p & q \end{bmatrix}.$$

Show that:

(a) $PT = P = TP$,   (b) $T^2 = \frac{1}{2}(P + T)$,

(c) $P^2 = P$,   (d) $T^n = P + (\frac{1}{2})^{n-1}(T - P)$

and if $S = \frac{1}{4}I + \frac{1}{2}T + \frac{1}{4}P$, then

(e) $ST = TS = T^2$ and   (f) $S^2 = \frac{1}{16}(I + 6T + 9P)$.

14. Referring to Exercise 7 in Chapter 1, assume the following probabilities apply to the demand for tickets:

No. of tickets demanded	1	2	3	4
Probability	.3	.4	.2	.1

The matrix of conditional costs in the aforementioned exercise was

$$\begin{bmatrix} 3 & 3 & 3 & 3 \\ -2 & 6 & 6 & 6 \\ -7 & 1 & 9 & 9 \\ -12 & -4 & 4 & 12 \end{bmatrix}.$$

Multiply this matrix by the above vector of probabilities to obtain the expected cost for each decision alternative. Does it make any difference whether you pre-multiply or post-multiply the matrix by the vector?

15. The expected cash flows resulting from an investment with an economic life of three years are $1,000, $2,000 and $10,000 for the three years, respectively. If the present value of the investment is greater than the required outlay of $5,000 when the expected cash flows are discounted at an interest rate of 10%, the project will be accepted. At 10%, the discount factors for the three years are 0.9091, 0.8264 and 0.7513. Treating the cash flows and the time discount factors as vectors, multiply them to obtain the present value of the investment. Will the investment be accepted?

16. (a) Consider the linear production function

$$q_t = b_1 x_{1t} + b_2 x_{2t}$$

where x_{1t} and x_{2t} are factor inputs in period t, b_1 and b_2 are technological coefficients and q_t is output in period t. Letting $q' = \{q_t\}$ be the row vector of outputs in years $t = 1, \ldots, n$, $X = \{x_{it}\}$ be the matrix of inputs, $i = 1, 2$, with each row corresponding to a particular input and each column to a particular year, and $b' = \{b_i\}$ be the row vector expressing the technological relationship between inputs and outputs, write the matrix expression for q in terms of X and b.

(b) Suppose $n = 3$ and the following observations are given:

$$X = \begin{bmatrix} 320 & 360 & 290 \\ 700 & 780 & 800 \end{bmatrix} \quad \text{and} \quad b' = [3 \quad 5].$$

Evaluate q', the vector of outputs.

REFERENCES

Bierman, H., Jr. (1966). The bond refunding decision as a Markov process. *Management Science*, **12**, 545–555.

Deming, W. E., and G. J. Glasser (1968). A Markovian analysis of the life of newspaper subscriptions. *Management Science*, **14**, B, 283–293.

Derman, C. (1963). Optimal replacement and maintenance under Markovian deterioration with probability bounds on failure. *Management Science*, **9**, 478–481.

CHAPTER 3

OTHER MATRIX OPERATIONS

---

1. LINEAR TRANSFORMATIONS

In the previous chapter we have seen that the product Ax of a matrix A and a vector x is itself a vector. Suppose this product vector is called y. Then $y = Ax$. By the nature of the multiplying process we see that every element of y is a sum of elements of x, each multiplied by an element of A. For example,

$$\begin{bmatrix} y_1 \\ y_2 \end{bmatrix} = y = Ax = \begin{bmatrix} a_{11} & a_{12} & a_{13} \\ a_{21} & a_{22} & a_{23} \end{bmatrix} \begin{bmatrix} x_1 \\ x_2 \\ x_3 \end{bmatrix}$$

gives $\qquad\qquad y_1 = a_{11}x_1 + a_{12}x_2 + a_{13}x_3$

and $\qquad\qquad y_2 = a_{21}x_1 + a_{22}x_2 + a_{23}x_3.$

Thus, so long as not all of the terms a_{11}, a_{12} and a_{13} are zero, the element y_1 is a weighted sum of the elements of x, with the weights being given by the elements in the first row of A. Such a weighted sum is called a *linear combination* of the elements of x; it is linear because the elements of x are not raised to powers other than unity, and no term contains more than one element of x. Thus in the product $y = Ax$ the elements of y are linear combinations of those of x; i.e. by the multiplication process the vector x has been transformed into the vector y. The matrix A in this situation is said to represent the *linear transformation* of x into y. It is the operational means by which the elements of x are transformed into the elements of y.

The idea of a linear transformation of one set of variables into another (that is, of one vector into another) arises in a variety of circumstances. A problem in determining production requirements will serve as an illustration.

[*41*]

Illustration. Parts requirements for each of three types of toys are given in Table 1. Suppose we let x_i represent the number of toys of type i, for

TABLE 1.   PARTS REQUIREMENTS FOR TOYS
(NUMBER OF PARTS)

Parts	Type of Toy		
	Toy 1	Toy 2	Toy 3
Part 1 (Wheels)	4	6	8
Part 2 (Axles)	2	2	3
Part 3 (Bodies)	1	1	1

$i = 1$, 2 and 3, requested in a sales order; and let y_j represent the total number of parts of variety j, for $j = 1$, 2 and 3, needed to assemble the toys so as to fill the sales order. Then the total parts requirements can be stated in equation form as

$$y_1 = 4x_1 + 6x_2 + 8x_3$$
$$y_2 = 2x_1 + 2x_2 + 3x_3 \qquad\qquad (1)$$
$$y_3 = x_1 + x_2 + x_3.$$

Using matrix notation these equations are

$$\begin{bmatrix} y_1 \\ y_2 \\ y_3 \end{bmatrix} = \begin{bmatrix} 4 & 6 & 8 \\ 2 & 2 & 3 \\ 1 & 1 & 1 \end{bmatrix} \begin{bmatrix} x_1 \\ x_2 \\ x_3 \end{bmatrix} \qquad \text{or} \quad y = Ax, \qquad (2)$$

so defining A. Thus the vector of toys is transformed into the vector of required parts by pre-multiplication by the matrix A; this matrix is said to represent the linear transformation of the vector x into the vector y, and the linear transformation is written as

$$y = Ax.$$

A common use of the general linear transformation $y = Ax$ is that characteristics of y can be derived from those of x. To continue the above illustration, consider the raw material requirements of plastic and steel for producing each of the parts as presented in Table 2. Then, if m_k for $k = 1$ and 2 represents the total amount of material of type k needed to make y_1 wheels, y_2 axles and y_3 bodies, we have the equations

$$m_1 = .5y_1 + 0y_2 + 3y_3$$
and
$$m_2 = 0y_1 + 1y_2 + 1y_3. \qquad\qquad (3)$$

TABLE 2.   RAW MATERIAL REQUIREMENTS FOR PARTS
(IN POUNDS)

Material	Type of Part		
	Part 1 (Wheels)	Part 2 (Axles)	Part 3 (Bodies)
Material 1 (Plastic)	.5	0	3
Material 2 (Steel)	0	1	1

Suppose we wish to find the raw material requirements for the parts needed for the sales order. These requirements could be derived by substituting into equations (3) the values for y_1, y_2 and y_3 given by equations (1). However, writing (3) in matrix notation as

$$\begin{bmatrix} m_1 \\ m_2 \end{bmatrix} = \begin{bmatrix} .5 & 0 & 3 \\ 0 & 1 & 1 \end{bmatrix} \begin{bmatrix} y_1 \\ y_2 \\ y_3 \end{bmatrix} \tag{4}$$

enables this substitution to be carried out in matrix form. Substituting (2) into (4) gives

$$\begin{bmatrix} m_1 \\ m_2 \end{bmatrix} = \begin{bmatrix} .5 & 0 & 3 \\ 0 & 1 & 1 \end{bmatrix} \begin{bmatrix} 4 & 6 & 8 \\ 2 & 2 & 3 \\ 1 & 1 & 1 \end{bmatrix} \begin{bmatrix} x_1 \\ x_2 \\ x_3 \end{bmatrix} \tag{5}$$

and carrying out the matrix multiplication on the right-hand side leads to the result

$$\begin{bmatrix} m_1 \\ m_2 \end{bmatrix} = \begin{bmatrix} 5 & 6 & 7 \\ 3 & 3 & 4 \end{bmatrix} \begin{bmatrix} x_1 \\ x_2 \\ x_3 \end{bmatrix} . \tag{6}$$

The effort involved in this derivation is less than that of making the direct algebraic substitutions of equations (1) into (3), and would be considerably less in a situation concerned with more variables than those of this illustration. The major effort required, that of reducing (5) to (6) by performing the matrix multiplication, is solely arithmetical—effort that is nowadays easily provided by high-speed computers, even when dealing with large numbers of variables.

The illustration just given exemplifies the general result that if $y = Ax$ and $x = Bw$ then $y = ABw$. This is true for any vectors x, y and w and any matrices A and B.

2. THE TRANSPOSE OF A MATRIX

Illustration. In first defining a matrix (Section 1.2) we considered the average price of automobiles of four different ages in three different years. The prices were arrayed as a 4×3 matrix

$$A = \begin{bmatrix} 1,881 & 2,120 & 2,445 \\ 1,512 & 1,676 & 1,825 \\ 1,261 & 1,397 & 1,484 \\ 1,054 & 1,144 & 1,218 \end{bmatrix}$$

where rows represent ages and columns represent years. This choice of rows is quite arbitrary as far as presentation of the prices of automobiles is concerned, and we could just as well have interchanged rows and columns with no loss of information, having rows for years and columns for ages. The array of the prices would then be

$$B = \begin{bmatrix} 1,881 & 1,512 & 1,261 & 1,054 \\ 2,120 & 1,676 & 1,397 & 1,144 \\ 2,445 & 1,825 & 1,484 & 1,218 \end{bmatrix}.$$

Although the elements of this matrix are the same as those of the first, according to the definition of matrix equality the two matrices are not the same—in fact, they are not even of the same order. They are related through the rows of one being the columns of the other, and whereas the first has order 4×3 the order of the second is 3×4. When matrices are related in this fashion each is said to be the *transpose* of the other; for example, B is said to be the transpose of A, and A is the transpose of B.

In general, the transpose of a matrix A is the matrix whose columns are the rows of A, with order retained, from first to last. The transpose is written as A'. The rows of A' are the same as the columns of A, and if A is $r \times c$, the order of A' is $c \times r$. If a_{ij} is the term in the ith row and the jth column of A it is also the term in the jth row and the ith column of A'. Hence if

$$A = \{a_{ij}\}$$

its transpose is $\qquad A' = \{a_{ij}\}' = \{a_{ji}\},$

and if we define a'_{ij} as in $A' = \{a'_{ij}\}$ then $a'_{ij} = a_{ji}$, for $i = 1, 2, \ldots, c$ and $j = 1, 2, \ldots, r$.

A matter of notation arises here. $A_{r \times c}$ has r rows and c columns. But combining this notation with that of the transpose $A'_{r \times c}$ is ambiguous, for it does not clearly indicate whether A is of order $r \times c$ or whether A' has that order, thereby implying that A has order $c \times r$. For clarity, one of the equivalent forms $(A_{r \times c})'$ or $(A')_{c \times r}$ must be used whenever it is necessary to have subscript notation for the order of a transposed matrix.

Properties and consequences of the transpose operation are now discussed.

a.   A reflexive operation

The transpose operation is reflexive; that is, the transpose of a transposed matrix is the matrix itself, i.e. $(A')' = A$. This may be shown as follows:

$$(A')' = \{a'_{ij}\}' = \{a_{ji}\}' = \{a'_{ji}\} = \{a_{ij}\} = A.$$

Example.

If
$$A = \begin{bmatrix} 3 & 1 \\ -4 & 2 \end{bmatrix}, \qquad A' = \begin{bmatrix} 3 & -4 \\ 1 & 2 \end{bmatrix},$$

then
$$(A')' = \begin{bmatrix} 3 & 1 \\ -4 & 2 \end{bmatrix} = A.$$

b.   Vectors

The transpose of a row vector is a column vector and vice versa. For example, the transpose of

$$x = \begin{bmatrix} 1 \\ 6 \\ 4 \end{bmatrix} \qquad \text{is} \qquad x' = [1 \quad 6 \quad 4].$$

This is consistent with the notation already introduced, of denoting a row vector by a superscript prime. It indicates that a row vector is the transpose of the column vector of the same elements and distinguishes it from that column vector.

c.   Sums

The transpose of a sum of matrices is the sum of the transposed matrices. For, if

$$A + B = C = \{c_{ij}\} = \{a_{ij} + b_{ij}\},$$

then
$$(A + B)' = C' = \{c'_{ij}\} = \{c_{ji}\} = \{a_{ji} + b_{ji}\} = \{a_{ji}\} + \{b_{ji}\}$$

and so
$$(A + B)' = A' + B'.$$

d.   Products

The transpose of a product matrix is the product of the transposed matrices taken in reverse order,[1] i.e. $(AB)' = B'A'$.

Example.

$$AB = \begin{bmatrix} 1 & 0 & -1 \\ 2 & -1 & 3 \end{bmatrix} \begin{bmatrix} 1 & 1 & 1 \\ 0 & 2 & 4 \\ 3 & 0 & 7 \end{bmatrix} = \begin{bmatrix} -2 & 1 & -6 \\ 11 & 0 & 19 \end{bmatrix}$$

$$B'A' = \begin{bmatrix} 1 & 0 & 3 \\ 1 & 2 & 0 \\ 1 & 4 & 7 \end{bmatrix} \begin{bmatrix} 1 & 2 \\ 0 & -1 \\ -1 & 3 \end{bmatrix} = \begin{bmatrix} -2 & 11 \\ 1 & 0 \\ -6 & 19 \end{bmatrix} = (AB)'.$$

Consideration of order and conformability for multiplication confirms this result. If A is $r \times s$ and B is $s \times t$ the product $P = AB$ is $r \times t$; i.e., $A_{r \times s}B_{s \times t} = P_{r \times t}$. But A' is $s \times r$ and B' is $t \times s$ and the only product to be derived from these is $B'_{t \times s}A'_{s \times r} = Q_{t \times r}$ say. That $Q = B'A'$ is the transpose of $P = AB$ is also apparent from the definition of multiplication: the ijth term of Q is the inner product of the ith row of B' and the jth column of A', which in turn is the inner product of the ith column of B and the jth row of A, and this by the definition of multiplication is the jith term of P. Hence $Q = P'$, or $B'A' = (AB)'$.

A direct extension of the result for the transpose of the product of two matrices is $(ABCD)' = D'C'B'A'$, and so on.

e.   Symmetric matrices

Illustration. When two variables x_1 and x_2 bear some relationship to each other they are said to be correlated, and the degree of correlation between variables is measured by the *correlation coefficient*. Correlation coefficients range in value from -1.0 to 1.0, with negative values indicating an inverse relationship between the variables (if one increases the other tends to decrease) and positive values representing a direct relationship (if one increases the other tends to increase). The larger the absolute value of the correlation coefficient, the stronger the relationship between the variables, with the

---

[1] *Proof:* Let

$$AB = C = \{c_{ij}\} = \left\{\sum_k a_{ik}b_{kj}\right\}.$$

Then

$$(AB)' = C' = \{c'_{ij}\} = \{c_{ji}\}$$

$$= \left\{\sum_k a_{jk}b_{ki}\right\} = \left\{\sum_k a'_{kj}b'_{ik}\right\}$$

$$= \left\{\sum_k b'_{ik}a'_{kj}\right\} = B'A'.$$

extreme values -1.0 and 1.0 representing perfect linear relationships between the variables.

Darling and Lovell (1965), in a study of the factors influencing investment in inventories, computed the correlation coefficients between all pairs of the following variables: the change in total business inventories (ΔI), capacity utilization of manufacturers (C), and seasonally adjusted aggregate sales (S). The correlation coefficients can be displayed as in Table 3:

TABLE 3.   CORRELATION COEFFICIENTS

	ΔI	C	S
ΔI	1.00	0.27	0.11
C	0.27	1.00	0.69
S	0.11	0.69	1.00

and the corresponding matrix is

$$\begin{bmatrix} 1.00 & 0.27 & 0.11 \\ 0.27 & 1.00 & 0.69 \\ 0.11 & 0.69 & 1.00 \end{bmatrix}.$$

The correlation of capacity utilization with sales, 0.27, is identical to the correlation of sales with capacity utilization; therefore the elements in the matrix form a symmetric pattern around the diagonal. Thus 0.27 appears both as the element in the first row and second column and as the element in the second row and first column. This symmetry is true for all off-diagonal elements, and as a result the elements of any row of the matrix are the same as those of the corresponding column. Consequently the transpose of the matrix equals the matrix itself. Such a matrix is called a *symmetric matrix*; it is, by definition, square.

A square matrix is defined as symmetric when it equals its transpose; that is, A is symmetric when $A = A'$, in which case if it is of order r, $a_{ij} = a_{ji}$, for $i, j = 1, 2, \ldots, r$.

We state three results concerning symmetry.

(i) The product of two symmetric matrices exists only if the matrices are of the same order, and the product is not necessarily symmetric. Thus if $A = A'$ and $B = B'$, A and B are symmetric and hence square; they must be of the same order for their product AB to exist. Assuming they are of the same order, the transpose of their product is

$$(AB)' = B'A' = BA$$

but the product AB is not necessarily symmetric.

Example.

$$\text{With } A = \begin{bmatrix} 1 & 2 \\ 2 & 3 \end{bmatrix} \quad \text{and} \quad B = \begin{bmatrix} 3 & 7 \\ 7 & 6 \end{bmatrix}, \text{ then } AB = \begin{bmatrix} 17 & 19 \\ 27 & 32 \end{bmatrix}$$

and AB is clearly not symmetric.

(ii) The product of a matrix with its transpose is symmetric. For

$$(AA')' = (A')'A' = AA'$$

and

$$(A'A)' = A'(A')' = A'A.$$

Although both products AA' and $A'A$ are symmetric, they are not necessarily equal.

Example.

$$\text{If} \quad A = \begin{bmatrix} 1 & 0 & 1 \\ 2 & -1 & 3 \\ 0 & 0 & 4 \end{bmatrix},$$

$$AA' = \begin{bmatrix} 1 & 0 & 1 \\ 2 & -1 & 3 \\ 0 & 0 & 4 \end{bmatrix} \begin{bmatrix} 1 & 2 & 0 \\ 0 & -1 & 0 \\ 1 & 3 & 4 \end{bmatrix} = \begin{bmatrix} 2 & 5 & 4 \\ 5 & 14 & 12 \\ 4 & 12 & 16 \end{bmatrix},$$

$$\text{and} \quad A'A = \begin{bmatrix} 1 & 2 & 0 \\ 0 & -1 & 0 \\ 1 & 3 & 4 \end{bmatrix} \begin{bmatrix} 1 & 0 & 1 \\ 2 & -1 & 3 \\ 0 & 0 & 4 \end{bmatrix} = \begin{bmatrix} 5 & -2 & 7 \\ -2 & 1 & -3 \\ 7 & -3 & 26 \end{bmatrix}.$$

As noted, both AA' and $A'A$ are symmetric, but they are not equal.

(iii) A row vector post-multiplied by a column vector of the same order yields a scalar and is therefore symmetric; i.e., $x'y = (x'y)' = y'x$.

Example. If $x' = \begin{bmatrix} 1 & 2 & 3 \end{bmatrix}$ and $y' = \begin{bmatrix} 4 & 3 & 7 \end{bmatrix}$, $x'y = 31 = y'x$.

3. QUADRATIC FORMS

With vectors

$$x' = \begin{bmatrix} x_1 & x_2 & x_3 \end{bmatrix} \quad \text{and} \quad y' = \begin{bmatrix} y_1 & y_2 & y_3 \end{bmatrix}$$

where the subscripted x's and y's are scalars, let us consider the product

$x'Ay$ where A is some 3×3 matrix. For example,

$$x'Ay = [x_1 \quad x_2 \quad x_3] \begin{bmatrix} 1 & 2 & 3 \\ 4 & 7 & 6 \\ 2 & -2 & 5 \end{bmatrix} \begin{bmatrix} y_1 \\ y_2 \\ y_3 \end{bmatrix}$$

$$= [x_1 + 4x_2 + 2x_3 \quad 2x_1 + 7x_2 - 2x_3 \quad 3x_1 + 6x_2 + 5x_3] \begin{bmatrix} y_1 \\ y_2 \\ y_3 \end{bmatrix}$$

$$= x_1 y_1 + 4x_2 y_1 + 2x_3 y_1 + 2x_1 y_2 + 7x_2 y_2$$
$$- 2x_3 y_2 + 3x_1 y_3 + 6x_2 y_3 + 5x_3 y_3.$$

This is a second degree function of the first degree in each of the x's and y's, since each term contains the product of one x with one y. It is called a *bilinear form*.

Suppose, though, that y is replaced by x. Then we have

$$x'Ax = x_1^2 + 4x_2 x_1 + 2x_3 x_1 + 2x_1 x_2 + 7x_2^2$$
$$- 2x_3 x_2 + 3x_1 x_3 + 6x_2 x_3 + 5x_3^2, \quad (7)$$

which simplifies to

$$x'Ax = x_1^2 + (4 + 2)x_1 x_2 + (2 + 3)x_1 x_3 + 7x_2^2 + (-2 + 6)x_2 x_3 + 5x_3^2 \quad (8)$$
$$= x_1^2 + 7x_2^2 + 5x_3^2 + 6x_1 x_2 + 5x_1 x_3 + 4x_2 x_3. \quad (9)$$

This is a quadratic function of the x's, and is referred to as a *quadratic form*. Two properties are apparent from its development. First, in equation (7) we see that $x'Ax$ is the sum of products of all possible pairs of the x_i's, each multiplied by an element of A; thus the second term in equation (7), $4x_2 x_1$, is $x_2 x_1$ multiplied by the element of A in the second row and first column. Second, in simplifying (7) to (9) we see in (8) that the coefficient of $x_1 x_2$, for example, is the sum of two elements in A: the one in the first column and second row plus that in the second column and first row. These results are true generally.

If x is a vector of order n with elements x_i for $i = 1, 2, \ldots, n$, and if A is a square matrix of order n with elements a_{ij} for $i, j = 1, 2, \ldots, n$, then

$$x'Ax = \left[\sum_i x_i a_{i1} \quad \sum_i x_i a_{i2} \quad \cdots \quad \sum_i x_i a_{in} \right] x$$

$$= \sum_{j=1}^{n} \left(\sum_i x_i a_{ij} \right) x_j$$

$$= \sum_i \sum_j x_i x_j a_{ij} \quad \text{similar to equation (7),}$$

$$= \sum_i x_i^2 a_{ii} + \sum_i \sum_{j \neq i} x_i x_j a_{ij}$$

and as in (8) and (9) this is

$$x'Ax = \sum_{i}^{n} x_i^2 a_{ii} + \sum_{i=1}^{n} \sum_{j=i+1}^{n} x_i x_j (a_{ij} + a_{ji}). \tag{10}$$

Thus $x'Ax$ is the sum of squares of the elements of x, each square multiplied by the corresponding diagonal element of A, plus the sum of products of the elements of x, each product multiplied by the sum of the corresponding elements of A; i.e., the product of the ith and jth element of x is multiplied by $(a_{ij} + a_{ji})$.

Returning to the example, notice that just as (9) was derived from (8) so also can (9) be written in terms of (10) as

$$x'Ax = x_1^2 + 7x_2^2 + 5x_3 + x_1 x_2(1 + 5) + x_1 x_3(1 + 4) + x_2 x_3(0 + 4).$$

In this way we see that

$$x'Ax = x' \begin{bmatrix} 1 & 2 & 3 \\ 4 & 7 & 6 \\ 2 & -2 & 5 \end{bmatrix} x \quad \text{is the same as} \quad x'Bx = x' \begin{bmatrix} 1 & 1 & 1 \\ 5 & 7 & 0 \\ 4 & 4 & 5 \end{bmatrix} x,$$

where B is different from A. Note that the quadratic forms are identical although the associated matrices are not the same. In fact, there is no unique matrix A for which any particular quadratic form can be expressed as $x'Ax$. Many matrices can be so used. Each must have the same diagonal elements, and in each of them the sum of each pair of symmetrically placed off-diagonal elements a_{ij} and a_{ji} must be the same; for example, equation (9) can also be expressed as

$$x'Ax = x' \begin{bmatrix} 1 & 2{,}342 & -789 \\ -2{,}336 & 7 & 1.37 \\ 794 & 2.63 & 5 \end{bmatrix} x. \tag{11}$$

In particular, if we rewrite (9) as

$$x'Ax = x_1^2 + 7x_2^2 + 5x_3^2 + x_1 x_2(3 + 3) + x_1 x_3(2.5 + 2.5) + x_2 x_3(2 + 2)$$

we see that it can be expressed as

$$x'Ax = x' \begin{bmatrix} 1 & 3 & 2.5 \\ 3 & 7 & 2 \\ 2.5 & 2 & 5 \end{bmatrix} x \tag{12}$$

where A is now a symmetric matrix. As such it is unique; for any quadratic form there is a unique symmetric matrix A for which the quadratic form can be expressed as $x'Ax$. It can be found in any particular case by rewriting the quadratic $x'Ax$ where A is not symmetric as $x'[\frac{1}{2}(A + A')]x$, for $\frac{1}{2}(A + A')$

is symmetric. For example, if A is the matrix used in equation (11), then $\frac{1}{2}(A + A')$ is the symmetric matrix used in (12).

Taking A as symmetric with $a_{ij} = a_{ji}$, we see from equation (10) that the quadratic form $x'Ax$ can be expressed as

$$x'Ax = \sum_{i=1}^{n} x_i^2 a_{ii} + 2 \sum_{i=1}^{n} \sum_{j=i+1}^{n} x_i x_j a_{ij}.$$

For example, if A is symmetric,

$$x'Ax = x' \begin{bmatrix} a_{11} & a_{12} & a_{13} \\ a_{12} & a_{22} & a_{23} \\ a_{13} & a_{23} & a_{33} \end{bmatrix} x$$

$$= a_{11}x_1^2 + a_{22}x_2^2 + a_{33}x_3^2 + 2(a_{12}x_1x_2 + a_{13}x_1x_3 + a_{23}x_2x_3).$$

Illustration. For data from a business or economic study we often want to calculate a sample mean and variance. Suppose n observations are represented by the vector $x' = [x_1 \quad x_2 \quad x_3 \quad \cdots \quad x_n]$. Then the mean of the observations is $\bar{x} = \sum_{i=1}^{n} x_i/n$ and the sample variance s^2 is based on the sum of squares (SS) of deviations from the mean, $s^2 = (1/n)\text{SS}$, with

$$\text{SS} = \sum_{i=1}^{n} (x_i - \bar{x})^2 = \sum_{i=1}^{n} x_i^2 - n\bar{x}^2. \tag{13}$$

In matrix notation the first term on the right-hand side of (13) is

$$\sum_{i=1}^{n} x_i^2 = [x_1 \quad x_2 \quad \cdots \quad x_n] \begin{bmatrix} x_1 \\ x_2 \\ \cdot \\ \cdot \\ \cdot \\ x_n \end{bmatrix} = x'x,$$

and the second term is

$$n\bar{x}^2 = n\left(\frac{\Sigma x_i}{n}\right)^2 = (\Sigma x_i)\frac{1}{n}(\Sigma x_i)$$

$$= [x_1 \quad x_2 \quad \cdots \quad x_n] \begin{bmatrix} \frac{1}{n} & \frac{1}{n} & \cdots & \frac{1}{n} \\ & & & \\ \cdot & & \cdot & \\ \cdot & & \cdot & \\ \cdot & & \cdot & \\ \frac{1}{n} & \frac{1}{n} & \cdots & \frac{1}{n} \end{bmatrix} \begin{bmatrix} x_1 \\ x_2 \\ \cdot \\ \cdot \\ \cdot \\ x_n \end{bmatrix} = x'U_n x,$$

where U_n is the square matrix of order n whose every element equals $1/n$. Hence

$$SS = x'x - x'U_nx = x'(I - U_n)x$$

where $(I - U_n)$ is a symmetric matrix of order n, with every diagonal element equal to $(1 - 1/n)$ and all off-diagonal elements equal to $-1/n$.

A quadratic form $x'Ax$ which is positive for all values of x other than $x = 0$ (a null vector) is called a *positive definite quadratic form;* and if $x'Ax$ is either positive or zero for all x other than $x = 0$ (i.e. $x'Ax = 0$ for some $x \neq 0$) then it is a *positive semi-definite quadratic form.* Sometimes the description *non-negative definite* is used instead of positive semi-definite. When $x'Ax$ is positive definite the associated matrix A is called *positive definite*; when $x'Ax$ is positive semi-definite A is *positive semi-definite.* The terms *negative definite* and *negative semi-definite* are used in a corresponding manner.

In the previous illustration the sum of squares SS is an example of a positive semi-definite quadratic form since it can never be negative. (It is not always positive; if every x_i has the same value the sum of squares is zero.) Hence $x'(I - U_n)x$ is a positive semi-definite quadratic form and $I - U_n$ is a positive semi-definite matrix. These terms are used frequently in advanced mathematical statistics, mathematical programming, and related applications in business and economics.

4. PARTITIONING OF MATRICES

Illustration. Suppose a market research study involving three sales regions is carried out to discover the sales potential of a new product. There are two brands of the product (deluxe and standard) and four possible sets of prices. A sales test in a region consists of selling the two brands in four test marketing areas within the region, one area for each set of prices. In one study the number of units sold during a sales test in the first region at the four sets of prices is reported as

$$A_1 = \begin{bmatrix} 180 & 30 & 70 & 90 \\ 150 & 190 & 190 & 70 \end{bmatrix},$$

where rows represent the brands and columns represent the different sets of prices. Sales for the other two test regions are represented by matrices which we shall call A_2 and A_3. Similarly, in another study sales are represented by matrices B_1, B_2 and B_3. Using the matrices A_1, A_2, A_3 and B_1, B_2, B_3, the complete array of these sales data could be presented as a combined matrix

$$C = \begin{bmatrix} A_1 & A_2 & A_3 \\ B_1 & B_2 & B_3 \end{bmatrix}.$$

In terms of the detailed sales, C is a matrix of four rows and twelve columns. However, it is expressed here as consisting of six smaller matrices, each having two rows and four columns. In this form C is referred to as a *partitioned* matrix and the A's and B's as *sub-matrices* of C.

Example. Consider the matrix

$$B = \begin{bmatrix} 1 & 6 & 8 & 9 & 3 & 8 \\ 2 & 4 & 1 & 6 & 1 & 1 \\ 4 & 3 & 6 & 1 & 2 & 1 \\ 9 & 1 & 4 & 6 & 8 & 7 \\ 6 & 8 & 1 & 4 & 3 & 2 \end{bmatrix}.$$

On defining

$$B_{11} = \begin{bmatrix} 1 & 6 & 8 & 9 \\ 2 & 4 & 1 & 6 \\ 4 & 3 & 6 & 1 \end{bmatrix}, \qquad B_{12} = \begin{bmatrix} 3 & 8 \\ 1 & 1 \\ 2 & 1 \end{bmatrix},$$

$$B_{21} = \begin{bmatrix} 9 & 1 & 4 & 6 \\ 6 & 8 & 1 & 4 \end{bmatrix} \quad \text{and} \quad B_{22} = \begin{bmatrix} 8 & 7 \\ 3 & 2 \end{bmatrix},$$

B can be written in partitioned form as

$$\begin{bmatrix} B_{11} & B_{12} \\ B_{21} & B_{22} \end{bmatrix}.$$

B_{11}, B_{12}, B_{21} and B_{22} are the sub-matrices of B.

In this example B_{11} and B_{21} have the same number of columns and so do B_{12} and B_{22}. Likewise B_{11} and B_{12} have the same number of rows, as do B_{21} and B_{22}. This is the usual method of partitioning, as expressed in the general case for an $r \times c$ matrix:

$$A_{r \times c} = \begin{bmatrix} K_{p \times q} & L_{p \times (c-q)} \\ M_{(r-p) \times q} & N_{(r-p) \times (c-q)} \end{bmatrix}$$

where K, L, M and N are the sub-matrices with their orders shown as subscripts.

Partitioning is not restricted to dividing a matrix into just four sub-matrices; it can be divided into numerous rows and columns of matrices. Thus in the previous example if

$$B_{01} = \begin{bmatrix} 1 & 6 & 8 & 9 \\ 2 & 4 & 1 & 6 \end{bmatrix}, \qquad B_{02} = \begin{bmatrix} 3 & 8 \\ 1 & 1 \end{bmatrix},$$

$$B_{11}^* = [4 \quad 3 \quad 6 \quad 1] \quad \text{and} \quad B_{12}^* = [2 \quad 1]$$

with B_{21} and B_{22} as already given, B can be written in partitioned form as

$$B = \begin{bmatrix} B_{01} & B_{02} \\ B_{11}^* & B_{12}^* \\ B_{21} & B_{22} \end{bmatrix}.$$

In general a matrix A, of order $p \times q$, can be partitioned in r rows and c columns of sub-matrices as

$$A = \begin{bmatrix} A_{11} & A_{12} & \cdots & A_{1c} \\ A_{21} & A_{22} & \cdots & A_{2_c} \\ \cdot & \cdot & & \cdot \\ \cdot & \cdot & & \cdot \\ \cdot & \cdot & & \cdot \\ A_{r1} & A_{r2} & \cdots & A_{rc} \end{bmatrix}$$

where A_{ij} is the sub-matrix in the ith row and jth column of the matrix of sub-matrices. If the ith row of sub-matrices has p_i rows of elements and the jth column of sub-matrices has q_j columns, then A_{ij} has order $p_i \times q_j$, where

$$\sum_{i=1}^{r} p_i = p \quad \text{and} \quad \sum_{j=1}^{c} q_j = q, \quad \text{with} \quad r \leq p \quad \text{and} \quad c \leq q.$$

5. MULTIPLICATION OF PARTITIONED MATRICES

The greatest use of matrix partitioning is in matrix multiplication, for if two matrices A and B are partitioned so that their sub-matrices are appropriately conformable for multiplication, the product AB can be expressed in partitioned form having sub-matrices that are functions of the sub-matrices of A and B. For example, if

$$A = \begin{bmatrix} A_{11} & A_{12} \\ A_{21} & A_{22} \end{bmatrix} \quad \text{and} \quad B = \begin{bmatrix} B_{11} \\ B_{21} \end{bmatrix},$$

$$AB = \begin{bmatrix} A_{11} & A_{12} \\ A_{21} & A_{22} \end{bmatrix} \begin{bmatrix} B_{11} \\ B_{21} \end{bmatrix} = \begin{bmatrix} A_{11}B_{11} + A_{12}B_{21} \\ A_{21}B_{11} + A_{22}B_{21} \end{bmatrix},$$

provided the products $A_{11}B_{11}$, $A_{12}B_{21}$, $A_{21}B_{11}$ and $A_{22}B_{21}$ exist, and provided that $A_{11}B_{11}$ and $A_{12}B_{21}$ are conformable for addition, and $A_{21}B_{11}$ and $A_{22}B_{21}$ are also. This implies that A_{11} (and A_{21}) must have the same number of columns as B_{11} has rows, and A_{12} (and A_{22}) must have the same number of columns as B_{21} has rows.

When two matrices are appropriately partitioned the sub-matrices of their product are obtained by treating the sub-matrices of each of them as elements in a normal matrix product, and the individual elements of the product are derived in the usual way from the products of the sub-matrices. If

$$X = \begin{bmatrix} a_{11} & a_{12} & b_{11} & b_{12} & b_{13} \\ a_{21} & a_{22} & b_{21} & b_{22} & b_{23} \\ c_{11} & c_{12} & d_{11} & d_{12} & d_{13} \end{bmatrix}$$

is partitioned in an obvious way as

$$X = \begin{bmatrix} A & B \\ C & D \end{bmatrix}, \quad \text{and likewise} \quad Y = \begin{bmatrix} p_{11} & p_{12} \\ p_{21} & p_{22} \\ q_{11} & q_{12} \\ q_{21} & q_{22} \\ q_{31} & q_{32} \end{bmatrix} \quad \text{as} \quad Y = \begin{bmatrix} P \\ Q \end{bmatrix},$$

then $XY = \begin{bmatrix} AP + BQ \\ CP + DQ \end{bmatrix}$.

Illustration. Feller (1957, page 355) gives a special example of a matrix of transition probabilities $P = \{p_{ij}\}$, where p_{ij} is the probability of moving from state i to state j. In his example the P matrix can be partitioned as

$$P = \begin{bmatrix} A & 0 & 0 \\ 0 & B & 0 \\ U_1 & V_1 & T \end{bmatrix}.$$

In general it may be shown that

$$P^n = \begin{bmatrix} A^n & 0 & 0 \\ 0 & B^n & 0 \\ U_n & V_n & T^n \end{bmatrix},$$

where $U_n = U_1 A^{n-1} + T U_{n-1}$ and $V_n = V_1 B^{n-1} + T V_{n-1}$.

This representation of P^n provides a means of calculating P^n in terms of its sub-matrices.

6.   APPENDIX: VARIANCE-COVARIANCE MATRICES

Matrices important in statistical work are those whose elements are the variances and covariances of a set of random variables. Developing these matrices involves many of the topics discussed in this chapter.

Suppose we are dealing with two random variables x_1 and x_2, whose means are μ_1 and μ_2; i.e., using E to denote expected value, $E(x_1) = \mu_1$ and $E(x_2) = \mu_2$. Writing

$$x = \begin{bmatrix} x_1 \\ x_2 \end{bmatrix} \quad \text{and} \quad \mu = \begin{bmatrix} \mu_1 \\ \mu_2 \end{bmatrix}$$

we have

$$E(x) = E \begin{bmatrix} x_1 \\ x_2 \end{bmatrix} = \begin{bmatrix} \mu_1 \\ \mu_2 \end{bmatrix} = \mu,$$

and the transposed result

$$[E(x)]' = E(x') = \mu'.$$

Consider now the variances defined as

$$\sigma_1^2 = E(x_1 - \mu_1)^2 \quad \text{and} \quad \sigma_2^2 = E(x_2 - \mu_2)^2, \tag{14}$$

and the covariance defined as[1]

$$\sigma_{12} = E\{(x_1 - \mu_1)(x_2 - \mu_2)\}.$$

They may be arrayed in a matrix

$$V = \begin{bmatrix} \sigma_1^2 & \sigma_{12} \\ \sigma_{12} & \sigma_2^2 \end{bmatrix},$$

in which case

$$V = \begin{bmatrix} E(x_1 - \mu_1)^2 & E(x_1 - \mu_1)(x_2 - \mu_2) \\ E(x_1 - \mu_1)(x_2 - \mu_2) & E(x_2 - \mu_2)^2 \end{bmatrix}$$

$$= E\left\{ \begin{bmatrix} x_1 - \mu_1 \\ x_2 - \mu_2 \end{bmatrix} [x_1 - \mu_1 \quad x_2 - \mu_2] \right\}$$

$$= E[x - \mu][x - \mu]'$$

which can be rewritten as

$$V = E[x - \mu][x' - \mu'] = E[x - E(x)][x' - E(x')]. \tag{15}$$

---

[1] The curly brackets in this expression emphasize that the expectation operation applies to the whole product. For the sake of simplicity these brackets are omitted hereafter.

This is the matrix analogue of the definition for the variance of a single variable, as in (14). It applies generally to any number of variables in the vector x. Suppose $x_1, x_2, \ldots, x_n$ represent a set of n random variables with means $\mu_1, \mu_2, \ldots, \mu_n$, variances $\sigma_1^2, \sigma_2^2, \ldots, \sigma_n^2$ and covariances σ_{12}, $\sigma_{13}, \ldots, \sigma_{1n}, \sigma_{23}, \ldots, \sigma_{2n}, \ldots, \sigma_{n-1,n}$. If the random variables are represented by the vector $x' = [x_1 \; x_2 \; \cdots \; x_n]$ and their means by $\mu' = [\mu_1 \; \mu_2 \; \cdots \; \mu_n]$ then $E(x') = \mu'$, or $E(x) = \mu$. Furthermore, the variances and covariances can be written in matrix form as

$$\mathrm{Var}(x) = V = \begin{bmatrix} \sigma_1^2 & \sigma_{12} & \cdots & \sigma_{1n} \\ \sigma_{12} & \sigma_2^2 & \cdots & \sigma_{2n} \\ \cdot & & & \cdot \\ \cdot & & & \cdot \\ \cdot & & & \cdot \\ \sigma_{1n} & \sigma_{2n} & \cdots & \sigma_n^2 \end{bmatrix}.$$

This matrix is symmetric: $V' = V$. Its ith diagonal term is the variance of x_i, and its ijth term, for $i \neq j$, is the covariance between x_i and x_j. The matrix is known as the *variance-covariance matrix* of the elements of the vector x. From the basic definitions of variance and covariance, namely

$$\sigma_i^2 = E(x_i - \mu_i)^2$$

and

$$\sigma_{ij} = E(x_i - \mu_i)(x_j - \mu_j) \quad \text{for} \quad i \neq j,$$

we may write

$$V = E(x - \mu)(x' - \mu'). \tag{16}$$

If the means are zero, $\mu = 0$, and this becomes

$$V = E(xx'). \tag{17}$$

These forms are often utilized in multivariate analysis where they offer a most useful shorthand.

Matrix notation is also valuable in this context when we consider some linear transformation $y = Tx$ applied to the x's, and require the variance-covariance matrix of the elements of y. The vector of the means of the y variables is

$$E(y) = E(Tx) = TE(x) = T\mu$$

with

$$E(y') = (T\mu)' = \mu'T'.$$

The variance-covariance matrix of the y variables is, by analogy with (15)

$$\mathrm{Var}(y) = E[y - E(y)][y' - E(y')],$$

and on substituting $y = Tx$ and $E(y) = T\mu$ and their transposes this becomes

$$
\begin{aligned}
\text{Var}(y) &= E[Tx - T\mu][x'T' - \mu'T'] \\
&= ET[x - \mu][x' - \mu']T' \\
&= TE[x - \mu][x' - \mu']T' \\
&= TVT' \text{ from (16)}.
\end{aligned}
$$

This result,

$$
\text{Var}(y) = \text{Var}(Tx) = T \,\text{Var}(x)T', \tag{18}
$$

is the variance-covariance matrix of any set of linear combinations $y = Tx$, and it applies to any number of linear combinations of any number of variables.

A special case of this result occurs when y is a single variable, i.e., a simple linear function of the x's. T is then a row vector t', and $\text{var}(y) = t'Vt$ is also a scalar.[1] Since $\text{var}(y)$ is a variance, it is always non-negative and therefore so is $t'Vt$. V is therefore a positive semi-definite symmetric matrix. This is true for any variance-covariance matrix.

Example. A useful example of a transformation of variables is that which is often referred to as "normalizing" or "converting to standard scores." We illustrate for the case of two variables x_1 and x_2, with means μ_1 and μ_2, variances σ_1^2 and σ_2^2 and covariance σ_{12}. First, let

$$
y = \begin{bmatrix} y_1 \\ y_2 \end{bmatrix} = \begin{bmatrix} x_1 - \mu_1 \\ x_2 - \mu_2 \end{bmatrix} = x - \mu.
$$

Then

$$
E(y) = E(x - \mu) = 0
$$

and so, using (17)

$$
\text{Var}(y) = E(yy') = E(x - \mu)(x' - \mu') = \text{Var}(x) = V = \begin{bmatrix} \sigma_1^2 & \sigma_{12} \\ \sigma_{12} & \sigma_2^2 \end{bmatrix}.
$$

Now take

$$
T = \begin{bmatrix} 1/\sigma_1 & 0 \\ 0 & 1/\sigma_2 \end{bmatrix}
$$

and

$$
z = Ty = \begin{bmatrix} 1/\sigma_1 & 0 \\ 0 & 1/\sigma_2 \end{bmatrix}\begin{bmatrix} x_1 - \mu_1 \\ x_2 - \mu_2 \end{bmatrix} = \begin{bmatrix} (x_1 - \mu_1)/\sigma_1 \\ (x_2 - \mu_2)/\sigma_2 \end{bmatrix}.
$$

---

[1] Also, if x is a scalar, then t is too, and

$$
\text{var}(y) = \text{var}(tx) = t \,\text{var}(x)t = t^2 \,\text{var}(x).
$$

Then z is the vector of "normalized" x-variables, and from (18)

$$\begin{aligned}
\text{Var}(z) &= T \text{ Var}(y)T' \\
&= TVT' \\
&= \begin{bmatrix} 1/\sigma_1 & 0 \\ 0 & 1/\sigma_2 \end{bmatrix} \begin{bmatrix} \sigma_1^2 & \sigma_{12} \\ \sigma_{12} & \sigma_2^2 \end{bmatrix} \begin{bmatrix} 1/\sigma_1 & 0 \\ 0 & 1/\sigma_2 \end{bmatrix} \\
&= \begin{bmatrix} 1 & \sigma_{12}/\sigma_1\sigma_2 \\ \sigma_{12}/\sigma_1\sigma_2 & 1 \end{bmatrix} = \begin{bmatrix} 1 & \rho \\ \rho & 1 \end{bmatrix}
\end{aligned}$$

where $\rho = \sigma_{12}/\sigma_1\sigma_2$ is the correlation between x_1 and x_2; i.e., the variance-covariance matrix of the normalized variables is the correlation matrix. Note that the variables z_1 and z_2 have means equal to zero and variances equal to unity, and in this sense they are said to be normalized or standardized.

7. EXERCISES

1. If x and y are $n \times 1$ column vectors and A and B are $n \times n$ matrices, which of the following expressions are undefined? Of those that are defined, which are bilinear forms, quadratic forms or linear transformations?

   (a) $y = Ax$ (b) $x' = y'B'$

   (c) $xy = A'B$ (d) $x'Ay$

   (e) $x'Bx$ (f) $y'A'By$

   (g) yBx (h) $xy' = B'$

   (i) $y'B'Ax$ (j) $y + \begin{bmatrix} 30 \\ 40 \end{bmatrix} = AB'x$ for $n = 2$.

2. Use $A = \begin{bmatrix} 1 & 1 \\ 2 & 1 \end{bmatrix}$, $B = \begin{bmatrix} 3 & 6 \\ 4 & 8 \end{bmatrix}$, $x = \begin{bmatrix} 1 \\ 2 \end{bmatrix}$ and $y = \begin{bmatrix} 3 \\ 4 \end{bmatrix}$ in Exercise 1 to

   (a) check the validity of those equations that are defined;
   (b) calculate the value of the other defined expressions.

3. Use $x = \begin{bmatrix} x_1 \\ x_2 \end{bmatrix}$, $y = \begin{bmatrix} y_1 \\ y_2 \end{bmatrix}$ and A and B as given in Exercise 2 to write down

   (after multiplication) expressions (a), (b), (d), (e), (f) and (i) of Exercise 1.

4. For $A = \begin{bmatrix} 3 & 6 \\ 2 & 1 \end{bmatrix}$, $B = \begin{bmatrix} 1 & 0 & 3 & 2 \\ 0 & -1 & -1 & 1 \end{bmatrix}$, $x = \begin{bmatrix} 1 \\ 1 \\ 0 \\ -1 \end{bmatrix}$ and $y = \begin{bmatrix} 1 \\ -1 \end{bmatrix}$

demonstrate that:

(a) $A'B = \begin{bmatrix} 3 & -2 & 7 & 8 \\ 6 & -1 & 17 & 13 \end{bmatrix}$

(b) $[A + A']B = \begin{bmatrix} 6 & 8 \\ 8 & 2 \end{bmatrix} B = \begin{bmatrix} 6 & -8 & 10 & 20 \\ 8 & -2 & 22 & 18 \end{bmatrix} = AB + A'B$

(c) trace of BB' = trace of $B'B = 17$

(d) $x'B'Bx = 5 = (Bx)'Bx$

(e) $y'A'Ay = 10 = (Ay)'Ay$.

5. Demonstrate that for

$$B = \begin{bmatrix} 1/\sqrt{6} & 1/\sqrt{3} & 1/\sqrt{2} \\ -2/\sqrt{6} & 1/\sqrt{3} & 0 \\ 1/\sqrt{6} & 1/\sqrt{3} & -1/\sqrt{2} \end{bmatrix}, \quad y = \begin{bmatrix} y_1 \\ y_2 \\ y_3 \end{bmatrix} \quad \text{and} \quad x = \begin{bmatrix} x_1 \\ x_2 \\ x_3 \end{bmatrix}:$$

(a) $B'B = BB' = I$

(b) $x'B'Bx = (Bx)'Bx = x'(B'B)x = x_1^2 + x_2^2 + x_3^2$ (calculate all three ways)

(c) $x'B'By = x_1y_1 + x_2y_2 + x_3y_3$.

6. For $A = \begin{bmatrix} 3 & 0 & 3 & 2 \\ 0 & -1 & 5 & 0 \\ -1 & 6 & 0 & 8 \end{bmatrix}$ and $B = \begin{bmatrix} -1 & 0 & 8 & 3 \\ 0 & -2 & -7 & -1 \\ -1 & 3 & 0 & 2 \end{bmatrix}:$

find AA', BB', AB' and BA' and verify that

$$(A + B)(A + B)' = (A + B)(A' + B') = AA' + BA' + AB' + BB'$$

$$= AA' + (AB')' + AB' + BB'$$

$$= \begin{bmatrix} 150 & -27 & 46 \\ -27 & 14 & -37 \\ 46 & -37 & 185 \end{bmatrix}.$$

7. Show that $\begin{bmatrix} 3 & 8 & 4 \\ 8 & 7 & -1 \\ 4 & -1 & 2 \end{bmatrix} + \begin{bmatrix} 1 & -1 & 3 \\ -1 & 2 & 4 \\ 3 & 4 & 6 \end{bmatrix}$ is symmetric.

8. With $A = \begin{bmatrix} 1 & 0 & 3 \\ 2 & -1 & 0 \\ 0 & 4 & 1 \end{bmatrix}$ and $B = \begin{bmatrix} 3 & 3 & 1 \\ -4 & 2 & -1 \\ -2 & 0 & 0 \end{bmatrix}:$

(a) Partition A and B as follows:

$$A = \begin{bmatrix} A_{11} & A_{12} \\ A_{21} & A_{22} \end{bmatrix}, \qquad B = \begin{bmatrix} B_{11} & B_{12} \\ B_{21} & B_{22} \end{bmatrix}$$

where both A_{21} and B_{21} have order 1×2.

(b) Calculate AB both with and without the partitioning, to demonstrate the validity of multiplication of partitioned matrices.

(c) Calculate AB', showing that

$$B' = \begin{bmatrix} B'_{11} & B'_{21} \\ B'_{12} & B'_{22} \end{bmatrix}.$$

9. Partition the following matrices in any manner conformable for the products indicated. Calculate the products both with and without using the partitioned forms, to test their validity.

(a) $A = \begin{bmatrix} 1 & 1 & 0 \\ 0 & 1 & 1 \\ 1 & 1 & 1 \end{bmatrix}$, $\quad B = \begin{bmatrix} 2 & 1 \\ 1 & 3 \\ 0 & 1 \end{bmatrix}$, and the product AB.

(b) $C = \begin{bmatrix} 4 & 1 \\ -1 & 1 \end{bmatrix}$, $\quad D = \begin{bmatrix} 1 & 0 & 0 & 4 \\ 0 & 2 & 3 & -5 \end{bmatrix}$, and the product CD.

(c) $E = \begin{bmatrix} 1 & 0 \\ 3 & -1 \end{bmatrix}$, $\quad F = \begin{bmatrix} 0 & 1 & 3 \\ 4 & 4 & -2 \end{bmatrix}$, $\quad G = \begin{bmatrix} 1 & 2 & 3 \\ 4 & -3 & -2 \\ -1 & 0 & -1 \end{bmatrix}$,

and the product EFG.

10. Suppose three random variables x_1, x_2 and x_3 have the following characteristics:

(i) their means are 5, 10 and 8, respectively;

(ii) their variances are 6, 14 and 1, respectively;

(iii) the covariance between x_1 and x_2 is 3, that between x_1 and x_3 is 1 and that between x_2 and x_3 is 0;

(iv) three random variables which are linear combinations of the x's are

$$\begin{aligned} y_1 &= & x_1 + 3x_2 - 2x_3 \\ y_2 &= & 7x_1 - 4x_2 + x_3 \\ y_3 &= & -2x_1 - x_2 + 4x_3. \end{aligned}$$

(a) Write down the vector of means and the variance-covariance matrix of the x's.

(b) Write down the matrix T which represents the linear transformation of the x's into y's.

(c) Use T as obtained in (b) to derive the vector of means and the variance-covariance matrix of the y's.

(*d*) Derive the correlation matrix of the x's using a linear transformation for standardizing the x's.

Note: The appendix to this chapter is used in answering this question.

11. Given
$$(AB)' = B'A',$$
prove
$$(ABC)' = C'B'A'.$$

12. Hass (1968) considers a firm consisting of two divisions, with the first producing product x_1 and the second products y_1 and y_2. The revenue function for the firm is
$$\pi = (4 - x_1)x_1 + (3 - 0.5y_1)y_1 + (0.5x_1 + 3 - 2y_2)y_2.$$

(*a*) Let $z' = [x_1 \ y_1 \ y_2]$.
Find the row vector p' and the symmetric matrix A such that $\pi = p'z - z'Az$.

(*b*) Show by expanding the quadratic form and completing the squares that A is positive definite.

13. Suppose elements of the vectors w, x, y and z represent different types of raw materials, component parts, assemblies and finished products, respectively. Suppose the matrices A, B and C represent the requirements from one level of product to the next lower level, such that $y = Az$, $x = By$, $w = Cx$; i.e., the equations relating the need for component parts (x) to assemblies (y) is $x = By$.

(*a*) Write down the equation describing w as a function of z; i.e., derive the raw material requirements as a function of the finished products.

(*b*) If the amount of excess inventory at each level is represented by the vectors w^0, x^0, y^0 and z^0, write down the matrix equation describing net raw material requirements as a function of finished products.

14. [Notation for this exercise is not consistent with that of the rest of the book but is based, in part, on that of Cohen and Pogue (1967).]

Markowitz (1952) considers the expected return (in percent) over a period of time and the variance of return for a portfolio of n securities. If R_p is the return the expected return is
$$E(R_p) = \sum_{i=1}^{n} X_i A_i$$
where X_i and A_i are scalars: A_i is the expected return on security i and X_i is the proportion of the portfolio invested in security i. Further, if σ_{ij} represents the covariance between the returns for securities i and j when $i \neq j$ (and σ_{ij} is σ_i^2, the variance, when $i = j$) then the variance of the portfolio return is
$$\text{Var}(R_p) = \sum_{i=1}^{n} \sum_{j=1}^{n} X_i \sigma_{ij} X_j.$$
Defining the vectors $X = \{X_i\}$, $A = \{A_i\}$ and the matrix $B = \{\sigma_{ij}\}$ for $i, j = 1, 2, \ldots, n$, write down $E(R_p)$ and $\text{Var}(R_p)$ in terms of X, A and B.

15. (Exercise 14 continued.) If the expected returns and variance-covariance matrix for the three securities in a portfolio are

$$A' = \begin{bmatrix} .10 \\ .06 \\ .05 \end{bmatrix} \quad \text{and} \quad B = \begin{bmatrix} .04 & 0 & -.005 \\ 0 & .01 & 0 \\ -.005 & 0 & .0025 \end{bmatrix}$$

calculate the expected return and variance of the return, $E(R_p)$ and $\text{Var}(R_p)$, respectively, for

(a) $X' = [1 \quad 0 \quad 0]$ (b) $X' = [0 \quad 0 \quad 1]$
(c) $X' = [0 \quad 1 \quad 0]$ (d) $X' = [.1 \quad .5 \quad .4]$.

16. (Exercises 14 and 15 continued.) An efficient portfolio is defined by Markowitz (1952) as being one for which no other portfolio has both a larger expected return and the same or smaller variance *or* the same or larger expected return and a smaller variance. Based on the results of Exercise 15, which of the four alternatives there are not efficient?

17. Sharpe (1963) studies the portfolio selection problem by assuming that the return of various securities is related to an index of market performance, rather than assuming that the covariances among all pairs of securities are known, as does Markowitz (1952) (see Exercises 14–16 above). In presenting the Sharpe model we use a notation that is somewhat different from Sharpe, in order to conform to the general notation of this book. The correspondence between notations is as follows:

Sharpe's notation: X_i R_i A_i B_i C_i I Q_i R_p N
Notation used below: x_i r_i a_i b_i c_i d q_i p n

With this notation, Sharpe's model is that the random variable r_i, the return (in percent) from security i one time period from now, is

$$r_i = a_i + b_i d + c_i, \quad \text{for} \quad i = 1, 2, \ldots, n,$$

and that d, the value of the index of market performance one time period hence, is

$$d = a_{n+1} + c_{n+1}$$

where

$$a' = [a_1 \quad a_2 \quad \cdots \quad a_{n+1}]$$

and

$$b' = [b_1 \quad b_2 \quad \cdots \quad b_n]$$

are vectors of parameters;

$$c' = [c_1 \quad c_2 \quad \cdots \quad c_{n+1}]$$

is a vector of mutually independent random variables having zero mean and diagonal variance-covariance matrix Q, of order $n + 1$; with

$$Q = \begin{bmatrix} q_1 & & & & \\ & q_2 & & & \\ & & \cdot & & \\ & & & \cdot & \\ & & & & \cdot \\ & & & & & q_{n+1} \end{bmatrix},$$

so that q_i is the variance of c_i. Then, because $E(c_{n+1}) = 0$, $E(d) = a_{n+1}$; and $\mathrm{Var}(d) = \mathrm{Var}(c_{n+1}) = q_{n+1}$.

Now let x_i be the proportion of the total portfolio that is invested in the ith security, for $i = 1, \ldots, n$, with $\sum_{i=1}^{n} x_i = 1$. Then the return on the entire portfolio over the next time period is

$$p = \sum_{i=1}^{n} x_i r_i$$

and if we define

$$x_{n+1} = \sum_{i=1}^{n} x_i b_i \quad \text{and} \quad x' = [x_1 \ \ x_2 \ \ \cdots \ \ x_n \ \ x_{n+1}]$$

it can be shown that

$$E(p) = x'a \quad \text{and} \quad \mathrm{Var}(p) = x'Qx.$$

With this formulation consider a situation involving three securities ($n = 3$) with the following characteristics:

$$r_1 = 50 + .8d + c_1, \quad \text{and} \quad q_1 = 100;$$
$$r_2 = 40 + .1d + c_2, \quad \text{and} \quad q_2 = 60;$$
$$r_3 = 20 - .2d + c_3, \quad \text{and} \quad q_3 = 20;$$
$$d = 40 + c_4 \quad \text{and} \quad q_4 = 2500.$$

Write down Q and use the results given above to calculate the expected return and variance of the return for the following two portfolios:

(a) $x_1 = 1$, $x_2 = 0$, $x_3 = 0$.
(b) $x_1 = .5$, $x_2 = .1$, $x_3 = .4$.

18. (Exercise 17 continued.) For $q_i = 100$ and $r_i = 50 + .8d + c_i$, for $i = 1, 2$ and 3 calculate the expected return and variance for the two portfolios given above

(a) with $q_4 = 2500$;
(b) with $q_4 = \quad 25$.

19. (Exercise 17 continued.) Show that

$$E(p) = x'a \quad \text{and} \quad \mathrm{Var}(p) = x'Qx.$$

20. The Cobb-Douglas (1928) production function is one of the most widely used aggregate production functions in macroeconomics. If, in year t, q_t denotes the value of aggregate output, L_t and K_t denote the value of aggregate labor and capital inputs respectively (q_t, L_t and K_t are scalars), and if a_t is also a scalar, then the Cobb-Douglas production function takes the form

$$q_t = a_t L_t^\alpha K_t^\beta$$

where α and β are constants. The form of a_t is assumed to be $a_t = a_0 e^{gt}$, implying that if labor and capital inputs were held constant, output would grow at rate g, reflecting technological change.

Suppose we have observations for T years on L_t and K_t and know the values of a_0, g, α and β. Writing $q' = [\log_e q_1 \quad \cdots \quad \log_e q_T]$

and $$x' = [\log_e a_0 \quad g \quad \alpha \quad \beta],$$

determine the matrix A of the linear transformation $q = Ax$.

REFERENCES

Cobb, C. W., and P. H. Douglas (1928). A theory of production. *American Economic Review*, **18**, 139–165.

Cohen, K. J., and J. A. Pogue (1967). An empirical evaluation of alternative portfolio-selection models. *The Journal of Business*, **40**, 167–193.

Darling, P. G., and M. C. Lovell (1965). Factors influencing investment in inventories. In *The Brookings Quarterly Econometric Model of the U.S.A.* Rand McNally, Chicago, Chapter 4.

Feller, W. (1957). *An Introduction to Probability Theory and Its Applications*. Vol. I, Second Edition, Wiley, New York.

Hass, J. E. (1968). Transfer pricing in a decentralized firm. *Management Science*, **14**, 310–331.

Markowitz, H. M. (1952). Portfolio selection. *The Journal of Finance*, **12**, 77–91.

Sharpe, W. F. (1963). A simplified model for portfolio analysis. *Management Science*, **9**, 277–293.

CHAPTER 4

DETERMINANTS

We now discuss an operation applicable to a square matrix that leads to a scalar value known as the determinant of the matrix. Knowledge of this operation is necessary for our discussion of the counterpart of division in matrix algebra in Chapter 5, and it is also useful in succeeding chapters.

The literature of determinants is extensive but the development in this book is relatively brief and deals only with elementary methods of evaluation.

1. INTRODUCTION

A determinant is a scalar that is the sum of selected products of the elements of the matrix from which it is derived, each product being multiplied by $+1$ or -1 according to well-defined rules. Procedures for deriving these products are developed below, and a formal, rigorous definition is given in the appendix.

Determinants are defined only for square matrices—the determinant of a non-square matrix is undefined and does not exist. The determinant of a square matrix of order n is referred to as an *n-order determinant* and the customary notation for the determinant of the matrix A is $|A|$, where A is square. Some texts use the notation $\|A\|$, $[A]$ or $\det(A)$, but $|A|$ is more common and is used throughout this book. Obtaining the value of $|A|$ by adding the appropriate products of the elements of A (with the correct $+1$ or -1 factor included in the product) is variously referred to as *evaluating* the determinant, *expanding* the determinant or *reducing* the determinant. The procedure for evaluating a determinant will first be illustrated by a series of numerical examples.

The determinant of a 1×1 matrix is the value of its sole element. The value of a second-order determinant is the product of its diagonal terms minus the product of its off-diagonal terms. For example, the determinant of

$$A = \begin{bmatrix} 7 & 3 \\ 4 & 6 \end{bmatrix}$$

is written

$$|A| = \begin{vmatrix} 7 & 3 \\ 4 & 6 \end{vmatrix}$$

and is calculated as $\quad |A| = 7(6) - 3(4) = 30.$

This illustrates the general result for expanding a second-order determinant: it consists of the product (multiplied by $+1$) of the diagonal terms plus the product (multiplied by -1) of the off-diagonal terms. Hence in general

$$|A| = \begin{vmatrix} a_{11} & a_{12} \\ a_{21} & a_{22} \end{vmatrix} = a_{11}a_{22} - a_{12}a_{21}.$$

For brevity we often use the word determinant (or the symbol $|A|$) to refer to a determinant in its arrayed form as well as to the scalar value to which it reduces. Thus if

$$A = \begin{bmatrix} 9 & 3 \\ 7 & 2 \end{bmatrix}$$

the symbol $|A|$ might variously refer to the arrayed form

$$|A| = \begin{vmatrix} 9 & 3 \\ 7 & 2 \end{vmatrix}$$

and to its expanded value

$$|A| = 9(2) - 7(3) = -3.$$

This dual usage is very customary, and is seldom confusing. It is used regardless of the order of a determinant.

A third-order determinant can be expanded as a linear function of three second-order determinants derived from it. Their coefficients are elements of a row (or column) of the main determinant, each product being multiplied by $+1$ or -1. For example, the expansion of

$$|A| = \begin{vmatrix} 1 & 2 & 3 \\ 4 & 5 & 6 \\ 7 & 8 & 10 \end{vmatrix}$$

based on the elements of the first row, 1, 2 and 3, is

$$|A| = 1(+1)\begin{vmatrix} 5 & 6 \\ 8 & 10 \end{vmatrix} + 2(-1)\begin{vmatrix} 4 & 6 \\ 7 & 10 \end{vmatrix} + 3(+1)\begin{vmatrix} 4 & 5 \\ 7 & 8 \end{vmatrix}$$

$$= 1(50 - 48) - 2(40 - 42) + 3(32 - 35) = -3.$$

The determinant is computed by adding the signed products of each element of the chosen row (in this case the first row) with the determinant derived from $|A|$ by crossing out the row and column containing the element concerned. For example, the first element, 1, is multiplied by the determinant $\begin{vmatrix} 5 & 6 \\ 8 & 10 \end{vmatrix}$ which is obtained from $|A|$ by crossing out the first row and first column, and the element 2 is multiplied (apart from the factor of -1) by the determinant derived from $|A|$ by deleting the row and column containing that element—namely, the first row and second column, leaving $\begin{vmatrix} 4 & 6 \\ 7 & 10 \end{vmatrix}$.

Determinants obtained in this way are called *minors* of A; that is to say, $\begin{vmatrix} 5 & 6 \\ 8 & 10 \end{vmatrix}$ is the minor of the element 1 in A; and $\begin{vmatrix} 4 & 6 \\ 8 & 10 \end{vmatrix}$ is the minor of the element 2.

The $(+1)$ and (-1) factors are determined by the following rule: if A is written in the form $A = \{a_{ij}\}$, the product of a_{ij} and its minor in the expansion of the determinant $|A|$ is multiplied by $(-1)^{i+j}$. The element 1 in the example is the element a_{11}; thus the product of a_{11} and its minor is multiplied by $(-1)^{1+1} = +1$. Similarly, the product of the element 2, a_{12}, with its minor is multiplied by $(-1)^{1+2} = -1$.

The minor of an element of a square matrix of order n is necessarily a determinant of order $n - 1$. However, minors are not all of order $n - 1$. Deleting any r rows and any r columns from a square matrix of order n leaves a sub-matrix of order $n - r$; and the determinant of this sub-matrix is a *minor of order* $n - r$, or an $(n - r)$-*order minor*.

2. EXPANSION BY MINORS

Suppose we denote the minor of the element a_{11} by $|M_{11}|$, where M_{11} is a sub-matrix of A obtained by deleting the first row and first column. Then, in the above example $|M_{11}| = \begin{vmatrix} 5 & 6 \\ 8 & 10 \end{vmatrix}$. Similarly, if $|M_{12}|$ is the minor of

a_{12} then $|M_{12}| = \begin{vmatrix} 4 & 6 \\ 7 & 10 \end{vmatrix}$; and if $|M_{13}|$ is the minor of a_{13} then $|M_{13}| = \begin{vmatrix} 4 & 5 \\ 7 & 8 \end{vmatrix}$. With this notation, the expansion of $|A|$ given above is

$$|A| = a_{11}(-1)^{1+1}|M_{11}| + a_{12}(-1)^{1+2}|M_{12}| + a_{13}(-1)^{1+3}|M_{13}|.$$

This method of expanding a determinant is known as *expansion by the elements of a row (or column)* or as *expansion by minors*. It has been illustrated using elements of the first row, but it can also be applied to the elements of any row (or column). For example, the expansion of $|A|$ just considered, using elements of the second row, gives

$$|A| = 4(-1)\begin{vmatrix} 2 & 3 \\ 8 & 10 \end{vmatrix} + 5(+1)\begin{vmatrix} 1 & 3 \\ 7 & 10 \end{vmatrix} + 6(-1)\begin{vmatrix} 1 & 2 \\ 7 & 8 \end{vmatrix}$$
$$= -4(-4) + 5(-11) - 6(-6) = -3$$

as before, and using elements of the first column the expansion yields the same value:

$$|A| = 1(+1)\begin{vmatrix} 5 & 6 \\ 8 & 10 \end{vmatrix} + 4(-1)\begin{vmatrix} 2 & 3 \\ 8 & 10 \end{vmatrix} + 7(+1)\begin{vmatrix} 2 & 3 \\ 5 & 6 \end{vmatrix}$$
$$= 1(2) - 4(-4) + 7(-3) = -3.$$

The minors in these expansions are derived in exactly the same manner as above. For example, the minor of the element 4 is $|A|$ with the second row and first column deleted, and since 4 is a_{21} its product with its minor is multiplied by $(-1)^{2+1} = -1$. Other terms are obtained in a similar manner.

The foregoing example illustrates the expansion of the general third-order determinant

$$|A| = \begin{vmatrix} a_{11} & a_{12} & a_{13} \\ a_{21} & a_{22} & a_{23} \\ a_{31} & a_{32} & a_{33} \end{vmatrix}.$$

Expanding this by elements of the first row gives

$$|A| = a_{11}(+1)\begin{vmatrix} a_{22} & a_{23} \\ a_{32} & a_{33} \end{vmatrix} + a_{12}(-1)\begin{vmatrix} a_{21} & a_{23} \\ a_{31} & a_{33} \end{vmatrix} + a_{13}(+1)\begin{vmatrix} a_{21} & a_{22} \\ a_{31} & a_{32} \end{vmatrix}$$
$$= a_{11}a_{22}a_{33} - a_{11}a_{23}a_{32} - a_{12}a_{21}a_{33} + a_{12}a_{23}a_{31} + a_{13}a_{21}a_{32}$$
$$- a_{13}a_{22}a_{31}.$$

The reader should satisfy himself that expansion by the elements of any other row or column leads to the same result.

No matter by what row or column the expansion is made, the value of the determinant is the same. Note that once a row or column is decided on and the sign calculated for the product of the first element therein with its minor, the signs for the following products alternate from plus to minus and minus to plus.

The expansion of an n-order determinant by this method is an extension of the expansion of a third-order determinant as just given. Thus the determinant of the $n \times n$ matrix $A = \{a_{ij}\}$ for $i, j = 1, 2, \ldots, n$ is obtained as follows. Consider the elements of any one row (or column): multiply each element, a_{ij}, of this row (or column) by its minor, $|M_{ij}|$, the determinant derived from $|A|$ by crossing out the row and column containing a_{ij}; multiply the product by $(-1)^{i+j}$; add the signed products and their sum is the determinant $|A|$. This expansion is used recursively when n is large, i.e. each $|M_{ij}|$ is in turn expanded by the same procedure.

Example. Expansion by elements of the first row of

$$|A| = \begin{vmatrix} 1 & 2 & 3 & 0 \\ 0 & 4 & 5 & 6 \\ 1 & 0 & 2 & 3 \\ 4 & 0 & 2 & 3 \end{vmatrix}$$

gives

$$|A| = 1(-1)^2 |M_{11}| + 2(-1)^3 |M_{12}| + 3(-1)^4 |M_{13}| + 0(-1)^5 |M_{14}|.$$

The needed minors are

$$|M_{11}| = \begin{vmatrix} 4 & 5 & 6 \\ 0 & 2 & 3 \\ 0 & 2 & 3 \end{vmatrix} = 4(-1)^2 \begin{vmatrix} 2 & 3 \\ 2 & 3 \end{vmatrix} + 5(-1)^3 \begin{vmatrix} 0 & 3 \\ 0 & 3 \end{vmatrix}$$

$$+ 6(-1)^4 \begin{vmatrix} 0 & 2 \\ 0 & 2 \end{vmatrix} = 0,$$

$$|M_{12}| = \begin{vmatrix} 0 & 5 & 6 \\ 1 & 2 & 3 \\ 4 & 2 & 3 \end{vmatrix} = 0(1) \begin{vmatrix} 2 & 3 \\ 2 & 3 \end{vmatrix} + 5(-1) \begin{vmatrix} 1 & 3 \\ 4 & 3 \end{vmatrix} + 6(1) \begin{vmatrix} 1 & 2 \\ 4 & 2 \end{vmatrix} = 9$$

and

$$|M_{13}| = \begin{vmatrix} 0 & 4 & 6 \\ 1 & 0 & 3 \\ 4 & 0 & 3 \end{vmatrix} = 0(1) \begin{vmatrix} 0 & 3 \\ 0 & 3 \end{vmatrix} + 4(-1) \begin{vmatrix} 1 & 3 \\ 4 & 3 \end{vmatrix} + 6(1) \begin{vmatrix} 1 & 0 \\ 4 & 0 \end{vmatrix} = 36.$$

Hence $|A| = 1(1)0 + 2(-1)9 + 3(1)36 + 0(-1)|M_{14}| = 90.$

Expansion in similar manner by elements of some other row or column leads to the same result. Note that in this case $|M_{14}|$ is not calculated because the element a_{14} is zero. By judicious choice of the row (or column) upon which a determinant will be expanded, arithmetic efforts can be minimized. Methods that reduce the arithmetic still further are discussed in the next section.

When expanding an n-order determinant by elements of a row we may write

$$|A| = \sum_{j=1}^{n} a_{ij}(-1)^{i+j}|M_{ij}| \qquad \text{for any } i, \qquad (1)$$

and when expanding by elements of a column

$$|A| = \sum_{i=1}^{n} a_{ij}(-1)^{i+j}|M_{ij}| \qquad \text{for any } j. \qquad (2)$$

Thus, as shown in the previous example, a fourth-order determinant is first expanded as four signed products each involving a third-order minor, and each of these is expanded as a sum of three signed products involving a second-order determinant. Consequently a fourth-order determinant ultimately involves $4 \times 3 \times 2 = 24$ products of its elements, each product containing four elements. This leads us to the general statement that the determinant of a square matrix of order n is a sum of $n!$ signed products.[1] The determinant is referred to as an *n-order* determinant. Utilizing methods given in Aitken (1948), it can be shown that each product involves n elements of the matrix, that each product has one and only one element from each row and column and that all such products are included and none occur more than once.

This method of evaluating a determinant involves tedious calculations for determinants of order greater than three. Fortunately, easier methods exist, but because the method already discussed forms the basis of these easier methods it has been considered in detail. Furthermore, it is found useful in developing additional properties of determinants.

3. ELEMENTARY EXPANSIONS

a. Determinant of a transpose

As shown in the appendix, the determinant of the transpose of a matrix is the same as the determinant of the matrix: $|A'| = |A|$.

----

[1] $n!$, read as "factorial n," is the product of all integers 1 through n inclusive; e.g. $4! = 1(2)(3)(4) = 24$. Also, $0!$ is defined as unity.

Example.

$$\begin{vmatrix} 1 & -1 & 0 \\ 2 & 1 & 2 \\ 4 & 4 & 9 \end{vmatrix} = \begin{vmatrix} 1 & 2 \\ 4 & 9 \end{vmatrix} + \begin{vmatrix} 2 & 2 \\ 4 & 9 \end{vmatrix} = 1 + 10 = 11,$$

and

$$\begin{vmatrix} 1 & 2 & 4 \\ -1 & 1 & 4 \\ 0 & 2 & 9 \end{vmatrix} = \begin{vmatrix} 1 & 4 \\ 2 & 9 \end{vmatrix} - 2 \begin{vmatrix} -1 & 4 \\ 0 & 9 \end{vmatrix} + 4 \begin{vmatrix} -1 & 1 \\ 0 & 2 \end{vmatrix} = 1 + 18 - 8 = 11.$$

Thus the operation of transposing a matrix does not affect the value of the resulting determinant; i.e., a determinant is not altered if all its rows are interchanged with all its columns. Because of this, three properties of rows of a determinant now to be considered hold true for columns also, but for simplicity's sake they are presented only in terms of rows.

b. Interchanging two rows

Interchanging any two rows (adjacent or otherwise) of a determinant changes the sign of the determinant. Proof of this is given in the appendix.

Example.

$$\begin{vmatrix} 6 & 1 & 0 \\ 3 & -1 & 2 \\ 4 & 0 & -1 \end{vmatrix} = 6 \begin{vmatrix} -1 & 2 \\ 0 & -1 \end{vmatrix} - 1 \begin{vmatrix} 3 & 2 \\ 4 & -1 \end{vmatrix} = 6 + 11 = 17;$$

$$\begin{vmatrix} 4 & 0 & -1 \\ 3 & -1 & 2 \\ 6 & 1 & 0 \end{vmatrix} = 4 \begin{vmatrix} -1 & 2 \\ 1 & 0 \end{vmatrix} - 1 \begin{vmatrix} 3 & -1 \\ 6 & 1 \end{vmatrix} = -8 - 9 = -17.$$

Corollary. If two rows of a determinant are the same the determinant is zero. Interchanging the rows changes the sign as just discussed, but since the rows are the same the value of the determinant must be unaltered. Hence $|A| = -|A|$ so that $|A| = 0$.

c. Factorization

If λ (a scalar) is a factor of a row (i.e. is a factor of every element of the row), it is also a factor of the determinant. Suppose λ is a factor of the kth row. Then the determinant can be expanded by elements of that kth row, and so λ can be factored out from every term in the expansion; i.e. λ is a factor of the determinant. A consequence of this is that the determinant can be written with the factor λ removed from the kth row and expressed as a factor of the determinant.

Example.

$$|A| = \begin{vmatrix} 2 & 10 & 8 \\ 1 & 7 & 3 \\ 6 & 1 & 4 \end{vmatrix} = (2)\begin{vmatrix} 1 & 5 & 4 \\ 1 & 7 & 3 \\ 6 & 1 & 4 \end{vmatrix}.$$

Expanding the first form gives

$$|A| = 2(28 - 3) - 10(4 - 18) + 8(1 - 42) = -138$$

and expanding the second gives

$$|A| = 2[(28 - 3) - 5(4 - 18) + 4(1 - 42)] = -138.$$

When appropriate, this process can be carried out for several rows simultaneously.

Three useful corollaries can be derived from this property of factorization.

(i) **Corollary 1.** If one row of a determinant is a multiple of another row, the determinant is zero. Factoring out the multiple reduces the determinant to having two rows which are the same. Hence it is zero.

Example.

$$\begin{vmatrix} -3 & 6 & 12 \\ 2 & -4 & -8 \\ 7 & 5 & 9 \end{vmatrix} = -(1.5)\begin{vmatrix} 2 & -4 & -8 \\ 2 & -4 & -8 \\ 7 & 5 & 9 \end{vmatrix} = 0.$$

(ii) **Corollary 2.** If a determinant has a row of zeros the value of the determinant is zero. Zero is here a factor of one row and hence a factor of the determinant, which is therefore zero.

Example.

$$\begin{vmatrix} 0 & 0 \\ 3 & 7 \end{vmatrix} = 0.$$

(iii) **Corollary 3.** If A is an $n \times n$ matrix and λ is a scalar, the determinant of the matrix λA is $\lambda^n |A|$; i.e. $|\lambda A| = \lambda^n |A|$. In this case λ is a factor of each of the n rows of λA and when the λ factor is removed from each row the determinant $|A|$ remains.

Example.

$$\begin{vmatrix} 3 & 0 & 27 \\ -9 & 3 & 0 \\ 15 & 6 & -3 \end{vmatrix} = 3^3\begin{vmatrix} 1 & 0 & 9 \\ -3 & 1 & 0 \\ 5 & 2 & -1 \end{vmatrix} = -2,700.$$

d. Adding multiples of a row

Adding to one row of a determinant any multiple of another row does not affect the value of the determinant. For example,

$$|A| = \begin{vmatrix} 1 & 3 & 2 \\ 8 & 17 & 21 \\ 2 & 7 & 1 \end{vmatrix}$$

$$= 1(17 - 147) - 3(8 - 42) + 2(56 - 34) = 16;$$

and adding four times row 1 to row 2 does not affect the value of $|A|$:

$$|A| = \begin{vmatrix} 1 & 3 & 2 \\ 8+4 & 17+12 & 21+8 \\ 2 & 7 & 1 \end{vmatrix} = \begin{vmatrix} 1 & 3 & 2 \\ 12 & 29 & 29 \\ 2 & 7 & 1 \end{vmatrix}$$

$$= 1(29 - 203) - 3(12 - 58) + 2(84 - 58) = -174 + 138 + 52 = 16.$$

The generality of this result for a second-order determinant can be readily demonstrated by constructing from

$$|B| = \begin{vmatrix} a_1 & b_1 \\ a_2 & b_2 \end{vmatrix} \tag{3}$$

the determinant $|C|$ where C is B with λ times row 2 added to the first row, namely

$$|C| = \begin{vmatrix} a_1 + \lambda a_2 & b_1 + \lambda b_2 \\ a_2 & b_2 \end{vmatrix}$$

$$= a_1 b_2 - b_1 a_2 + \lambda a_2 b_2 - \lambda a_2 b_2$$

$$= \begin{vmatrix} a_1 & b_1 \\ a_2 & b_2 \end{vmatrix} + \lambda \begin{vmatrix} a_2 & b_2 \\ a_2 & b_2 \end{vmatrix}.$$

The determinant multiplying λ is zero since it has two rows which are the same, and therefore $|C| = |B|$ as in (3).

This property is used repeatedly in expanding determinants, adding both positive and negative multiples of rows to other rows. Thus in the previous example, where

$$|A| = \begin{vmatrix} 1 & 3 & 2 \\ 8 & 17 & 21 \\ 2 & 7 & 1 \end{vmatrix}, \tag{4}$$

adding (-8) times the first row to the second does not affect $|A|$ and gives

$$|A| = \begin{vmatrix} 1 & 3 & 2 \\ 0 & -7 & 5 \\ 2 & 7 & 1 \end{vmatrix}. \tag{5}$$

If we now add (-2) times the first row to the third row we get

$$|A| = \begin{vmatrix} 1 & 3 & 2 \\ 0 & -7 & 5 \\ 0 & 1 & -3 \end{vmatrix}. \tag{6}$$

Expansion by elements of the first column is now straightforward because two elements of the column are zero:

$$|A| = 1 \begin{vmatrix} -7 & 5 \\ 1 & -3 \end{vmatrix} = 16,$$

the same result as before. By repeated use of the process of adding a multiple of one row to another we can reduce one column (not necessarily the first) so that it has all elements except one equal to zero; expansion by elements of that column then involves only one minor, which is a determinant of order one less than the original one. This in turn can be reduced also, and so on until the original determinant is a multiple of a single 2×2 determinant. At each stage the process of adding multiples of one row to the other rows can be combined into one operation; e.g. in the previous example the steps represented by (5) and (6) can be made simultaneously and the form (6) written directly from (4).

The process can be started with any row; for example, consider basing the reduction of the foregoing example on its third row. If -21 times row 3 is added to row 2 and -2 times row 3 is added to row 1 we get

$$|A| = \begin{vmatrix} 1 & 3 & 2 \\ 8 & 17 & 21 \\ 2 & 7 & 1 \end{vmatrix} = \begin{vmatrix} -3 & -11 & 0 \\ -34 & -130 & 0 \\ 2 & 7 & 1 \end{vmatrix}$$

so that expansion by elements of the third column gives

$$|A| = \begin{vmatrix} -3 & -11 \\ -34 & -130 \end{vmatrix} = 390 - 374 = 16$$

as before. The choice of which row is used and which column is simplified is dependent on observing which might lead to the simplest arithmetic. For example, in expanding

$$D = \begin{vmatrix} 13 & 5 & 17 \\ 4 & 1 & 3 \\ 11 & 3 & 7 \end{vmatrix}$$

the arithmetic will be easier if multiples of the second row are added to the other rows to put zero in the first and third elements of the second column than if multiples of either the second or third row are added to the others.

The foregoing properties can be applied in endless variation in expanding determinants, with efficiency in perceiving a procedure that leads to a minimal amount of effort in any particular case being largely a matter of practice. The underlying method might be summarized as follows. Through adding multiples of one row to other rows of the determinant a column is reduced to having only one non-zero element. Expansion by elements of that column then involves only one non-zero product of an element with its minor, which is a determinant of order one less than the original determinant. Successive applications of this method reduce the determinant to a multiple of one of order 2×2. If at any stage these reductions lead to the elements of a row containing a common factor, this can be factored out as a factor of the determinant, and if they lead to a row of zeros or to two rows being identical, the determinant is zero.

e. Adding a row to a multiple of a row

We have seen that adding a multiple of a row to another row does not affect the value of a determinant. Thus for

$$|A| = \begin{vmatrix} a_1 & b_1 & c_1 \\ a_2 & b_2 & c_2 \\ a_3 & b_3 & c_3 \end{vmatrix} \quad \text{and} \quad |B| = \begin{vmatrix} a_1 + \lambda a_2 & b_1 + \lambda b_2 & c_1 + \lambda c_2 \\ a_2 & b_2 & c_2 \\ a_3 & b_3 & c_3 \end{vmatrix}$$

we have shown that $|B| = |A|$. Notice, however, that adding a row to a multiple of another row is not the same thing and leads to a different result. For example, adding row 2 to λ times row 1 gives the determinant

$$|C| = \begin{vmatrix} \lambda a_1 + a_2 & \lambda b_1 + b_2 & \lambda c_1 + c_2 \\ a_2 & b_2 & c_2 \\ a_3 & b_3 & c_3 \end{vmatrix} = \lambda |A|.$$

This is because $|C|$ is derived from $|A|$ by multiplying the first row of $|A|$ by λ and adding the second row to it. The first of these two steps changes $|A|$ to $\lambda |A|$, as shown above, and the second does not alter this value. Hence $|C| = \lambda |A|$.

Thus while adding a multiple of a row to a row does not affect a determinant, adding a row to a multiple of a row has the effect of multiplying the determinant by the factor involved.

Example.

$$\begin{vmatrix} -2 & 1 \\ 4 & 3 \end{vmatrix} = \begin{vmatrix} -2 & 1 \\ 0 & 5 \end{vmatrix} = -10,$$

but

$$\begin{vmatrix} 2(-2) + 4 & 2(1) + 3 \\ 4 & 3 \end{vmatrix} = \begin{vmatrix} 0 & 5 \\ 4 & 3 \end{vmatrix} = 2(-10).$$

4. ADDITION AND SUBTRACTION OF DETERMINANTS

In general, the sum of two determinants (or difference between them) cannot be written as a single determinant. The simplest demonstration of this is in the case of second-order determinants:

$$|A| + |B| = \begin{vmatrix} a_{11} & a_{12} \\ a_{21} & a_{22} \end{vmatrix} + \begin{vmatrix} b_{11} & b_{12} \\ b_{21} & b_{22} \end{vmatrix}$$

$$= a_{11}a_{22} - a_{12}a_{21} + b_{11}b_{22} - b_{12}b_{21}$$

cannot be written in determinantal form other than as $|A| + |B|$, and definitely not as $|A + B|$. The same applies to the difference $|A| - |B|$.

Both $|A| + |B|$ and $|A| - |B|$ have meaning even when A and B are square matrices of different orders, because the value of a determinant is a scalar. This is in contrast to the matrix expressions $A + B$ and $A - B$ which have meaning only when the matrices are conformable for addition (have the same order).

5. MULTIPLICATION OF DETERMINANTS

It can be shown that if A and B are square matrices of the same order, determinants of two partitioned matrices involving them are

$$\begin{vmatrix} O & A \\ -I & B \end{vmatrix} = |A|, \tag{7}$$

and

$$\begin{vmatrix} A & O \\ X & B \end{vmatrix} = |A|\,|B|, \tag{8}$$

where X is any matrix (of the same order as A and B) and O is a null matrix. (Proof of these results is given in the appendix to this chapter, Section 4.7e.)

Consider the determinant $\begin{vmatrix} A & O \\ -I & B \end{vmatrix}$. Adding A times the rows $[-I \quad B]$ to the rows $[A \quad O]$ changes these rows to be $[A \quad O] + A[-I \quad B] = [O \quad AB]$, and the determinant becomes $\begin{vmatrix} O & AB \\ -I & B \end{vmatrix}$. This operation does not alter the value of the determinant. Therefore

$$\begin{vmatrix} O & AB \\ -I & B \end{vmatrix} = \begin{vmatrix} A & O \\ -I & B \end{vmatrix}. \tag{9}$$

But by (7) the left-hand side of (9) is $|AB|$, and by (8) the right-hand side of (9) is $|A|\,|B|$. Therefore

$$|AB| = |A|\,|B|.$$

Furthermore, because $|A|\,|B| = |B|\,|A| = |BA|$, we have

$$|AB| = |BA| = |A|\,|B|.$$

Hence we have the important result that the determinant of the product of two square matrices (of the same order) is the product of the determinants of the individual matrices.

Example.

If
$$A = \begin{bmatrix} 1 & 3 \\ 2 & 8 \end{bmatrix} \quad \text{and} \quad B = \begin{bmatrix} 2 & 3 \\ 5 & 11 \end{bmatrix},$$

$$AB = \begin{bmatrix} 17 & 36 \\ 44 & 94 \end{bmatrix} \quad \text{and} \quad BA = \begin{bmatrix} 8 & 30 \\ 27 & 103 \end{bmatrix}.$$

Then $|A| = 2$, $|B| = 7$ and $|AB| = |A|\,|B| = |B|\,|A| = |BA| = 14$.

6. DIAGONAL EXPANSION

Evaluation of a determinant can sometimes be simplified by expressing its matrix as the sum of two matrices, one of which is a diagonal matrix; i.e., as $(A + D)$ where $A = \{a_{ij}\}$ for $i, j = 1, 2, \ldots, n$ and D is a diagonal matrix of order n. (Recall from Section 1.6 that a diagonal matrix is a square matrix having all non-diagonal elements zero.) We show that the determinant of $(A + D)$ can be expressed as a polynomial of the elements of D.

First we introduce abbreviations for minors. Writing

$$|A| = \begin{vmatrix} a_{11} & a_{12} & a_{13} & \cdots & a_{1n} \\ a_{21} & a_{22} & a_{23} & \cdots & a_{2n} \\ a_{31} & a_{32} & a_{33} & \cdots & a_{3n} \\ \cdot & \cdot & \cdot & & \cdot \\ \cdot & \cdot & \cdot & & \cdot \\ \cdot & \cdot & \cdot & & \cdot \\ a_{n1} & a_{n2} & a_{n3} & \cdots & a_{nn} \end{vmatrix},$$

minors will be denoted by just their diagonal elements; for example, $\begin{vmatrix} a_{11} & a_{12} \\ a_{21} & a_{22} \end{vmatrix}$

is written as $|a_{11} \quad a_{22}|$ and in similar fashion $\begin{vmatrix} a_{12} & a_{13} \\ a_{22} & a_{23} \end{vmatrix}$ is written as $|a_{12} \quad a_{23}|$.

Likewise $|a_{21} \quad a_{32}|$ denotes the second-order (2×2) minor having a_{21} and a_{32} as diagonal elements, and from $|A|$ we see that the elements in the same rows and columns as these are a_{22} and a_{31}, so that

$$|a_{21} \quad a_{32}| = \begin{vmatrix} a_{21} & a_{22} \\ a_{31} & a_{32} \end{vmatrix}.$$

Similarly, a third-order minor is

$$|a_{21} \quad a_{33} \quad a_{44}| = \begin{vmatrix} a_{21} & a_{23} & a_{24} \\ a_{31} & a_{33} & a_{34} \\ a_{41} & a_{43} & a_{44} \end{vmatrix}.$$

Suppose now that B is the sum of two 2×2 matrices

$$B = A + D \quad \text{with} \quad A = \begin{bmatrix} a_{11} & a_{12} \\ a_{21} & a_{22} \end{bmatrix} \quad \text{and} \quad D = \begin{bmatrix} x_1 & 0 \\ 0 & x_2 \end{bmatrix}.$$

Then the determinant of B is

$$|B| = |A + D| = \begin{vmatrix} a_{11} + x_1 & a_{12} \\ a_{21} & a_{22} + x_2 \end{vmatrix}.$$

By direct expansion

$$|B| = (a_{11} + x_1)(a_{22} + x_2) - a_{12}a_{21},$$

and on rearranging terms this becomes

$$|B| = x_1 x_2 + x_1 a_{22} + x_2 a_{11} + \begin{vmatrix} a_{11} & a_{12} \\ a_{21} & a_{22} \end{vmatrix}. \tag{10}$$

In similar fashion it can be shown for the 3×3 case that

$$\begin{vmatrix} a_{11} + x_1 & a_{12} & a_{13} \\ a_{21} & a_{22} + x_2 & a_{23} \\ a_{31} & a_{32} & a_{33} + x_3 \end{vmatrix}$$

$$= x_1 x_2 x_3 + x_1 x_2 a_{33} + x_1 x_3 a_{22} + x_2 x_3 a_{11}$$

$$+ x_1 \begin{vmatrix} a_{22} & a_{23} \\ a_{32} & a_{33} \end{vmatrix} + x_2 \begin{vmatrix} a_{11} & a_{13} \\ a_{31} & a_{33} \end{vmatrix} + x_3 \begin{vmatrix} a_{11} & a_{12} \\ a_{21} & a_{22} \end{vmatrix} + \begin{vmatrix} a_{11} & a_{12} & a_{13} \\ a_{21} & a_{22} & a_{23} \\ a_{31} & a_{32} & a_{33} \end{vmatrix}$$

which, using the abbreviated notation, can be written as

$$x_1 x_2 x_3 + x_1 x_2 a_{33} + x_1 x_3 a_{22} + x_2 x_3 a_{11}$$
$$+ x_1 \,|a_{22} \quad a_{33}| + x_2 \,|a_{11} \quad a_{33}| + x_3 \,|a_{11} \quad a_{22}| + |a_{11} \quad a_{22} \quad a_{33}|. \quad (11)$$

Considered as a polynomial in the x's we see that the coefficient of the product of all the x's is unity; the coefficients of the second-degree terms in the x's (the terms $x_1 x_2$, $x_1 x_3$ and $x_2 x_3$) are the diagonal elements of A; the coefficients of the first-degree terms in the x's are the second-order minors of A which have diagonal elements coincident with the diagonal of $|A|$; and the term independent of the x's is $|A|$ itself. The minors of A in these coefficients, namely those having diagonal elements coincident with the diagonal of $|A|$, are called the *principal minors* of $|A|$.

This method of expansion is known as *expansion by diagonal elements* or simply as *diagonal expansion*. It has particular merit when many of the principal minors of A are zero for then the expansion of $|A + D|$ by this method is greatly simplified.

Example.

If

$$|B| = \begin{vmatrix} 7 & 2 & 2 \\ 2 & 8 & 2 \\ 2 & 2 & 9 \end{vmatrix},$$

we can write $|B| = |A + D| = \left\| \begin{bmatrix} 2 & 2 & 2 \\ 2 & 2 & 2 \\ 2 & 2 & 2 \end{bmatrix} + \begin{bmatrix} 5 & 0 & 0 \\ 0 & 6 & 0 \\ 0 & 0 & 7 \end{bmatrix} \right\|.$

Every element of A is a 2 so that $|A|$ and all its 2×2 minors are zero. Consequently $|B|$ evaluated by (11) consists of only the first four terms:

$$|B| = 5(6)7 + 5(6)2 + 5(7)2 + 6(7)2 = 424.$$

Evaluating a determinant in this manner is also useful when all elements of the diagonal matrix D are the same, i.e. when the x_i's are equal. The expansion (11) then becomes

$$x^3 + x^2(a_{11} + a_{22} + a_{33}) + x(|a_{11} \quad a_{22}| + |a_{11} \quad a_{33}| + |a_{22} \quad a_{33}|) + |A|. \tag{12}$$

Suppose we let $tr_1(A)$ represent the trace of A (the sum of the diagonal elements; see Section 1.6), and $tr_2(A)$ represent the sum of the principal second-order minors of A. Then expression (12) can be written as

$$x^3 + x^2 tr_1(A) + x tr_2(A) + |A|. \tag{13}$$

The general diagonal expansion of a determinant of order n,

$$|A + D| = \begin{vmatrix} a_{11} + x_1 & a_{12} & a_{13} & \cdots & a_{1n} \\ a_{21} & a_{22} + x_2 & a_{23} & \cdots & a_{2n} \\ a_{31} & a_{32} & a_{33} + x_3 & \cdots & a_{3n} \\ \cdot & \cdot & \cdot & \cdot & \cdot \\ \cdot & \cdot & \cdot & \cdot & \cdot \\ \cdot & \cdot & \cdot & \cdot & \cdot \\ a_{n1} & a_{n2} & a_{n3} & \cdots & a_{nn} + x_n \end{vmatrix},$$

consists of the sum of all possible products of the x_i taken r at a time for $r = n, n - 1, \ldots, 2, 1, 0$, each product being multiplied by its complementary principal minor of order $n - r$ in A. By *complementary principal minor* in A is meant the principal minor whose diagonal elements are other than those associated in $|A + D|$ with the x's of the particular product concerned; for example, the complementary principal minor associated with the product $x_1 x_3 x_6$ is $|a_{22} \quad a_{44} \quad a_{55} \quad a_{77} \quad a_{88} \quad \cdots \quad a_{nn}|$. When the x's are all equal the expansion becomes $\sum_{i=0}^{n} x^{n-i} tr_i(A)$ where $tr_i(A)$ is the sum of the principal minors of order i of A and, by definition, $tr_0(A) = 1$ and $tr_n(A) = |A|$.

Example. Suppose

$$|A + D| = \begin{vmatrix} a + b & a & a & a \\ a & a + b & a & a \\ a & a & a + b & a \\ a & a & a & a + b \end{vmatrix}.$$

By diagonal expansion

$$|A + D| = b^4 + b^3 tr_1(A) + b^2 tr_2(A) + b tr_3(A) + |A|$$

where A is the 4×4 matrix whose every element is a. Thus $|A|$ and all minors of order 2 or more are zero. Hence

$$|A + D| = b^4 + b^3(4a) = (4a + b)b^3.$$

7. APPENDIX

a. Determinant of a transpose

A', the transpose of A, is A with the rows of A being the columns of A'. Therefore, in the expansion of $|A'|$ by elements of its first column, as in equation (2), the elements are the same as those in the first row of A, and the signs, $(-1)^{i+1}$ for $i = 1, \ldots, n$ are the same as $(-1)^{1+j}$ for $j = 1, \ldots, n$. However, in this expansion of $|A'|$ the minors used will be the transposes of those used in expanding $|A|$ by elements of its first row. But this will be the only difference between $|A'|$ and $|A|$. Furthermore, it will be the only difference in all the successive expansions of the minors themselves, until both $|A'|$ and $|A|$ are in terms of second-order minors that are transposes of one another. Then, because

$$\begin{vmatrix} a & c \\ b & d \end{vmatrix} = \begin{vmatrix} a & b \\ c & d \end{vmatrix} = ad - bc,$$

the second-order minors in $|A'|$ and $|A|$ have the same value. Hence, because all the signs and elements that make up the coefficients of these minors are the same, it follows that $|A'| = |A|$.

b. Interchanging rows

For $A = \{a_{ij}\}$, $i, j = 1, \ldots, n$, the expansion of $|A|$ by elements of its pth row is, from equation (1),

$$|A| = \sum_{j=1}^{n} a_{pj}(-1)^{p+j}|M_{pj}|, \tag{14}$$

where $|M_{pj}|$ is the minor of a_{pj} in A. Consider interchanging the pth row of A with the $(p + 1)$th row, calling the resulting matrix B. Then, in the expansion of $|B|$ by elements of its $(p + 1)$th row, these elements are a_{pj} for $j = 1, \ldots,$ n. But because they are in the $(p + 1)$th row of $|B|$ the signs in this expansion will be $(-1)^{(p+1)+j}$. Furthermore, it will be found that if the row and column containing a_{pj} are deleted from $|B|$, exactly the same minor is left as if these deletions are made from $|A|$. Hence the minor of a_{pj} in B is the same as the minor of a_{pj} in A, namely $|M_{pj}|$. Therefore, in expanding B by elements of its $(p + 1)$th row,

$$
\begin{aligned}
|B| &= \sum_{j=1}^{n} a_{pj}(-1)^{p+1+j}|M_{pj}| \\
&= (-1)\sum_{j=1}^{n} a_{pj}(-1)^{p+j}|M_{pj}| \\
&= (-1)|A|, \qquad \text{by (14).}
\end{aligned}
$$

Thus interchanging two adjacent rows of $|A|$ changes its sign.

We must now consider interchanging non-adjacent rows, those numbered p and $p + q$, for example. Shifting row p to row $p + q$ can be achieved by q interchanges of adjacent rows. What was row $p + q$ is then row $p + q - 1$, and shifting it to row p can be achieved with $q - 1$ interchanges of adjacent rows. There are therefore $2q - 1$ such interchanges and so the determinant is multiplied by $(-1)^{2q-1} = -1$; i.e. the sign of the determinant is changed. Thus interchanging any two rows of a determinant (adjacent or otherwise) changes its sign.

c. Formal definition

As indicated at the end of Section 2, the determinant of a square matrix is the sum of $n!$ signed products of the elements of the matrix, each product containing one and only one element from every row and column of the matrix. Furthermore, expansion by minors as in equations (1) and (2) yields the requisite products with their correct signs when the minors are successively expanded at each stage by the same procedure. A formal definition of a determinant, adapted from Ferrar (1941), is now given. Its equivalence to expansion by minors is proved in Searle (1966), Section 3.2.

The determinant of a square matrix A of order n, $A = \{a_{ij}\}$, $i, j = 1$, $2, \ldots, n$, is the sum of all possible products of n elements of A such that (i) each product has one and only one element from every row and column of A, and (ii) the sign of a product is $(-1)^p$ for $p = \sum_{i=1}^{n} n_i$ where, by writing the product with its i subscripts in natural order $a_{1j_1}a_{2j_2}\cdots a_{ij_i}\cdots a_{nj_n}$, with the j subscripts j_i, $i = 1, 2, \ldots, n$, being the first n integers in some order, n_i is defined as the number of j's less than j_i that follow j_i in this order.

Example. In the expansion of

$$
|A| = \begin{vmatrix} a_{11} & a_{12} & a_{13} \\ a_{21} & a_{22} & a_{23} \\ a_{31} & a_{32} & a_{33} \end{vmatrix}
$$

the j_1, j_2, j_3 array in the product $a_{12}a_{23}a_{31}$ is $j_1 = 2$, $j_2 = 3$ and $j_3 = 1$. Now n_1 is the number of j's in this array that are less than j_1 and follow it, i.e. that are less than 2 and follow it. There is only one, namely $j_3 = 1$; therefore $n_1 = 1$. Likewise $n_2 = 1$ because $j_2 = 3$ is followed by only one j less than 3, and finally $n_3 = 0$. Thus

$$p = n_1 + n_2 + n_3 = 1 + 1 + 0 = 2$$

and the sign of the product $a_{12}a_{23}a_{31}$ in $|A|$ is therefore $(-1)^2 = +1$.

d. The Laplace expansion

In the expansion of

$$|A| = \begin{vmatrix} a_{11} & a_{12} & a_{13} & a_{14} \\ a_{21} & a_{22} & a_{23} & a_{24} \\ a_{31} & a_{32} & a_{33} & a_{34} \\ a_{41} & a_{42} & a_{43} & a_{44} \end{vmatrix}$$

the minor of a_{11} is $|a_{22} \quad a_{33} \quad a_{44}|$. An extension of this, easily verified, is that the coefficient of $|a_{11} \quad a_{22}|$ is $|a_{33} \quad a_{44}|$; namely, the coefficient of

$$\begin{vmatrix} a_{11} & a_{12} \\ a_{21} & a_{22} \end{vmatrix} = a_{11}a_{22} - a_{12}a_{21}$$

in the expansion of $|A|$ is

$$\begin{vmatrix} a_{33} & a_{34} \\ a_{43} & a_{44} \end{vmatrix} = a_{33}a_{44} - a_{34}a_{43}.$$

Likewise the coefficient of $|a_{11} \quad a_{24}|$ is $|a_{32} \quad a_{43}|$: the coefficient of

$$|a_{11} \quad a_{24}| = \begin{vmatrix} a_{11} & a_{14} \\ a_{21} & a_{24} \end{vmatrix} = a_{11}a_{24} - a_{21}a_{14}$$

in the expansion of $|A|$ is

$$|a_{32} \quad a_{43}| = \begin{vmatrix} a_{32} & a_{33} \\ a_{42} & a_{43} \end{vmatrix} = a_{32}a_{43} - a_{33}a_{42}.$$

Each determinant just described as the coefficient of a particular minor in the expansion of $|A|$ is the *complementary minor* in A of that particular minor: it is the determinant obtained from $|A|$ by deleting from it all the rows and columns containing the particular minor. This is simply an extension of the procedure for finding the coefficient of an individual element in $|A|$

as derived in the expansion by elements of a row or column discussed earlier. In that case the particular minor is a single element and its coefficient in $|A|$ is $|A|$ amended by deletion of the row and column containing the element concerned. A sign factor is also involved, namely $(-1)^{i+j}$ for the coefficient of a_{ij} in $|A|$. In the extension to coefficients of minors the sign factor is -1 raised to the power of the sum of the subscripts of the diagonal elements of the chosen minor: for example, the sign factor for the coefficient of $|a_{32} \quad a_{43}|$ as just given is $(-1)^{3+2+4+3} = +1$. The complementary minor multiplied by this sign factor can be appropriately referred to as the coefficient of the particular minor concerned. Furthermore, just as the expansion of a determinant is the sum of products of elements of a row (or column) with their coefficients, so also is it the sum of products of all minors of order m that can be derived from any set of m rows, each multiplied by its coefficient as just defined. This generalization of the method of expanding a determinant by elements of a row to expanding it by minors of a set of rows was first established by the eighteenth-century mathematician Laplace and so bears his name. Aitken (1948) and Ferrar (1941) are two places where proof of the procedure is given; we shall be satisfied here with a general statement of the method and an example illustrating its use.

The Laplace expansion of a determinant $|A|$ of order n can be obtained as follows. (i) Consider any m rows of $|A|$. They contain $n!/[m!(n-m)!]$ minors of order m (where $n!$ is the expression defined in the footnote at the end of Section 4.2). (ii) Multiply each of these minors, $|M|$ say, by its complementary minor and by a sign factor, where (a) the complementary minor of $|M|$ is the $(n-m)$-order minor derived from $|A|$ by deleting the m rows and columns containing M, and (b) the sign factor is $(-1)^\mu$ where μ is the sum of the subscripts of the diagonal elements of M, A being defined as $A = \{a_{ij}\}$, $i, j = 1, 2, \ldots, n$. (iii) The sum of all such products is $|A|$.

Example. We will expand

$$|A| = \begin{vmatrix} 1 & 2 & 3 & 0 & 0 \\ -1 & 2 & -1 & 0 & 0 \\ 2 & 0 & 1 & 4 & 5 \\ 1 & 0 & 4 & 2 & 3 \\ 0 & 2 & 1 & 2 & 3 \end{vmatrix}$$

using the Laplace expansion based on the first two rows ($m = 2$). There are ten minors of order 2 in these two rows; however, seven of them are zero because they involve a column of zeros. Hence $|A|$ can be expanded as the sum of three products involving the three 2×2 non-zero minors in the first two rows, namely as

$$|A| = (-1)^{1+1+2+2} \begin{vmatrix} 1 & 2 \\ -1 & 2 \end{vmatrix} \begin{vmatrix} 1 & 4 & 5 \\ 4 & 2 & 3 \\ 1 & 2 & 3 \end{vmatrix}$$

$$+ (-1)^{1+1+2+3} \begin{vmatrix} 1 & 3 \\ -1 & -1 \end{vmatrix} \begin{vmatrix} 0 & 4 & 5 \\ 0 & 2 & 3 \\ 2 & 2 & 3 \end{vmatrix}$$

$$+ (-1)^{1+2+2+3} \begin{vmatrix} 2 & 3 \\ 2 & -1 \end{vmatrix} \begin{vmatrix} 2 & 4 & 5 \\ 1 & 2 & 3 \\ 0 & 2 & 3 \end{vmatrix}$$

The sign factors in these terms have been derived by envisaging A as $\{a_{ij}\}$.

Consequently the first 2×2 minor, $\begin{vmatrix} 1 & 2 \\ -1 & 2 \end{vmatrix}$, is $\begin{vmatrix} a_{11} & a_{12} \\ a_{21} & a_{22} \end{vmatrix}$, leading to $(-1)^{1+1+2+2}$ as its sign factor; likewise for the other terms. Simplification of the whole expression gives

$$|A| = 4 \begin{vmatrix} 1 & 4 & 5 \\ 3 & 0 & 0 \\ 1 & 2 & 3 \end{vmatrix} - 2(2) \begin{vmatrix} 4 & 5 \\ 2 & 3 \end{vmatrix} + (-8) \begin{vmatrix} 2 & 4 & 5 \\ 1 & 0 & 0 \\ 0 & 2 & 3 \end{vmatrix} = -24 - 8 + 16$$

and hence $|A| = -16$. Expansion by any other method leads to the same result.

A convenient use of the Laplace expansion is when A can be partitioned into sub-matrices one of which is a null matrix. An example of this is seen in the next section.

Certain other methods of evaluating determinants are based on extending the Laplace expansion, using it recurrently not only to expand a determinant by minors and their complementary minors, but also to expand these minors themselves by the Laplace expansion. Many of these extended expansions are identified by the names of their originators, for example Cauchy, Binet-Cauchy and Jacoby. Description of them is to be found in such places as Aitken (1948) and Ferrar (1941).

e. Multiplication of determinants

Proofs of results (7) and (8) quoted in Section 4.5 are given here. Both rely upon the Laplace expansion just discussed.

Suppose A and B are square, of order n. Then, by the Laplace expansion,

$$|D| = \begin{vmatrix} O & A \\ -I & B \end{vmatrix} = |A|\,|-I|\,(-1)^p$$

where p is the sum of the subscripts of the diagonal elements of A after writing D as $\{d_{ij}\}$ for i and $j = 1, 2, \ldots, 2n$. In this way

$$p = (1 + n + 1) + (2 + n + 2) + \cdots + (n + n + n)$$
$$= 2(1 + 2 + \cdots + n) + n(n)$$

and utilizing the fact that $1 + 2 + \cdots + n = n(n + 1)/2$ then gives p as

$$p = 2n(n + 1)/2 + n^2 = n(2n + 1).$$

And, since I is of order n, $|-I| = (-1)^n$ and so

$$|A|\,|-I|\,(-1)^p = |A|\,(-1)^{n+n(2n+1)} = |A|\,(-1)^{2n(n+1)} = |A|,$$

thus verifying (7).

Similarly,

$$\begin{vmatrix} A & O \\ X & B \end{vmatrix} = |A|\,|B|$$

because Laplace expansion of the left-hand side by the first n rows gives the right-hand side. Thus is (8) proven.

8. EXERCISES

1. Show that both

(a) $\begin{vmatrix} 1 & 5 & -5 \\ 3 & 2 & -5 \\ 6 & -2 & -5 \end{vmatrix}$ and $\begin{vmatrix} -3 & 2 & -6 \\ -3 & 5 & -7 \\ -2 & 3 & -4 \end{vmatrix}$ equal -5;

(b) $\begin{vmatrix} 2 & 6 & 5 \\ -2 & 7 & -5 \\ 2 & -7 & 9 \end{vmatrix}$ and $\begin{vmatrix} 2 & -1 & 9 \\ -1 & 7 & 2 \\ 3 & -21 & 2 \end{vmatrix}$ equal 104;

(c) $\begin{vmatrix} 1 & 6 & 4 & 3 \\ 2 & 8 & 5 & 4 \\ 3 & 8 & 7 & 5 \\ 4 & 9 & 7 & 7 \end{vmatrix}$ and $\begin{vmatrix} 1 & -1 & -1 & -1 \\ -1 & 1 & -1 & -1 \\ -1 & -1 & 1 & -1 \\ -1 & -1 & -1 & 1 \end{vmatrix}$ equal -16;

(d) $\begin{vmatrix} 21 & 6 & 3 & 9 \\ 12 & 16 & 36 & 4 \\ 13 & 10 & 19 & 5 \\ 1 & 93 & 81 & 6 \end{vmatrix}$ and $\begin{vmatrix} 4 & 6 & 8 & 1 & 2 \\ -1 & -7 & 2 & 3 & 1 \\ 2 & -8 & 12 & 7 & 4 \\ 7 & 9 & 17 & 27 & -5 \\ 8 & 3 & 6 & 2 & 37 \end{vmatrix}$ are zero.

2.

Show and explain why $\begin{vmatrix} 1 & 0 & 0 \\ 2 & 3 & 0 \\ 4 & 5 & 6 \end{vmatrix} = 18,$ but $\begin{vmatrix} 0 & 0 & 1 \\ 0 & 3 & 2 \\ 6 & 5 & 4 \end{vmatrix} = -18,$

whereas $\begin{vmatrix} 1 & 0 & 0 & 0 \\ 2 & 3 & 0 & 0 \\ 4 & 5 & 6 & 0 \\ 7 & 8 & 9 & 10 \end{vmatrix} = \begin{vmatrix} 0 & 0 & 0 & 1 \\ 0 & 0 & 3 & 2 \\ 0 & 6 & 5 & 4 \\ 10 & 9 & 8 & 7 \end{vmatrix} = 180.$

3.

Show that $\begin{vmatrix} 1 & 1 & 1 & 1 \\ 2 & 3 & -1 & 5 \\ 4 & 9 & 1 & 25 \\ 8 & 27 & -1 & 125 \end{vmatrix} = 432.$

4. Expand:

(a) $\begin{vmatrix} 6 - \lambda & 3 & 1 \\ 2 & 4 - \lambda & 2 \\ 1 & 5 & 7 - \lambda \end{vmatrix}$ as $-\lambda^3 + 17\lambda^2 - 77\lambda + 78;$

(b) $\begin{vmatrix} 2 - \lambda & -2 & 3 \\ 10 & -4 - \lambda & 5 \\ 5 & -4 & 6 - \lambda \end{vmatrix}$ as $-(\lambda - 1)^2(\lambda - 2);$

(c) $\begin{vmatrix} -1 - \lambda & -2 & 1 \\ -2 & 2 - \lambda & -2 \\ -1 & -2 & -1 - \lambda \end{vmatrix}$ as $-\lambda^3 + 10\lambda + 12.$

5. Reduce the following determinants using expansions by minors and/or Laplace expansions, or other methods of expansion given in this chapter. Wherever possible utilize one result to obtain succeeding ones (e.g., d is the transpose of b). When expansion by minors is unnecessary—as it is in most cases—state why, and give the value of the determinant.

(a) $\begin{vmatrix} 1 & 4 & 1 & 8 \\ 2 & 8 & 2 & 16 \\ 1 & 0 & 0 & 12 \\ 7 & 8 & 1 & 2 \end{vmatrix}$ (b) $\begin{vmatrix} 2 & 3 & 7 \\ 4 & 1 & 8 \\ 1 & 0 & 2 \end{vmatrix}$

(c) $\begin{vmatrix} 1 & 4 & 1 \\ 2 & 6 & 3 \\ -1 & -2 & 7 \end{vmatrix}$ (d) $\begin{vmatrix} 2 & 4 & 1 \\ 3 & 1 & 0 \\ 7 & 8 & 2 \end{vmatrix}$

(e) $\begin{vmatrix} 2 & 3 & 7 & 0 & 0 & 0 \\ 4 & 1 & 8 & 0 & 0 & 0 \\ 1 & 0 & 2 & 0 & 0 & 0 \\ 17 & -1 & -3 & 1 & 4 & 1 \\ 6 & 2 & 1 & 2 & 6 & 3 \\ 7 & 8 & 12 & -1 & -2 & 7 \end{vmatrix}$

(f) $\begin{vmatrix} 2 & 8 & 2 \\ 2 & 6 & 3 \\ -1 & -2 & 7 \end{vmatrix}$ (g) $\begin{vmatrix} 1 & 2 & 0 \\ 3 & 1 & 0 \\ 1 & 7 & 0 \end{vmatrix}$

(h) $\begin{vmatrix} 1 & 0 & 7 \\ 2 & 8 & 1 \end{vmatrix}$ (i) $\begin{vmatrix} 3 & 1 & 3 \\ 2 & 8 & 6 \\ -1 & 7 & -1 \end{vmatrix}$

(j) $\begin{vmatrix} 5 & 5 & 1 \\ 3 & 1 & 0 \\ 7 & 8 & 2 \end{vmatrix}$ (k) $\begin{vmatrix} 1 & 3 & 2 \\ 1 & 7 & 2 \\ 1 & 4 & 2 \end{vmatrix}$

(l) $\begin{vmatrix} 1 & 0 & 0 \\ 0 & 2 & 0 \\ 0 & 0 & -1 \end{vmatrix}$ (m) $\begin{vmatrix} 0 & 0 & -1 \\ 0 & 2 & 0 \\ 1 & 0 & 0 \end{vmatrix}$

(n) $\begin{vmatrix} 3 & 20 & 62 \\ 2 & 22 & 67 \\ 0 & 4 & 16 \end{vmatrix} = \begin{vmatrix} \begin{bmatrix} 2 & 3 & 7 \\ 4 & 1 & 8 \\ 1 & 0 & 2 \end{bmatrix} \begin{bmatrix} 2 & 8 & 2 \\ 2 & 6 & 3 \\ -1 & -2 & 7 \end{bmatrix} \end{vmatrix}$

(o) $\begin{vmatrix} 7 & 1 & 3 & 0 \\ 3 & 1 & 4 & 1 \\ 2 & 7 & 8 & 4 \\ 1 & 4 & 1 & 4 \end{vmatrix}$

6.

Show that
$$\begin{vmatrix} a+b+c & a+b & a & a \\ a+b & a+b+c & a & a \\ a & a & a+b+c & a+b \\ a & a & a+b & a+b+c \end{vmatrix}$$
$$= c^2(4a + 2b + c)(2b + c).$$

7.

Evaluate
$$\begin{vmatrix} 7 & 6 & 6 & 6 & 6 & 6 \\ 6 & 7 & 6 & 6 & 6 & 6 \\ 6 & 6 & 7 & 6 & 6 & 6 \\ 6 & 6 & 6 & 7 & 6 & 6 \\ 6 & 6 & 6 & 6 & 7 & 6 \\ 6 & 6 & 6 & 6 & 6 & 7 \end{vmatrix}$$
and
$$\begin{vmatrix} 3 & 21 & 18 & 9 \\ 2 & 13 & 13 & 5 \\ 4 & 29 & 31 & 17 \\ -2 & -8 & 2 & 6 \end{vmatrix}.$$

8. Explain why the determinant of a diagonal matrix is the product of its diagonal elements. Is the same true for a triangular matrix? Give examples.

9. Without expanding the determinants, suggest values of x that satisfy the following equations:

(a) $\begin{vmatrix} x & x & x \\ 2 & -1 & 0 \\ 7 & 4 & 5 \end{vmatrix} = 0;$

(b) $\begin{vmatrix} 1 & x & x^2 \\ 1 & 2 & 4 \\ 1 & -1 & 1 \end{vmatrix} = 0;$

(c) $\begin{vmatrix} 4 & x & x \\ x & 4 & x \\ x & x & 4 \end{vmatrix} = 0;$

(d) $\begin{vmatrix} x & 4 & 4 \\ 4 & x & 4 \\ 4 & 4 & x \end{vmatrix} = 0.$

10. A typical transportation problem involves the shipment of goods from production plants to warehouses, with more than one plant often supplying the needs of one warehouse. Suppose we have a situation of two plants with 10 and 25 units of product available, having to supply three warehouses whose needs are 15, 15 and 5 units, respectively. If we let x_{ij} represent the number of units shipped from plant i to warehouse j the amount of product that is available at each plant necessitates the following two equations:
$$x_{11} + x_{12} + x_{13} = 10;$$
$$x_{21} + x_{22} + x_{23} = 25.$$

Furthermore, the warehouse requirements demand that the following equations must be satisfied:
$$x_{11} + x_{21} = 15;$$
$$x_{12} + x_{22} = 15;$$
$$x_{13} + x_{23} = 5.$$

Defining

$$x' = [x_{11} \ \ x_{12} \ \ x_{13} \ \ x_{21} \ \ x_{22} \ \ x_{23}] \quad \text{and} \quad y' = [10 \ \ 25 \ \ 15 \ \ 15 \ \ 5],$$

the above five equations can be written as $Ax = y$. With this formulation

(a) write down the matrix A;

(b) take any five square sub-matrices of A, one each of order 5, 4, 3, 2 and 1, and show that the determinant of each sub-matrix is either $+1$, -1 or zero. (This is a general characteristic of transportation problems of this nature, which leads to their having integer solutions for the x's.)

REFERENCES

Aitken, A. C. (1948). *Determinants and Matrices.* Fifth Edition; Oliver and Boyd, Edinburgh.

Ferrar, W. L. (1941). *Algebra, a Text-Book of Determinants, Matrices and Algebraic Forms.* First Edition; Oxford University Press.

Searle, S. R. (1966). *Matrix Algebra for the Biological Sciences.* Wiley, New York.

CHAPTER 5

THE INVERSE OF A MATRIX

The arithmetic operations of addition, subtraction and multiplication in matrix algebra have already been dealt with, but division has not. In its usual sense division does not exist in matrix algebra; the concept of "dividing" by a matrix A is replaced by the concept of multiplying by a matrix called the inverse of A.

1. INTRODUCTION

The inverse of the matrix A is a matrix whose product with A is the identity matrix. It is usually denoted by the symbol A^{-1}, often read as "A to the (power of) minus one," as "the inverse of A" or as "A, inverse." We illustrate the importance of inverses by considering the problem of solving linear equations.

Illustration. Suppose an economist determines the cost of a product to be \$186 for 300 production units. He calculates the cost per unit as:

$$186/300 = 0.62.$$

In doing so he is effectively solving a simple equation $300x = 186$, where x is a scalar. One way of doing this is to multiply both sides of the equation by $1/300$, which gives $(1/300)300x = (1/300)186$, i.e. $x = 186/300 = 0.62$. And in general, for scalars a, b and x where a and b are known values and a is not zero, the equation

$$ax = b$$

can be solved by multiplying both sides of it by $1/a$ to give $(1/a)ax = (1/a)b$ or

$$x = (1/a)b = a^{-1}b. \tag{1}$$

An extension of this problem is that of solving not just a single equation, but several equations simultaneously, i.e. of solving a set of simultaneous (linear) equations.

Illustration. An important belief in both economics and business is that a competitive market mechanism forces the price of a product to a level which equates supply with demand. Suppose the demand curve for automobiles over some time period can be written

$$x_1 = 12{,}000 - 0.2x_2$$

where x_1 is the price of an automobile and x_2 is the corresponding quantity; and suppose the supply curve is

$$x_1 = 300 + 0.1x_2.$$

Rewriting these equations as

$$x_1 + 0.2x_2 = 12{,}000$$

and
$$10x_1 - x_2 = 3{,}000, \tag{2}$$

they could be solved by substitution, giving, for example

$$10(12{,}000 - 0.2x_2) - x_2 = 3{,}000$$

and hence $x_2 = 39{,}000$ and $x_1 = 4{,}200$. Thus the equilibrium price is \$4,200 and at that price the quantity demanded equals the quantity supplied, namely, 39,000 automobiles.

This method of solving equations is satisfactory for equations as simple as these, but were there to be many equations and unknowns, say eight or ten, obtaining their solution by these means would be arduous. However, with the aid of matrices the solution can be put in a form similar to (1) that is applicable no matter how many equations and unknowns there are. If (1) were to be considered in matrix form x would be a column vector of solutions, a^{-1} would be what is known as the inverse of a matrix, for which the notation A^{-1} is used, and b would be a column vector. To achieve this the equations are first written as $Ax = b$; they are then solved as $x = A^{-1}b$ where A^{-1} is the inverse of A, and this form of solution applies whether we are solving two simultaneous linear equations in two unknowns or a hundred equations in a hundred unknowns.

Continuing the illustration, equations (2) can be written in matrix form as

$$\begin{bmatrix} 1 & 0.2 \\ 10 & -1 \end{bmatrix} \begin{bmatrix} x_1 \\ x_2 \end{bmatrix} = \begin{bmatrix} 12{,}000 \\ 3{,}000 \end{bmatrix}.$$

Let us pre-multiply both sides of this equation by

$$A^{-1} = \frac{1}{-3} \begin{bmatrix} -1 & -0.2 \\ -10 & 1 \end{bmatrix}.$$

(We shall later show how A^{-1} is computed.) Then, on carrying out the multiplication the equations become

$$\begin{bmatrix} 1 & 0 \\ 0 & 1 \end{bmatrix} \begin{bmatrix} x_1 \\ x_2 \end{bmatrix} = \begin{bmatrix} 4,200 \\ 39,000 \end{bmatrix}.$$

This is simply

$$\begin{bmatrix} x_1 \\ x_2 \end{bmatrix} = \begin{bmatrix} 4,200 \\ 39,000 \end{bmatrix},$$

namely the solution $x_1 = 4,200$ and $x_2 = 39,000$ obtained previously. Hence we have found that the equations

$$\begin{bmatrix} 1 & 0.2 \\ 10 & -1 \end{bmatrix} \begin{bmatrix} x_1 \\ x_2 \end{bmatrix} = \begin{bmatrix} 12,000 \\ 3,000 \end{bmatrix}$$

have the solution

$$\begin{bmatrix} x_1 \\ x_2 \end{bmatrix} = \frac{1}{-3} \begin{bmatrix} -1 & -0.2 \\ -10 & 1 \end{bmatrix} \begin{bmatrix} 12,000 \\ 3,000 \end{bmatrix} = \begin{bmatrix} 4,200 \\ 39,000 \end{bmatrix}.$$

The matrix A^{-1} used in obtaining this result has been so constructed that its product with the original matrix is an identity matrix, i.e.,

$$\frac{1}{-3} \begin{bmatrix} -1 & -0.2 \\ -10 & 1 \end{bmatrix} \begin{bmatrix} 1 & 0.2 \\ 10 & -1 \end{bmatrix} = \begin{bmatrix} 1 & 0 \\ 0 & 1 \end{bmatrix}.$$

Let us now re-write the illustration, writing the components of the problem in symbols:

$$A = \begin{bmatrix} 1 & 0.2 \\ 10 & -1 \end{bmatrix}, \quad x = \begin{bmatrix} x_1 \\ x_2 \end{bmatrix} \quad \text{and} \quad b = \begin{bmatrix} 12,000 \\ 3,000 \end{bmatrix}.$$

We have then solved the equation

$$Ax = b \tag{3}$$

by pre-multiplying it by A^{-1} to get

$$A^{-1}Ax = A^{-1}b \tag{4}$$

and because

$$A^{-1}A = I \tag{5}$$

we have

$$x = A^{-1}b. \tag{6}$$

Since I is the "one" of matrix algebra the matrix A^{-1} is called the *reciprocal of A*, or *inverse of A*. We define it for the moment by equation (5), namely as a matrix which, when post-multiplied by A, gives the identity matrix I.

Notice the solution has not been derived from equation (3) by dividing both sides of (3) by A; at no stage has there been division by A, for division is undefined in matrix algebra and has no meaning. Instead, we obtain the solution by pre-multiplying both sides of (3) by a matrix A^{-1} to get (4). Then, because A^{-1} is defined as in (5) we find that (4) simplifies to the solution (6).

The remainder of this chapter is devoted first to sharpening the definition of A^{-1} as given by the equation $A^{-1}A = I$, and then to specifying how A^{-1} is obtained from A. When this is known, any set of simultaneous linear equations that has a unique solution can be put in the form $Ax = b$ and solved as $x = A^{-1}b$, no matter how many such equations there are. Computational difficulties may sometimes arise, but the form of the solution and the procedure for obtaining it remain the same.

2.   PRODUCTS EQUAL TO I

The concept of a matrix inverse has been introduced by considering the solution of a set of simultaneous linear equations. However, we have not yet discussed the basic question: "Is there a matrix R such that the product RA equals I?" There are three answers to this question, depending on the characteristics of A: (i) in some cases R exists and is unique for a given A; (ii) sometimes numerous matrices R exist for a particular A, i.e. R exists but is not unique; and (iii) in some instances R does not exist at all. Examples of these three situations are now given.

i. If $\qquad A_1 = \begin{bmatrix} 2 & 5 \\ 3 & 8 \end{bmatrix} \quad$ and $\quad R_1 = \begin{bmatrix} 8 & -5 \\ -3 & 2 \end{bmatrix}$

$R_1A_1 = I$ and, as shown later, R_1 is the unique matrix for this particular A such that $RA = I$.

ii. For $\qquad A_2 = \begin{bmatrix} 1 & 1 \\ -1 & 0 \\ 3 & -1 \end{bmatrix}$

there is an infinite number of matrices R such that $RA_2 = I$; for example,

$$R_2 = \begin{bmatrix} 1 & 3 & 1 \\ 2 & 5 & 1 \end{bmatrix} \quad \text{and} \quad R_2^* = \begin{bmatrix} 4 & 15 & 4 \\ 7 & 25 & 6 \end{bmatrix}.$$

iii. When $\qquad A_3 = \begin{bmatrix} 0 & 3 & 7 \\ 0 & 2 & 5 \end{bmatrix}$

there is no matrix R such that $RA_3 = I$, for the leading element of RA_3 will always be zero and so the product matrix can never equal I.

One might notice a further characteristic of these three situations:

i. $A_1R_1 = I$ as well as $R_1A_1 = I$;

ii. $A_2R_2 \neq I$ even though $R_2A_2 = I$;

iii. For $S = \begin{bmatrix} 4 & 8 \\ 5 & -7 \\ -2 & 3 \end{bmatrix}$, $A_3S = I$ even though there is no matrix R such

that $RA_3 = I$.

The A^{-1} established in considering examples of solving linear equations was a matrix such that $A^{-1}A = I$. But we have just illustrated how $RA = I$ does not necessarily yield a matrix R that is unique for a given A, nor one for which AR necessarily equals I even if RA does. Only in the case of A_1 did R have both these properties, namely, that it be unique for a given A and that both pre- and post-multiplication by A lead to the identity matrix. Analogous to the familiar reciprocal of an ordinary number in scalar algebra, these properties seem appropriate for the inverse (reciprocal) of a matrix. Accordingly we will derive a matrix from A which has these properties. The derived matrix will be the inverse of A and instead of satisfying only $A^{-1}A = I$ it will have the two properties (a) $A^{-1}A = AA^{-1} = I$ and (b) A^{-1} unique for given A.

Property (a), that $A^{-1}A = AA^{-1}$, implies that the two products $A^{-1}A$ and AA^{-1} both exist. As explained in Section 2.5d, this occurs only when A and A^{-1} are both square matrices of the same order. This means that an A^{-1} such that property (a) is true exists only when A is square. What we are about to describe as the inverse of a matrix can therefore exist only if that matrix is square (and its inverse will also be square, of the same order). In contrast, rectangular matrices do not have inverses, although some of them do have a restricted kind of inverse of the type discussed in case (ii) above (see appendix, Section 5.9).

We have just established that the inverse A^{-1} can exist only when A is square; and the determinant of A exists only when A is square. Therefore the inverse can exist only when A has a determinant. We now derive the inverse in a manner that utilizes the determinant. This form of derivation has been chosen because it is particularly suitable for describing how the elements of the inverse of a matrix can be obtained directly from those of the matrix itself, although it is tedious except for small-ordered matrices. It is therefore seldom used for calculating the inverse of a matrix in practical situations. Nevertheless, it is a method that presents with clarity the underlying form of the elements of an inverse and, furthermore, it readily leads to demonstrating properties (a) and (b) just discussed.

3.   DERIVATION OF THE INVERSE

We saw in equation (1) of Chapter 4 how the determinant of a matrix $A = \{a_{ij}\}$ for $i, j, = 1, 2, \ldots, n$ could be expanded as

$$|A| = \sum_{j=1}^{n} a_{ij}(-1)^{i+j}|M_{ij}| \qquad \text{for any } i,$$

where $|M_{ij}|$ is the minor of a_{ij} and is the determinant obtained from $|A|$ by crossing out the ith row and jth column. The product $(-1)^{i+j}|M_{ij}|$ is known as the *cofactor* of a_{ij}. It is the minor of a_{ij} with sign attached, and so the cofactor is often called the *signed minor*. If we write μ_{ij} for the cofactor of a_{ij} we have

$$\mu_{ij} = (-1)^{i+j}|M_{ij}|.$$

Example. Consider again the matrix used in the market equilibrium illustration at the beginning of this chapter:

$$A = \begin{bmatrix} 1 & 0.2 \\ 10 & -1 \end{bmatrix}. \tag{7}$$

Its determinant is $|A| = -1 - 2 = -3$; the cofactors of its elements are

$$\mu_{11} = (-1)^{1+1}(-1) = -1, \qquad \mu_{12} = (-1)^{1+2}(10) = -10,$$
$$\mu_{21} = (-1)^{2+1}(0.2) = -0.2 \quad \text{and} \quad \mu_{22} = (-1)^{2+2}(1) = 1.$$

Now consider the matrix of cofactors

$$\begin{bmatrix} \mu_{11} & \mu_{12} \\ \mu_{21} & \mu_{22} \end{bmatrix} = \begin{bmatrix} -1 & -10 \\ -0.2 & 1 \end{bmatrix} \tag{8}$$

and its transpose
$$\begin{bmatrix} \mu_{11} & \mu_{21} \\ \mu_{12} & \mu_{22} \end{bmatrix} = \begin{bmatrix} -1 & -0.2 \\ -10 & 1 \end{bmatrix}.$$

The product of this transpose with A is

$$\begin{bmatrix} \mu_{11} & \mu_{21} \\ \mu_{12} & \mu_{22} \end{bmatrix}\begin{bmatrix} a_{11} & a_{12} \\ a_{21} & a_{22} \end{bmatrix} = \begin{bmatrix} -1 & -0.2 \\ -10 & 1 \end{bmatrix}\begin{bmatrix} 1 & 0.2 \\ 10 & -1 \end{bmatrix} = \begin{bmatrix} -3 & 0 \\ 0 & -3 \end{bmatrix}$$

and since $-3 = |A|$ this is

$$\begin{bmatrix} \mu_{11}a_{11} + \mu_{21}a_{21} & \mu_{11}a_{12} + \mu_{21}a_{22} \\ \mu_{12}a_{11} + \mu_{22}a_{21} & \mu_{12}a_{12} + \mu_{22}a_{22} \end{bmatrix} = \begin{bmatrix} |A| & 0 \\ 0 & |A| \end{bmatrix}. \tag{9}$$

This result serves two purposes. First, it leads very readily to the inverse of A. For, on multiplying both sides of (9) by the scalar $1/|A|$, the reciprocal of the determinant of A, we get an identity matrix:

$$\frac{1}{|A|}\begin{bmatrix} \mu_{11} & \mu_{21} \\ \mu_{12} & \mu_{22} \end{bmatrix}\begin{bmatrix} a_{11} & a_{12} \\ a_{21} & a_{22} \end{bmatrix} = \frac{1}{|A|}\begin{bmatrix} |A| & 0 \\ 0 & |A| \end{bmatrix} = I.$$

Thus, by the preliminary definition of A^{-1} given above, namely $A^{-1}A = I$, we have

$$A^{-1} = \frac{1}{|A|}\begin{bmatrix} \mu_{11} & \mu_{21} \\ \mu_{12} & \mu_{22} \end{bmatrix} = \frac{1}{|A|}\begin{bmatrix} \mu_{11} & \mu_{12} \\ \mu_{21} & \mu_{22} \end{bmatrix}';$$

i.e.

$$A^{-1} = \frac{1}{-3}\begin{bmatrix} -1 & -0.2 \\ -10 & 1 \end{bmatrix} \tag{10}$$

is the inverse of A given in (7). It has been derived by replacing every element of A by its cofactor, transposing the resulting matrix and multiplying by the reciprocal of the determinant. This is the formal definition of an inverse of a square matrix (if it exists): it is the matrix of cofactors, transposed and multiplied by the reciprocal of the determinant.

Equation (9) also incorporates the general result that a determinant $|A|$ is the sum of products of the elements of a row (or column) with their cofactors:

$$|A| = \sum_{j=1}^{n} a_{ij}\mu_{ij} = \sum_{i=1}^{n} a_{ij}\mu_{ij}; \tag{11}$$

this is the result that gives rise to the diagonal elements $|A|$ in (9). Likewise the 0's in (9) arise from another general result, that for any determinant the sum of products of the elements of one row (or column) with the cofactors of the elements of another row (or column) is zero:[1]

$$\sum_{j=1}^{n} a_{ij}\mu_{hj} = 0 \quad \text{for} \quad i \neq h \quad \text{and} \quad \sum_{i=1}^{n} a_{ij}\mu_{ik} = 0 \quad \text{for} \quad j \neq k. \tag{12}$$

---

[1] (11) is true because, by the definition of μ_{ij}, it is identical to the expansion of $|A|$ by minors given in Section 4.2. To see that (12) is true observe that $\sum_{j=1}^{n} a_{ij}\mu_{hj}$ for $i \neq h$ represents the expansion of a determinant having $a_{i1}, a_{i2}, \ldots, a_{in}$ as a row, with the other $n - 1$ rows being those of $|A|$ from which the cofactors $\mu_{h1}, \mu_{h2}, \ldots, \mu_{hn}$ have come. But since these cofactors are cofactors of the elements $a_{h1}, a_{h2}, \ldots, a_{hn}$, these rows consist of all rows of A except $(a_{h1}, a_{h2}, \ldots, a_{hn})$. Therefore they contain $(a_{i1}, a_{i2}, \ldots, a_{in})$ as a row; i.e. the determinant form of $\sum_{j=1}^{n} a_{ij}\mu_{hj}$ has two rows the same and hence is zero. Similarly, the second expression in (12) is a determinant having two columns the same and so it, too, is zero; thus (12) is shown true.

The procedure for deriving A^{-1} illustrated above is more clearly demonstrated for a 3×3 matrix. Consider the matrix

$$A = \begin{bmatrix} 1 & 2 & 3 \\ 4 & 5 & 6 \\ 7 & 8 & 10 \end{bmatrix}. \tag{13}$$

Derivation of its inverse involves the cofactors of all elements of the matrix. We begin with the first column. The cofactors of its elements are, respectively,

$$(-1)^{1+1}\begin{vmatrix} 5 & 6 \\ 8 & 10 \end{vmatrix}, \qquad (-1)^{2+1}\begin{vmatrix} 2 & 3 \\ 8 & 10 \end{vmatrix} \qquad \text{and} \qquad (-1)^{3+1}\begin{vmatrix} 2 & 3 \\ 5 & 6 \end{vmatrix},$$

$$= 2, 4 \text{ and } -3. \tag{14}$$

Similarly the cofactors of elements of the second column are

$$(-1)\begin{vmatrix} 4 & 6 \\ 7 & 10 \end{vmatrix}, \qquad (+1)\begin{vmatrix} 1 & 3 \\ 7 & 10 \end{vmatrix} \qquad \text{and} \qquad (-1)\begin{vmatrix} 1 & 3 \\ 4 & 6 \end{vmatrix},$$

$$= 2, -11 \text{ and } 6; \tag{15}$$

and the cofactors of the elements of the third column are

$$(+1)\begin{vmatrix} 4 & 5 \\ 7 & 8 \end{vmatrix}, \qquad (-1)\begin{vmatrix} 1 & 2 \\ 7 & 8 \end{vmatrix} \qquad \text{and} \qquad (+1)\begin{vmatrix} 1 & 2 \\ 4 & 5 \end{vmatrix},$$

$$= -3, 6 \text{ and } -3. \tag{16}$$

Now consider the matrix of cofactors

$$\begin{bmatrix} 2 & 2 & -3 \\ 4 & -11 & 6 \\ -3 & 6 & -3 \end{bmatrix}$$

obtained by replacing in A every element by its cofactor just as was done in (8); i.e. the columns of the new matrix are the values in (14), (15) and (16). Let us transpose this newly formed matrix and multiply it by a scalar, the reciprocal of the determinant. As in (10), the resulting matrix is the inverse of A:

$$A^{-1} = \frac{1}{-3}\begin{bmatrix} 2 & 4 & -3 \\ 2 & -11 & 6 \\ -3 & 6 & -3 \end{bmatrix}. \tag{17}$$

Pre-multiplying (13) by (17) shows that $A^{-1}A = I$:

$$\frac{1}{-3}\begin{bmatrix} 2 & 4 & -3 \\ 2 & -11 & 6 \\ -3 & 6 & -3 \end{bmatrix}\begin{bmatrix} 1 & 2 & 3 \\ 4 & 5 & 6 \\ 7 & 8 & 10 \end{bmatrix} = \frac{1}{-3}\begin{bmatrix} -3 & 0 & 0 \\ 0 & -3 & 0 \\ 0 & 0 & -3 \end{bmatrix}$$

$$= \begin{bmatrix} 1 & 0 & 0 \\ 0 & 1 & 0 \\ 0 & 0 & 1 \end{bmatrix} = I. \tag{18}$$

Hence (17) is the inverse of A.

The diagonal matrix with non-zero elements equal to $|A|$ (-3 in this case) occurs in (18) for exactly the same reason as in the previous 2×2 example. The (-3)'s and 0's stem from (11) and (12). For example, adding the products of the cofactors of the elements of the first column of (18) with these elements gives the determinant, in accord with (11):

$$2(1) + 4(4) - 3(7) = -3 = |A|. \tag{19}$$

And multiplying the cofactors in (14) with the elements of the second and third columns of A and summing gives zeros, in accord with (12):

$$2(2) + 4(5) - 3(8) = 0,$$

and

$$2(3) + 4(6) - 3(10) = 0. \tag{20}$$

These sums of products, (19) and (20), are the elements of the first row of the matrix product in (18). Similar results hold for the cofactors of the elements of other columns of A, as shown in (15) and (16), leading to the second and third rows of (18).

The matrix A^{-1} of (17) is such that $A^{-1}A = I$, as shown in (18). So far this is the sole property of A^{-1} that has been shown. We have not shown that for A^{-1} so obtained AA^{-1} also equals I, nor have we shown that it is the unique matrix for which $A^{-1}A = AA^{-1} = I$. These are properties (a) and (b) mentioned earlier. We shall discuss them shortly, but first let us reconstruct the derivation of A^{-1}.

Initially we had A as

$$A = \begin{bmatrix} a_{11} & a_{12} & a_{13} \\ a_{21} & a_{22} & a_{23} \\ a_{31} & a_{32} & a_{33} \end{bmatrix}$$

and then formed a new matrix by replacing each element of A by its cofactor:

$$\begin{bmatrix} \mu_{11} & \mu_{12} & \mu_{13} \\ \mu_{21} & \mu_{22} & \mu_{23} \\ \mu_{31} & \mu_{32} & \mu_{33} \end{bmatrix}.$$

This was transposed, giving

$$\begin{bmatrix} \mu_{11} & \mu_{21} & \mu_{31} \\ \mu_{12} & \mu_{22} & \mu_{32} \\ \mu_{13} & \mu_{23} & \mu_{33} \end{bmatrix}, \tag{21}$$

and multiplied by the scalar $1/|A|$ to produce

$$\frac{1}{|A|}\begin{bmatrix} \mu_{11} & \mu_{21} & \mu_{31} \\ \mu_{12} & \mu_{22} & \mu_{32} \\ \mu_{13} & \mu_{23} & \mu_{33} \end{bmatrix} = A^{-1},$$

the inverse of A. Pre-multiplying A by this matrix gives the identity matrix because in the matrix product each diagonal element is a sum of products of elements of a column of A with their cofactors in $|A|$ and therefore equal to $|A|$ by equation (11); the scalar $1/|A|$ reduces them to 1. Moreover, the off-diagonal elements of the product matrix are sums of products of elements of columns of A with cofactors in $|A|$ of elements of other columns, and therefore equal zero by equation (12). Hence $A^{-1}A = I$. (The product AA^{-1} is considered later.)

The matrix (21), namely A with its elements replaced by their cofactors and then transposed, is called the *adjoint*, or sometimes the *adjugate* matrix of A. Thus the inverse, A^{-1}, can be described as the adjoint of A multiplied by the scalar $1/|A|$.

Example. The determinant of

$$A = \begin{bmatrix} 2 & 5 \\ 3 & 9 \end{bmatrix} \quad \text{is} \quad |A| = \begin{vmatrix} 2 & 5 \\ 3 & 9 \end{vmatrix} = 18 - 15 = 3.$$

The adjoint matrix is $\begin{bmatrix} 9 & -5 \\ -3 & 2 \end{bmatrix}$, and so the inverse is

$$A^{-1} = \frac{1}{3}\begin{bmatrix} 9 & -5 \\ -3 & 2 \end{bmatrix}.$$

4.   CONDITIONS FOR EXISTENCE OF THE INVERSE

The inverse A^{-1} of a matrix A, such that $A^{-1}A = I$, has been developed in the preceding paragraph as the adjoint matrix of A multiplied by the scalar $1/|A|$. The conditions imposed on A in order for A^{-1} to exist are two:

i. A^{-1} *can* exist only when A is square—as discussed in Section 5.2.

ii. A^{-1} *does* exist only if $|A|$ is non-zero. (If $|A|$ is zero the scalar factor $1/|A|$ in A^{-1} is not defined and A^{-1} does not exist. $|A|$ must therefore be non-zero for A^{-1} to exist.)

A square matrix is said to be *singular* when its determinant is zero and *nonsingular* when its determinant is non-zero. Singularity is therefore a property of square matrices only, not of rectangular matrices, and it is only nonsingular matrices that have inverses. Just as conformability is necessary for the existence of a matrix product, so is nonsingularity necessary for the existence of a matrix inverse. In both instances the necessary condition is sometimes not mentioned explicitly but it must always be satisfied.

5. PROPERTIES OF THE INVERSE

If A is a square, nonsingular matrix its inverse, A^{-1}, has the following properties. Proofs follow.

Properties

i. The inverse commutes with A, both products being the identity matrix: $A^{-1}A = AA^{-1} = I$.

ii. The inverse of A is unique. $SA = AS = I$ if and only if $S = A^{-1}$.

iii. The determinant of the inverse of A is the reciprocal of the determinant of A: $|A^{-1}| = 1/|A|$.

iv. The inverse matrix is nonsingular.

v. The inverse of A^{-1} is A: $(A^{-1})^{-1} = A$.

vi. The inverse of a transpose is the transpose of the inverse:

$$(A')^{-1} = (A^{-1})'.$$

vii. If A is symmetric so is its inverse: if $A' = A$, then $(A^{-1})' = A^{-1}$.

viii. The inverse of a product is the product of the inverses taken in reverse order, provided the inverses exist: if A^{-1} and B^{-1} exist,

$$(AB)^{-1} = B^{-1}A^{-1}.$$

ix. If A is such that its inverse equals its transpose, A is said to be an *orthogonal* matrix and $AA' = I$.

The first two of the foregoing have been referred to earlier as properties (*a*) and (*b*). Proofs of these properties are now indicated; examples are left to the reader.

Proofs.

i. It has been shown that for A^{-1} as developed, $A^{-1}A = I$. This is because the elements of the product matrix $A^{-1}A$ are sums of products of elements of columns of A multiplied by cofactors of elements of the same and different columns of A. In like fashion the product AA^{-1} has elements that are sums of products of elements of rows of A multiplied by cofactors of elements of the same and different rows of A. Hence $AA^{-1} = I$.

ii. Suppose that A^{-1} is not a unique inverse of A and that S is another inverse different from A^{-1}, such that $SA = I$. Then post-multiplying $SA = I$ by A^{-1} gives

$$SAA^{-1} = IA^{-1} = A^{-1}$$

and because, from (i), $AA^{-1} = I$ this means

$$SI = A^{-1};$$

i.e., $S = A^{-1}$, thus proving that A^{-1} is the unique inverse of A.

iii and iv. It was shown in Section 4.5 that, if A and B are two square matrices of the same order, $|A| |B| = |AB|$. Therefore

$$|A| |A^{-1}| = |AA^{-1}| = |I| = 1, \quad \text{and so} \quad |A^{-1}| = 1/|A|.$$

The nonsingularity of A^{-1} follows from this.

v, vi and vii. Consider the identity $I = A^{-1}A$. Pre-multiplying it by $(A^{-1})^{-1}$ leads to the result $(A^{-1})^{-1} = A$; transposing it and pre-multiplying by $(A')^{-1}$ leads to $(A')^{-1} = (A^{-1})'$. For $A' = A$ the latter becomes $A^{-1} = (A^{-1})'$.

viii. Provided A and B are both square, of the same order and nonsingular we may write

$$B^{-1}A^{-1}AB = B^{-1}(A^{-1}A)B = B^{-1}IB = B^{-1}B = I,$$

and so, by post-multiplying by $(AB)^{-1}$, we have $(AB)^{-1} = B^{-1}A^{-1}$. The reversal rule that applies to transposing a product therefore applies also to inverting it.

ix. If $A^{-1} = A'$, then $AA' = A'A = I$. The reader might satisfy himself that this is true for $A = \frac{1}{15} \begin{bmatrix} 5 & -14 & 2 \\ -10 & -5 & -10 \\ 10 & 2 & -11 \end{bmatrix}$. As already stated, such a matrix is said to be *orthogonal*.

6. SOME USES OF INVERSES

Uses of the inverse matrix are so many and varied that only a selection can be given. Four have been chosen for illustration.

a. Equations

Repeated reference has already been made to solving equations. A further illustration is a problem in production capacity.

Illustration. Suppose a manufacturer of n different products has production capacity in n different departments. We let b_i represent the total available capacity in department i and a_{ij} represent the amount of the capacity in department i that is required for each unit of product j. Then if x_j represents the number of units of product j which are produced, the following equations indicate how the available capacity could be used with no surplus or deficit:

$$
\begin{bmatrix}
a_{11} & a_{12} & \cdots & a_{1n} \\
a_{21} & a_{22} & \cdots & a_{2n} \\
\cdot & & & \cdot \\
\cdot & & & \cdot \\
\cdot & & & \cdot \\
a_{n1} & a_{n2} & \cdots & a_{nn}
\end{bmatrix}
\begin{bmatrix}
x_1 \\ x_2 \\ \cdot \\ \cdot \\ \cdot \\ x_n
\end{bmatrix}
=
\begin{bmatrix}
b_1 \\ b_2 \\ \cdot \\ \cdot \\ \cdot \\ b_n
\end{bmatrix}.
$$

Hence, if A denotes the matrix of requirements, x the vector of production quantities and b the vector of capacities, $Ax = b$. The production quantities which satisfy this relationship can be obtained by solving this equation for x using the inverse of A, i.e. $x = A^{-1}b$.

b.   Input-output analysis

Input-output analysis was developed by Leontief (1936) as a vehicle for analyzing the interdependencies of the various sectors of the economy. In 1949 the Department of Commerce published the first officially compiled input-output table for the United States, for the year 1947. A table for 1958 has appeared recently and newer tables are presently being compiled. We show here the basic matrix algebra of input-output analysis.

Consider an economy with n sectors, each producing a single commodity.

Let $x_i =$ gross output of commodity i,

$\quad a_{ij} =$ amount of commodity i used in producing one unit of commodity j,

and $y_i =$ final demand for commodity i.

Now, gross product for each commodity must equal the sum of the amounts of that commodity used in producing all the commodities, plus the final demand for the commodity itself; in symbols

$$
x_i = \sum_{j=1}^{n} a_{ij} x_j + y_i \qquad \text{for} \quad i = 1, 2, \ldots, n.
$$

In matrix notation

$$
x = Ax + y
$$

where the vectors x and y and the matrix A are

$$
x = \{x_i\}, \qquad y = \{y_i\} \quad \text{and} \quad A = \{a_{ij}\} \qquad \text{for} \quad i, j = 1, 2, \ldots, n.
$$

The matrix A is called the input-output coefficient matrix; all its elements are, by definition, non-negative, as will be the elements of x if those of y are non-negative also.

Assuming the a_{ij}'s remain constant over some period of time, we can determine the gross output of all sectors (x) required for any level of final demand (y) by solving the above equation for x. Rearranging terms,

$$(I - A)x = y$$

and, on pre-multiplying by $(I - A)^{-1}$, provided $(I - A)$ is nonsingular, we have

$$x = (I - A)^{-1}y. \tag{22}$$

Illustration. Consider a grossly simplified four-sector economy in which there are two industries (agriculture and manufacturing), one primary factor of production (labor) and a governmental sector that consumes output from both industries and utilizes labor as well. In this simple example the government produces nothing for the economy and its consumption from the industry and labor sectors represents the final demand for the commodities produced by these sectors. In its production processes each industry uses some output of the other industry as well as labor, and the labor force requires output from both industries as well as labor services to sustain itself. Labor can be freely imported and exported, so there is never any unemployment or excess demand for labor, and capital stock and inventories are held constant through time. Suppose we observe the physical flow of goods among the four sectors of the economy for a particular period and compile an input-output table for the economy as shown in Table 1. Adding across each row gives the total output of each industry and the total labor force employed. Since output has been measured in physical units, adding down columns would be meaningless. Yet, any column taken as a whole is meaningful; it shows the inputs a given sector requires in order to produce its total output; in effect, it describes the production function of that sector.

TABLE 1. INPUT-OUTPUT TABLE

Production Sector	Consumption Sector				Total
	Agriculture	Manufacturing	Labor	Final Demand by Government	
Agriculture (tons)	600	400	1,400	600	3,000
Manufacturing (machines)	1,500	800	700	1,000	4,000
Labor (employees)	900	4,800	700	600	7,000

For example, the second column describes the basic production process being currently utilized in manufacturing. To produce 4,000 machines, the manufacturing industry requires 400 tons of agricultural output, 800 machines and 4,800 employees.

Were this hypothetical economy to advance technologically and/or were the relative prices of the three factors to change over time, we would expect the production processes implied by the input-output table to change as well; but suppose prices are relatively stable and that technology changes slowly. Further, suppose that the governmental sector now wishes to use (a) 1,000 tons of agricultural products, (b) 1,200 machines and (c) 800 employees in the next time period. Given that these governmental desires represent final demand, what must be the size of the labor force, and what will be the levels of gross output by each of the industrial sectors? That is, given that the new value of the y vector is

$$y = \begin{bmatrix} 1,000 \\ 1,200 \\ 800 \end{bmatrix},$$

what will the corresponding x be? We can use equation (22) to answer these questions if we can derive the input-output coefficient matrix A from the input-output table for the economy.

Recall that the element a_{ij} of A is defined as the amount of commodity i used in producing one unit of commodity j. If we divide the ijth element of the input-output table by the sum of row j, the result is a_{ij}; that is, if we divide the amount of commodity i sold to sector j by the total output of sector j, we get the amount of commodity i used in producing one unit of commodity j. For example, $a_{12} = \dfrac{400}{4,000} = 0.1$: it takes 0.1 ton of agricultural output to produce one machine. In this way the input-output coefficient matrix is obtained:

$$A = \begin{bmatrix} \dfrac{600}{3,000} & \dfrac{400}{4,000} & \dfrac{1,400}{7,000} \\[2mm] \dfrac{1,500}{3,000} & \dfrac{800}{4,000} & \dfrac{700}{7,000} \\[2mm] \dfrac{900}{3,000} & \dfrac{4,800}{4,000} & \dfrac{700}{7,000} \end{bmatrix} = \begin{bmatrix} 0.2 & 0.1 & 0.2 \\ 0.5 & 0.2 & 0.1 \\ 0.3 & 1.2 & 0.1 \end{bmatrix}$$

and

$$(I - A)^{-1} = \frac{1}{0.264} \begin{bmatrix} 0.60 & 0.33 & 0.17 \\ 0.48 & 0.66 & 0.18 \\ 0.84 & 0.99 & 0.59 \end{bmatrix}.$$

Before using (22) to answer the questions raised earlier, these calculations can be verified by substituting the initial government consumption (column 4 of Table 1) as the final demand vector y in equation (22). The resulting x should be the gross outputs shown in the final column of Table 1. Thus, letting

$$y = \begin{bmatrix} 600 \\ 1,000 \\ 600 \end{bmatrix}$$

in equation (22) we have

$$x = (I - A)^{-1}y = \frac{1}{0.264} \begin{bmatrix} 0.60 & 0.33 & 0.17 \\ 0.48 & 0.66 & 0.18 \\ 0.84 & 0.99 & 0.59 \end{bmatrix} \begin{bmatrix} 600 \\ 1,000 \\ 600 \end{bmatrix} = \begin{bmatrix} 3,000 \\ 4,000 \\ 7,000 \end{bmatrix};$$

we see our calculations are correct.

Now, to determine the level of agricultural and manufacturing activity and the labor force required under the new demand levels, we calculate x for the new value of y:

$$x = (I - A)^{-1}y = \frac{1}{0.264} \begin{bmatrix} 0.60 & 0.33 & 0.17 \\ 0.48 & 0.66 & 0.18 \\ 0.84 & 0.99 & 0.59 \end{bmatrix} \begin{bmatrix} 1,000 \\ 1,200 \\ 800 \end{bmatrix} = \begin{bmatrix} 4287.88 \\ 5363.64 \\ 9469.70 \end{bmatrix}.$$

Thus, in order to meet the new demand levels, approximately 4,288 tons of agricultural output and 5,364 machines will have to be produced and 9,470 employees will be required.

Three general comments can be made concerning input-output analysis. First, unlike Table 1, most input-output tables are expressed in dollar transactions rather than physical units. This avoids the problem of defining units and making units comparable. In Table 1 we had manufacturers producing machines, but it is unreasonable to assume that machines consumed by manufacturers, the agriculture sector, the labor force and the government are all identical. One way to aggregate various commodities is by weighting each according to its dollar value. Hence, published input-output tables are usually expressed in terms of dollar transactions. In Exercise 15 at the end of this chapter, the reader is asked to restate and solve the problem emanating from Table 1 in terms of dollar values.

Second, it seems inappropriate to treat the demand for goods and services of the household sector of an economy in the same fashion that the demand for a manufacturing sector would be treated. Recent tables have therefore considered household demand as a component of final demand (y) rather than intermediate demand (Ax), and have thus avoided the implication that there is a "production function" for the labor force, as column 3 in Table 1 implies.

Third, the elements of $(I - A)^{-1}$ contain economic meaning. Let $B = (I - A)^{-1}$. The ijth element of B, b_{ij}, is the amount of commodity i that the economy is required to produce in order to deliver one unit of commodity j as final output. For instance, in the above illustration, in order to produce one machine as a final product, the agricultural sector must supply $\dfrac{0.330}{0.264} =$ 1.25 tons of output. Why so much, given that it takes only 0.1 (a_{12}) units of agricultural output to produce one machine? The reason is that labor and machines are required as well; and each of these in turn requires agricultural output for their production, as does the agricultural output itself. Thus, while in the total of 1.25 tons only 0.1 ton is required directly, 1.15 tons are required indirectly to obtain one machine.

Equation (22) may also apply to incremental changes in final demand rather than to absolute values. Thus if y^* is the vector of such changes, $x^* = (I - A)^{-1}y^*$ is the vector of changes in gross output that are needed to satisfy the changes in demand, y^*. Hence, in the above example, if

$$y^* = \begin{bmatrix} 400 \\ 200 \\ 200 \end{bmatrix}, \qquad x^* = \begin{bmatrix} 1{,}287.88 \\ 1{,}363.64 \\ 2{,}469.70 \end{bmatrix}.$$

In this instance, we see that y^* is equal to the changes in value of the y vectors used earlier and, correspondingly, x^* is equal to the changes in the x vector. However, any values, positive or negative, may be used for elements of y^*. Furthermore, one may not always want the complete vector x^*, but only certain elements of it. In that case just the appropriate rows of $(I - A)^{-1}$ are used. Thus if the effect of the changes y^* on labor alone were required, just the last row of $(I - A)^{-1}$ would be used to derive x_3^*:

$$x_3^* = \frac{1}{0.264} [0.84 \quad 0.99 \quad 0.59] \begin{bmatrix} 400 \\ 200 \\ 200 \end{bmatrix} = 2{,}469.7$$

as before, that is, 2,470 additional laborers must be obtained in order for the economy to meet the altered demand.

For varying national and international economic situations, economic planners are often able to predict changes in the vector of final demand for the national economy. A detailed and accurate $(I - A)^{-1}$ matrix then enables the consequences of the changed demand to be assessed. Consider, for example, how useful a matrix like this would be to the Council of Economic

Advisers and the Board of Governors of the Federal Reserve System in developing fiscal and monetary policies to meet such changes while attempting, at the same time, to maintain full employment and stable prices.

c. Algebraic simplifications

There are numerous ways in which inverse matrices occur in matrix algebra in some sense analogous to division in scalar algebra. One is as follows. For x being a scalar other than unity, the result

$$1 + x + x^2 + \cdots + x^{n-1} = \frac{x^n - 1}{x - 1}$$

has as its matrix counterpart

$$I + X + X^2 + \cdots + X^{n-1} = (X - I)^{-1}(X^n - I)$$

where $(X - I)$ is a nonsingular matrix.

Illustration. Suppose a 6×6 transition probability matrix can be written as a partitioned matrix (Section 3.4)

$$P = \begin{bmatrix} I & 0 \\ C & B \end{bmatrix},$$

where C and B take particular forms based on the description of the six states, and I refers to all the trapping states (a trapping state is one in which the probability of remaining there is unity). The matrix of probabilities after two transitions is

$$P^2 = \begin{bmatrix} I & 0 \\ C & B \end{bmatrix}^2 = \begin{bmatrix} I & 0 \\ C + BC & B^2 \end{bmatrix},$$

since B and C are, by definition, conformable for the product BC; and after n transitions

$$P^n = \begin{bmatrix} I & 0 \\ C & B \end{bmatrix}^n = \begin{bmatrix} I & 0 \\ (I + B + B^2 + \cdots + B^{n-1})C & B^n \end{bmatrix}$$

which, by use of the result given above, is

$$P^n = \begin{bmatrix} I & 0 \\ (B - I)^{-1}(B^n - I)C & B^n \end{bmatrix}.$$

For large n this form of P^n is more readily calculated than is P^n directly. Also, it shows the relationship of P^n to B and C and provides easy consideration of the limiting form of P^n for infinitely large n.

d. Determinants of partitioned matrices

Evaluating the determinant of a partitioned matrix is sometimes aided by the availability of inverse matrices. Suppose a square matrix M is partitioned as

$$M = \begin{bmatrix} A & B \\ C & D \end{bmatrix},$$

such that D is square and nonsingular; i.e., its inverse, D^{-1}, exists. We seek to evaluate the determinant $|M|$. Recalling equation (7) of Section 4.5, we see that for any matrix X

$$\begin{vmatrix} I & 0 \\ X & I \end{vmatrix} = |I| \, |I| = 1.$$

Since this result is true for any matrix X it is true for $X = -D^{-1}C$, so that

$$\begin{vmatrix} I & 0 \\ -D^{-1}C & I \end{vmatrix} = 1.$$

Using this result to multiply $|M|$ by 1 we get

$$|M| = |M| \, (1) = \begin{vmatrix} A & B \\ C & D \end{vmatrix} \begin{vmatrix} I & 0 \\ -D^{-1}C & I \end{vmatrix}.$$

Then, since for any two square matrices P and Q of the same order $|P| \, |Q| = |PQ|$ (see Section 4.5) we have

$$|M| = \begin{vmatrix} \begin{bmatrix} A & B \\ C & D \end{bmatrix} \begin{bmatrix} I & 0 \\ -D^{-1}C & I \end{bmatrix} \end{vmatrix} = \begin{vmatrix} A - BD^{-1}C & B \\ O & D \end{vmatrix}.$$

On using a Laplace expansion[1] of the right-hand side this gives

$$|M| = |D| \, |A - BD^{-1}C|,$$

i.e.
$$\begin{vmatrix} A & B \\ C & D \end{vmatrix} = |D| \, |A - BD^{-1}C|; \tag{23}$$

and with A^{-1} existing it can be shown in similar fashion that

$$\begin{vmatrix} A & B \\ C & D \end{vmatrix} = |A| \, |D - CA^{-1}B|. \tag{24}$$

---

[1] See Section 4.7d.

Example. The following partitioned matrix,

$$Q = \begin{bmatrix} 21 & 37 & 1 & -1 & 2 \\ 18 & 12 & 3 & 5 & 7 \\ \hline -4 & 6 & 2 & 0 & 0 \\ 6 & 0 & 0 & 3 & 0 \\ 8 & 4 & 0 & 0 & 4 \end{bmatrix} = \begin{bmatrix} A & B \\ C & D \end{bmatrix},$$

has determinant

$$|Q| = |D| \, |A - BD^{-1}C|$$

$$= 24 \left| \begin{bmatrix} 21 & 37 \\ 18 & 12 \end{bmatrix} - \begin{bmatrix} 1 & -1 & 2 \\ 3 & 5 & 7 \end{bmatrix} \begin{bmatrix} \frac{1}{2} & 0 & 0 \\ 0 & \frac{1}{3} & 0 \\ 0 & 0 & \frac{1}{4} \end{bmatrix} \begin{bmatrix} -4 & 6 \\ 6 & 0 \\ 8 & 4 \end{bmatrix} \right|$$

$$= 24 \begin{vmatrix} 21 & 32 \\ 0 & -4 \end{vmatrix} = -2{,}016.$$

7. INVERSION BY PARTITIONING

Obtaining the inverse of a matrix can often be simplified by partitioning it into four sub-matrices in a manner such that those at the top left and bottom right are square; i.e.

$$M = \begin{bmatrix} A & B \\ C & D \end{bmatrix} \tag{25}$$

where A and D are square. If the corresponding partitioned form of M^{-1} is

$$M^{-1} = \begin{bmatrix} X & Y \\ Z & W \end{bmatrix}, \tag{26}$$

post-multiplying (25) by (26) gives

$$\begin{bmatrix} A & B \\ C & D \end{bmatrix} \begin{bmatrix} X & Y \\ Z & W \end{bmatrix} = I = \begin{bmatrix} I & 0 \\ 0 & I \end{bmatrix}.$$

Hence

$$AX + BZ = I, \tag{27}$$

$$AY + BW = 0, \tag{28}$$

$$CX + DZ = 0 \tag{29}$$

and

$$CY + DW = I. \tag{30}$$

For the moment it is supposed that both A and D are nonsingular. Then from (29)

$$Z = -D^{-1}CX, \qquad (31)$$

which substituted in (27) leads to

$$X = (A - BD^{-1}C)^{-1}. \qquad (32)$$

Similarly from (28) $\qquad Y = -A^{-1}BW \qquad (33)$

and so from (30) $\qquad W = (D - CA^{-1}B)^{-1}. \qquad (34)$

These results demand inverting four matrices. Expressions more convenient from a computational point of view can, however, be derived. First note that the identity $M^{-1}M = I$ also holds, so that from (25) and (26)

$$\begin{bmatrix} X & Y \\ Z & W \end{bmatrix} \begin{bmatrix} A & B \\ C & D \end{bmatrix} = \begin{bmatrix} I & 0 \\ 0 & I \end{bmatrix}$$

and hence $\qquad XA + YC = I, \qquad (35)$

$$XB + YD = 0, \qquad (36)$$

$$ZA + WC = 0, \qquad (37)$$

and $\qquad ZB + WD = I. \qquad (38)$

From (36) $\qquad Y = -XBD^{-1}$

and from (38) $\qquad W = D^{-1} - ZBD^{-1}.$

Using these results and X and Z from (32) and (31) we then have expressions that require inverting only two matrices:

$$X = (A - BD^{-1}C)^{-1},$$
$$Y = -XBD^{-1},$$
$$Z = -D^{-1}CX,$$
$$W = D^{-1} - ZBD^{-1}.$$

These expressions can be calculated by proceeding in the following sequence:

Procedure a.

i. D^{-1} $\qquad\qquad$ ii. $D^{-1}C$

iii. $BD^{-1}C$ $\qquad\qquad$ iv. $A - BD^{-1}C$

[v.] $X = (A - BD^{-1}C)^{-1}$ $\qquad\qquad$ vi. BD^{-1}

[vii.] $Y = -XBD^{-1}$ $\qquad\qquad$ [viii.] $Z = -D^{-1}CX$

ix. ZBD^{-1} $\qquad\qquad$ [x.] $W = D^{-1} - ZBD^{-1}.$

The steps enclosed in square brackets are those which yield the elements of M^{-1}. Note that this procedure, evolved from equations (31), (32), (36) and (38), requires the existence of D^{-1} but not A^{-1}. In a similar manner equations (33), (34), (35) and (37) can be used to derive an alternative procedure that requires the existence of A^{-1} but not D^{-1}:

Procedure b.

    i. A^{-1} ii. $A^{-1}B$

    iii. $CA^{-1}B$ iv. $D - CA^{-1}B$

    [v.] $W = (D - CA^{-1}B)^{-1}$ vi. CA^{-1}

    [vii.] $Z = -WCA^{-1}$ [viii.] $Y = -A^{-1}BW$

    ix. YCA^{-1} [x.] $X = A^{-1} - YCA^{-1}$.

The choice of which of these procedures shall be used in any particular case will depend on any properties pertaining to M and its sub-matrices that may simplify the calculations. For example, if

$$M = \begin{bmatrix} A & B \\ C & D \end{bmatrix} = \left[\begin{array}{cc:ccc} 1 & 2 & 1 & 2 & 3 \\ 3 & 4 & 9 & 4 & 1 \\ \hdashline 6 & 3 & 7 & 0 & 0 \\ 1 & 0 & 0 & 8 & 0 \\ 4 & 2 & 0 & 0 & 9 \end{array} \right]$$

procedure **a** is preferable because it involves D^{-1} and $(A - BD^{-1}C)^{-1}$, which are relatively easy to calculate since D is diagonal and A is only 2×2. Procedure **b** would require inverting A, which is easy enough, but $(D - CA^{-1}B)$ has also to be inverted—a non-diagonal 3×3. The difference in effort between the two procedures is small in this example, but it would not be were D of order 30 say, and diagonal, and A only 4×4 say. The nature of M should therefore be carefully studied before embarking on the inversion.

These procedures simplify a little when M is symmetric, $M = M'$, for then $A = A'$, $C = B'$, $D = D'$ and $X = X'$, $Z = Y'$ and $W = W'$. That is,

$$M = M' = \begin{bmatrix} A & B \\ B' & D \end{bmatrix} \text{ and } M^{-1} = (M^{-1})' = \begin{bmatrix} X & Y \\ Y' & W \end{bmatrix}. \text{ The steps for}$$

calculating M^{-1} are then as follows:

Procedure aa.

i. D^{-1} ii. $BD^{-1} = (D^{-1}B')'$

iii. $BD^{-1}B'$ iv. $A - BD^{-1}B'$

[v.] $X = (A - BD^{-1}B')^{-1}$ [vi.] $Y = -XBD^{-1}$

vii. $Y'BD^{-1}$ [viii.] $W = D^{-1} - Y'BD^{-1}$.

Compared to procedure **a** there are two less steps, resulting from the symmetry of M. In a similar manner procedure **b** becomes

Procedure bb.

i. A^{-1} ii. $B'A^{-1} = (A^{-1}B)'$

iii. $B'A^{-1}B$ iv. $D - B'A^{-1}B$

[v.] $W = (D - B'A^{-1}B)^{-1}$ [vi.] $Y = -A^{-1}BW$

vii. $YB'A^{-1}$ [viii.] $X = A^{-1} - YB'A^{-1}$.

It is an interesting exercise to show algebraically that the sub-matrices of M^{-1} as given by procedures **a** and **b** are equivalent, as are those of procedures **aa** and **bb**.

8. INVERTING MATRICES ON HIGH-SPEED COMPUTERS

Our development of the matrix inverse has been entirely in terms of determinants. This procedure has been described in detail because it demonstrates so clearly the derivation of each element of the inverse. But it is not an efficient procedure for inverting matrices of large order, since it results in excessive computing. Alternative methods are available, however, and even though they involve considerable effort (especially if carried out on a desk calculator) they are preferable to calculating determinants. However, the decision as to which of these alternative methods to use has nowadays largely become the concern of computer programmers, numerical analysts and other personnel associated with computer installations. As a result, the presence of computers has almost entirely removed this decision from the concern of those having a specific matrix to invert. These conditions render it unnecessary to discuss the mechanics of the different methods available for inverting a matrix using a computer.[1]

Two matters only will be discussed in connection with inverting matrices by computers. The first is the general procedure used in solving a set of linear equations with the aid of a desk calculator—a procedure which is essentially

---

[1] The reader who wishes to pursue the subject in detail might turn to Dwyer (1951), Hildebrand (1956), Bodewig (1959), Fadeeva (1958) or Wilkinson (1963), to mention only a few of the many sources available.

the foundation of any method of inverting a matrix. The second is the subject of rounding error—since both desk calculators and computers have limited capacity in the number of decimal digits that they carry for any particular value entered or generated therein, there is always a problem of rounding errors, the cumulative effect of which can sometimes lead to erroneous results. We now briefly discuss these two topics.

a. Solving equations

Suppose we have a matrix

$$A = \begin{bmatrix} 2 & 5 & 7 \\ 3 & 9 & 15 \\ 5 & 16 & 20 \end{bmatrix}$$

whose inverse we wish to find by a method other than evaluating $|A|$ and the co-factor of every element. One method would be to imagine having the equations

$$A \begin{bmatrix} x \\ y \\ z \end{bmatrix} = \begin{bmatrix} a \\ b \\ c \end{bmatrix} \tag{39}$$

which could be solved explicitly for x, y and z by customary procedures of successive elimination. (Both here and for the remainder of the chapter the symbols x, y, z and a, b, c are scalar quantities.)

Pre-multiplying by A^{-1} gives the solution as

$$\begin{bmatrix} x \\ y \\ z \end{bmatrix} = A^{-1} \begin{bmatrix} a \\ b \\ c \end{bmatrix}, \tag{40}$$

so long as A^{-1} exists.

Hence if the solutions of (39) for x, y and z obtained explicitly in terms of a, b and c were written down in matrix form, the resulting matrix would be A^{-1}. Let us carry through this process.

Equations (39) are

$$2x + 5y + 7z = a, \tag{41}$$

$$3x + 9y + 15z = b \tag{42}$$

and

$$5x + 16y + 20z = c. \tag{43}$$

Dividing (41) by 2 gives

$$x + 2.5y + 3.5z = 0.5a, \tag{44}$$

and subtracting 3 times this from (42) and 5 times it from (43) gives

$$1.5y + 4.5z = b - 1.5a \tag{45}$$

and
$$3.5y + 2.5z = c - 2.5a. \tag{46}$$

Dividing (45) by 1.5 and multiplying (46) by 2 yields

$$y + 3z = b/1.5 - a \tag{47}$$

and
$$7y + 5z = 2c \quad - 5a. \tag{48}$$

Subtract 7 times (47) from (48) and we obtain

$$-16z = 2c - 5a - 7(b/1.5 - a)$$

so that
$$z = (2a - 7b/1.5 + 2c)/(-16)$$
$$= (-3a + 7b - 3c)/24. \tag{49}$$

Substituting z into (47) leads to

$$y = (-15a - 5b + 9c)/24 \tag{50}$$

and putting both y and z in (44) gives

$$x = (60a - 12b - 12c)/24. \tag{51}$$

The last three results can be assembled in matrix form as

$$\begin{bmatrix} x \\ y \\ z \end{bmatrix} = \tfrac{1}{24} \begin{bmatrix} 60 & -12 & -12 \\ -15 & -5 & 9 \\ -3 & 7 & -3 \end{bmatrix} \begin{bmatrix} a \\ b \\ c \end{bmatrix} \tag{52}$$

and from comparison with (40) we conclude that

$$A^{-1} = \tfrac{1}{24} \begin{bmatrix} 60 & -12 & -12 \\ -15 & -5 & 9 \\ -3 & 7 & -3 \end{bmatrix}.$$

Multiplication with A will show that this is so. This whole procedure is, of course, very tedious and would be even more so were A to be of order larger than three. Though tedious, it is not difficult in concept, and sustained effort combined with arithmetic accuracy will always yield correct results.

The arithmetic processes just outlined are the foundation of most computer methods of inverting a matrix, although no dummy variables such as x, y, z, a, b and c are used, only the arithmetic operations which determine their coefficients at each stage. Several names are familiarly associated with procedures for organizing (and hence programming) these calculations, for example, Gauss, Doolittle, Fox and Crout. The methods which bear these names are largely, though not entirely, variants of each other directed towards improving the efficiency of organizing the calculations rather than being distinctly different methods of calculation. The reader who follows the arithmetic just described will therefore have a basic understanding of the calculations involved in most computer methods.

Other benefits also accrue from using computers to invert matrices. Since individual steps of the calculations are essentially the same for both large and small matrices, although more numerous for large matrices, computers can readily be made to cope with matrices of any order (limited solely by the size of the computer). Further, the accuracy of the computer far exceeds that of a desk calculator operated by *homo sapiens*, who needs to repeat specific steps of a calculation in order to check on accuracy. Generally speaking, a computer makes no errors (although it is a high-speed idiot and will carry out erroneous instructions just as quickly as it will correct ones) and requires no repetitive operations for the sake of accuracy. The greatest advantage of computers, though, is their speed, for it makes feasible tasks that might otherwise never be undertaken. For example, modern computers can invert a matrix of order 100 in a matter of minutes, whereas, in contrast, instances are known of desk calculator work requiring six weeks for the inversion of a 40×40 matrix.

In general, then, for reasons of size, accuracy and speed, computers are a most useful tool for inverting matrices, and their existence has eliminated the chore of the bothersome arithmetic involved.

b. Rounding error

Although the accuracy of high-speed computers has just been put forward as one of their advantages, there is nevertheless one problem concerning accuracy which must be appreciated—the problem of rounding error. It does, of course, arise with desk calculators also, since in both computers and calculators there is a limited number of decimal digits that can be carried in each value handled by the machine. Even though modern computers can handle numbers containing very many digits, the rounding error problem still exists. No matter how large the computer, nor how great its capacity, it can never contain the value $4\frac{2}{3}$ as a decimal number with complete, 100 per cent accuracy. Some computers will simply truncate their numbers, taking $4\frac{2}{3}$, for example, as 4.666 . . . to as many sixes as capacity will allow; others round off their numbers taking $4\frac{2}{3}$ as 4.6667, again with as many sixes as capacity allows. Understandably this can lead to error when large numbers of multiplication and division operations take place as in inverting a large matrix. This creates a problem, especially since the situations in which error will occur cannot be specified with certainty, and a mathematical discussion of the cumulative effects of rounding errors becomes very involved in anything but simple situations (see Wilkinson, 1963). At best one can usually say that trouble does not occur very often. At worst one must be cognizant of the possibility of trouble of this sort on any occasion that a computer is used.

Computer methods for inverting a matrix are not based on computing the determinant of the matrix and the cofactors of each element. But suppose they were. And suppose the determinant of a matrix was truly zero: there would then be no inverse. However, rounding errors incurred while evaluating the determinant might lead to its being given a non-zero value.

Example. The determinant

$$|M| = \begin{vmatrix} 0.12 & -0.21 & 0.12 \\ 0.23 & 0.17 & 0.19 \\ 0.15 & 0.31 & 0.11 \end{vmatrix}$$

expanded by elements of its first row is

$$0.12 \begin{vmatrix} 0.17 & 0.19 \\ 0.31 & 0.11 \end{vmatrix} + 0.21 \begin{vmatrix} 0.23 & 0.19 \\ 0.15 & 0.11 \end{vmatrix} + 0.12 \begin{vmatrix} 0.23 & 0.17 \\ 0.15 & 0.31 \end{vmatrix}$$

$$= 0.12(0.0187 - 0.0589) + 0.21(0.0253 - 0.0285)$$
$$+ 0.12(0.0713 - 0.0255)$$
$$= 0.12(-0.0402) + 0.21(-0.0032) + 0.12(0.0458)$$
$$= -0.004824 - 0.000672 + 0.005496 = 0. \tag{53}$$

Suppose this determinant were evaluated on a computer of very limited capacity, one capable of carrying only three decimal digits, meaning by this that at every stage numbers would be rounded off to the nearest integer in the third decimal place, a 5 in the fourth place being rounded up (a common procedure in computers). Were exactly the same method of calculation used as just shown expression (53) would become, as a result of the rounding-off procedure,

$$0.12(0.019 - 0.059) + 0.21(0.025 - 0.029) + 0.12(0.071 - 0.026)$$
$$= 0.12(-0.040) + 0.21(-0.004) + 0.12(0.045).$$

The exact value of this is

$$-0.00480 - 0.00084 + 0.00540,$$

but because of rounding off its computed value would be

$$-0.005 - 0.001 + 0.005$$
$$= -0.001,$$

which is not the correct value of the determinant, namely zero.

The above example demonstrates how rounding error can lead to computers producing erroneous results. It is, of course, an oversimplification of the general problem, which is one that becomes quite difficult to follow when more numerous arithmetic operations are involved. Furthermore, the example might suggest that rounding error problems can readily become acute and perhaps frequently do so. For several reasons this is not the case at all. First of all, any respectable computer can efficiently handle many more than three decimal digits, and rounding errors always originate in the least significant one. By accumulation through numerous arithmetic operations they many, however, affect the last two, three or even four digits and thus a computer capable of handling eleven significant digits might reliably give results correct to seven or eight digits, which, of course, is sufficient for many problems. A second reason that rounding error is not a frequent problem is that computational methods for inverting matrices have been devised to minimize its effect (see Dwyer, 1951, and Wilkinson, 1963, for example); and third, computer programming techniques have been developed for handling numbers with very many decimals (for example, floating decimal arithmetic and multiple-word length operations). The result is that computing establishments seldom return an inverted matrix that is "bugged" by rounding error. But such an event *can* occur and those who rely on computers for obtaining their inverse matrices should be aware of the possibility and the nature of the consequences.

Consider what would have happened in the foregoing example if our "three-decimal" computer, having found $|M| = -0.001$, had proceeded to derive M^{-1} using the method of cofactors. Rounding off all arithmetic values to three decimal places would have resulted in

$$\frac{1}{-0.001} \begin{bmatrix} -0.040 & 0.060 & -0.060 \\ 0.004 & -0.005 & 0.005 \\ 0.045 & -0.069 & 0.068 \end{bmatrix} = \begin{bmatrix} 40 & -60 & 60 \\ -4 & 5 & -5 \\ -45 & 69 & -68 \end{bmatrix}.$$

This is the kind of result that may sometimes occur: rounding errors can cause true zero values to be computed as very small, non-zero quantities, and division by these small quantities leads to an alleged inverse which may have some unduly large elements. This might be just the clue that something is in error.

It is important to know the possible consequences of such errors, to scrutinize results carefully and even to check them by using a desk calculator to derive an element or two of the product of the matrix with its inverse. Total prevention of rounding errors is impossible, but one useful method of reducing them is to see that the elements of a matrix that has to be inverted are all of approximately the same order of magnitude. This can often be achieved by pre- or post-multiplying the matrix by a diagonal matrix before inverting it in order to amend the elements of a row or column to bring them into line with the others. It is necessary, of course, to remember to appropriately multiply the inverse matrix obtained from the computer by the inverse of the diagonal matrix to derive the inverse of the original matrix.

Example. Consider the matrix:

$$A = \begin{bmatrix} 2 & 1,365 \\ 3 & 2,050 \end{bmatrix}.$$

Elements of the second column are much larger than those of the first, which leads to expressing A as a product of two matrices B and D, where D is diagonal, with

$$B = \begin{bmatrix} 2 & 2.73 \\ 3 & 4.10 \end{bmatrix} \quad \text{and} \quad D = \begin{bmatrix} 1 & 0 \\ 0 & 500 \end{bmatrix}.$$

Then, since $A = BD$,

$$A^{-1} = D^{-1}B^{-1},$$

or

$$\begin{bmatrix} 2 & 1,365 \\ 3 & 2,050 \end{bmatrix}^{-1} = \begin{bmatrix} 1 & 0 \\ 0 & 500 \end{bmatrix}^{-1} \begin{bmatrix} 2 & 2.73 \\ 3 & 4.10 \end{bmatrix}^{-1}$$

$$= \begin{bmatrix} 1 & 0 \\ 0 & 0.002 \end{bmatrix} \frac{1}{0.01} \begin{bmatrix} 4.10 & -2.73 \\ -3 & 2 \end{bmatrix} = \begin{bmatrix} 410 & -273 \\ -0.6 & 0.4 \end{bmatrix}.$$

9.   APPENDIX: LEFT AND RIGHT INVERSES

In introducing the idea of an inverse matrix in Section 2 it was pointed out that A^{-1} could exist only for A being square, for this is the only condition under which both of the products $A^{-1}A$ and AA^{-1} exist. Thereafter inverse matrices have been considered only in terms of square matrices. However, in regard to rectangular matrices the following question can be asked: Does there exist, for any rectangular matrix $A_{r \times c}$ $(r \neq c)$, a matrix $B_{c \times r}$ for which $AB = I_r$ and $BA = I_c$? The answer is "no," there is no such matrix B, as is now shown.

Proof. There is no loss of generality in assuming $r < c$ and in writing A and B in partitioned form as

$$A = [X_{r \times r} \quad Y_{r \times (c-r)}] \quad \text{and} \quad B = \begin{bmatrix} Z_{r \times r} \\ W_{(c-r) \times r} \end{bmatrix}.$$

Then $BA = \begin{bmatrix} ZX & ZY \\ WX & WY \end{bmatrix}$ and in order for this to equal I_c it is necessary that

$ZX = I_r$, which occurs only if X is nonsingular and $Z = X^{-1}$. It is also necessary that $WX = 0$, which implies $W = 0$ because X is nonsingular. As a result $BA = \begin{bmatrix} I & ZY \\ 0 & 0 \end{bmatrix}$, which can never be an identity matrix and therefore no B of the required form exists. Thus for a rectangular matrix $A_{r \times c}$ there is no matrix B such that $AB = I_r$ and $BA = I_c$.

However, for certain rectangular matrices, $A_{r \times c}$ for $r \neq c$, there are matrices $B_{c \times r}$ for which $AB = I_r$ (but $BA \neq I_c$); B in such cases is called the *right inverse* of A. And for other matrices $A_{r \times c}$ there are matrices $D_{c \times r}$ for which $DA = I_c$ (but $AD \neq I_r$); in these cases D is called the *left inverse* of A. As has just been shown, no matrix can be both the left and right inverse of the same rectangular matrix. As is shown in Chapter 6, no rectangular matrix has both a left and a right inverse. It will have neither of them or one or the other, but not both. Only square (nonsingular) matrices have both a left and a right inverse, and they are the same, namely the (unique) inverse that has already been described.

Example.

$$L = \begin{bmatrix} 1 & 3 & 1 \\ 2 & 5 & 1 \end{bmatrix} \text{ is the left inverse of } A = \begin{bmatrix} 1 & 1 \\ -1 & 0 \\ 3 & -1 \end{bmatrix}$$

because $LA = I_2$. Likewise A is the right inverse of L. But there is no right inverse of A (no matrix B for which $AB = I_3$) nor is there a left inverse of L (no matrix D for which $DL = I_3$).

10. EXERCISES

1. Show that if

   (a) $A = \begin{bmatrix} 6 & 13 \\ 5 & 12 \end{bmatrix}$, $|A| = 7$, $A^{-1} = \tfrac{1}{7} \begin{bmatrix} 12 & -13 \\ -5 & 6 \end{bmatrix}$,

   and $AA^{-1} = A^{-1}A = I$;

   (b) $B = \begin{bmatrix} 3 & -4 \\ 7 & 14 \end{bmatrix}$, $|B| = 70$, $B^{-1} = \tfrac{1}{70} \begin{bmatrix} 14 & 4 \\ -7 & 3 \end{bmatrix}$,

   and $BB^{-1} = B^{-1}B = I$.

2. Demonstrate the reversal rule for the inverse of a product of two matrices, using A and B given in Exercise 1.

3. Show that if

(a)
$$A = \begin{bmatrix} -1 & 3 & 0 \\ 0 & 2 & 1 \\ 1 & 0 & 4 \end{bmatrix}, \quad |A| = -5, \quad A^{-1} = \tfrac{1}{5}\begin{bmatrix} -8 & 12 & -3 \\ -1 & 4 & -1 \\ 2 & -3 & 2 \end{bmatrix},$$

and $A^{-1}A = AA^{-1} = I$;

(b)
$$B = \begin{bmatrix} 10 & 6 & -1 \\ 6 & 5 & 4 \\ -1 & 4 & 17 \end{bmatrix}, \quad |B| = 25, \quad B^{-1} = \begin{bmatrix} 2.76 & -4.24 & 1.16 \\ -4.24 & 6.76 & -1.84 \\ 1.16 & -1.84 & 0.56 \end{bmatrix},$$

and $B^{-1}B = BB^{-1} = I$;

(c)
$$C = \tfrac{1}{10}\begin{bmatrix} 0 & -6 & 8 \\ -10 & 0 & 0 \\ 0 & -8 & -6 \end{bmatrix}, \quad |C| = 1, \quad C^{-1} = C',$$

and $CC' = C'C = I$.

4. For each of the following matrices derive the determinant and inverse. Check each inverse by multiplication.

(a) $\begin{bmatrix} 7 & 3 \\ 8 & 9 \end{bmatrix}$ (b) $\begin{bmatrix} 6 & 31 \\ 8 & 29 \end{bmatrix}$ (c) $\begin{bmatrix} -7 & -7 \\ 3 & 1 \end{bmatrix}$

(d) $\begin{bmatrix} 1 & 5 & -5 \\ 3 & 2 & -5 \\ 6 & -2 & -5 \end{bmatrix}$ (e) $\begin{bmatrix} -3 & 2 & -6 \\ -3 & 5 & -7 \\ -2 & 3 & -4 \end{bmatrix}$ (f) $\begin{bmatrix} 2 & 1 & 3 \\ -5 & 1 & 0 \\ 1 & 4 & -2 \end{bmatrix}$

(g) $\begin{bmatrix} -1 & -2 & 1 \\ -2 & 2 & -2 \\ 1 & -2 & -1 \end{bmatrix}$ (h) $\begin{bmatrix} 1 & -2 & 1 \\ -2 & 4 & -2 \\ 1 & -2 & 1 \end{bmatrix}$ (i) $\begin{bmatrix} 7 & 4 & -1 \\ 4 & 7 & -1 \\ -4 & -4 & 4 \end{bmatrix}$

5. Write down the inverse of $\begin{bmatrix} 1 & 0 & 6 & 8 \\ 0 & 1 & 5 & 4 \\ 0 & 0 & -1 & 0 \\ 0 & 0 & 0 & -1 \end{bmatrix}$.

6. For $A = \begin{bmatrix} 6 & -1 & 4 \\ 2 & 5 & -3 \\ 2 & 5 & 2 \end{bmatrix}$

    i. calculate the transpose of A^{-1} and the inverse of A';
   ii. calculate the inverse of A^{-1}.

7. Find the inverse of $\begin{bmatrix} 3 & 0 & 0 & 0 \\ 0 & 7 & 0 & 0 \\ 0 & 0 & 1 & 0 \\ 0 & 0 & 0 & 5 \end{bmatrix}$.

8. Write each of the following sets of equations in matrix and vector form. Solve the equations, after obtaining the necessary inverse matrices (if they exist). If an inverse does not exist, explain why.

(a)
$$3x + 4y - 2z = 4$$
$$-x - y + 3z = 6$$
$$x - 7y + z = -2$$

(b)
$$-3x + 4y = 2$$
$$x + 7z = -8$$
$$-2x - 3y - 8z = -11$$

(c)
$$2w + 8y - z = 0$$
$$8x - 3z = -4$$
$$4y + \tfrac{3}{2}z = 42$$
$$\tfrac{3}{8}z = 7\tfrac{1}{2}$$

(d)
$$x + 3y - z = 7$$
$$4x - y + 2z = 8\tfrac{1}{2}$$
$$5x + 2y + z = 10.$$

9. Calculate the inverse of the following matrices. Check each inverse by calculating a product that is an identity matrix.

(a) $\begin{bmatrix} 1 & -4 & 1 \\ 2 & 7 & 0 \\ 0 & 1 & 0 \end{bmatrix}$ (b) $\begin{bmatrix} \tfrac{1}{2} & \tfrac{1}{4} & \tfrac{1}{8} \\ \tfrac{1}{4} & \tfrac{1}{8} & 0 \\ 4 & 0 & 1 \end{bmatrix}$ (c) $\begin{bmatrix} 1 & 1 & 0 \\ 0 & 1 & 1 \\ 1 & 0 & 1 \end{bmatrix}$.

10. In Exercise 9 denote each matrix in turn by A and solve the equations

$$A \begin{bmatrix} x \\ y \\ z \end{bmatrix} = \begin{bmatrix} 1 \\ 1 \\ 1 \end{bmatrix}$$

for x, y and z.

11. The media selection problem is a marketing problem in which a decision maker must choose how much of a limited budget to spend on each of several advertising media (e.g. radio, television and newspapers). Suppose a marketing manager wants to reach exactly 30,000 housewives, 20,000 representatives of homes having more than \$10,000 annual income, and 20,000 persons age 25–45. These categories are not mutually exclusive sets, of course, but the marketing manager knows that twice the dollars spent in media one (m_1) plus 3 times the dollars spent in media two (m_2) plus 5 times the dollars spent in media three (m_3) equals the number (x) of housewives who will be reached, on the average, by the

advertising. The other two categories are reached according to the following equations, respectively:

$$y = \text{no. of people earning over } \$10,000 = 7m_1 + 1m_2 + 0m_3$$
$$z = \text{no. of people aged 25–45} \qquad = 4m_1 + 5m_2 + 1m_3.$$

(a) If the marketing manager's requirements are met exactly, how many dollars does he spend on each medium?

(b) How many additional dollars must he spend to reach 10% more women, 20% more people earning over $10,000 and 30% more age 25–45?

12. Suppose a manufacturer produces three products: 1, 2 and 3, using three production processes: stamping, assembling and painting. The capacity (in man-hours per period) of the three production processes is 40, 40 and 80, respectively, while the man-hours required in each process to produce one unit of finished product are contained in

$$A = \{a_{ij}\} = \begin{bmatrix} 2 & 2 & 1 \\ 1 & 4 & 1 \\ 1 & 6 & 4 \end{bmatrix}$$

where a_{ij} represents the number of man-hours required from process i to produce one unit of product j.

(a) In matrix form write down the three equations representing the equality between capacity used and available capacity for each production process.

(b) Assuming each production process is used to capacity, what volumes of the three products are produced?

(c) Suppose product 1 is much more profitable than products 2 and 3; find the maximum amount of product 1 that can be produced. Then try to use the remaining production capacity to maximize output of product 2; and use any further remaining capacity to produce product 3. Are all three production processes used to capacity under this schedule? How many different products are produced?

13. Consider aggregating all industrial and service functions in an economy into two sectors: durable goods production and nondurable goods production. Each sector produces goods for the other, as well as for final consumption. Suppose we have the following sales period data (in millions of dollars):

Sales from:	Sales to:		
	Durable	Nondurable	Consumers
Durable	24	90	6
Nondurable	12	45	93

(a) What is the total sales volume for each industry?

(b) What is the Gross National Product (GNP), defined as the sum of the values of final goods produced for consumption purposes? What is the "value added" by each industry? (Recall that the sum of values added over all sectors equals GNP.)

(c) Derive the economy's input-output coefficient matrix.

(d) What would be the dollar value of each industry's sales if consumers purchased $10 million of durables and $90 million of nondurables?

(e) What value of output must the durable goods industry produce in order for the economy to deliver one dollar's worth of nondurables to the consumers?

14. Consider a vertically integrated steel firm with coal mines, ore mines, blast furnaces that produce iron ingots and rolling mills that produce steel sheets. Assume that the following input-output coefficient matrix accurately reflects the production processes involved:

Inputs	Outputs			
	Coal	Ore	Ingots	Sheets
Coal (tons)	0.1	0.1	0.5	0.1
Ore (tons)	0.0	0.0	1.0	0.0
Ingots	0.0	0.0	0.0	1.0
Sheets	0.0	0.0	0.0	0.0

Suppose the firm wishes to sell 3,000 tons of coal, no ore, 6,000 ingots and 12,000 rolled sheets to the rest of the economy in which it operates, with no purchases of any inputs from outside the company.

(a) How much of each product will have to be produced?

(b) Where is all the coal that is being produced utilized?

(c) Interpret the (2, 4) element in $(I - A)^{-1}$ and relate this to the original production function for steel sheets implied by the input-output coefficient matrix.

15. Rework, in dollar amounts, the problem given in Table 1 of Section 5.6b. Assume that the price of agricultural output per ton is $1,000, a machine costs $2,000 and labor costs are $4,000 per man-year. [Begin by converting the input-output table to dollar amounts; then proceed to determine A, $(I - A)^{-1}$ and x.]

16. Consider the following model of income determination, where all symbols are scalars:

$Y = C + I + G + X - M,$ where Y = net national product

$C = a + bY,$ C = consumption expenditures

$I = c + dY,$ I = net investment expenditures

$M = e + fC,$ M = imports

X = exports

and G = government expenditures.

Assume that G and X are determined exogenously and that $a, b, \ldots, f$ are known constants.

(a) Given values for G and X, is it possible to derive values for Y, C, I and M? Give reasons for your answer.

(b) With $x = \begin{bmatrix} Y \\ C \\ I \\ M \end{bmatrix}$ and $y = \begin{bmatrix} G + X \\ a \\ c \\ e \end{bmatrix}$ write down the equation that relates x

to y, and solve for x. [*Hint:* See equation (22) of the text.]

(c) Suppose government spending were to increase by a dollar. What is the algebraic form of the resultant increase in GNP? What economic term is used to describe this effect? By how much would consumption, investment and imports change?

17. Livingstone (1968) uses matrix algebra in simultaneous allocation of costs among various departments of a corporation. As a simplified example, an accountant may want to allocate the costs of two service departments (repairs and buildings) to an operating department, in order to calculate average costs of production. Before costs can be assigned to the operating departments, it is necessary to allocate costs from each service department to the other. Assume 80% of the cost of repairs provided by the repair department are to be allocated to the building department and 20% to the operating department. Also, assume 30% of the total costs of the building department are to be allocated to the repair department and 70% to the operating department. The directly identifiable costs (before any allocations) incurred by the building department are $6,000 and those incurred by the repair department are $2,000.

Compute the total allocated costs of the repair and building departments and the allocation of these costs to the various departments, using matrix algebra to solve the two simultaneous equations involved.

18. Show that the following statements are true:

(a) If A and B are symmetric, $[(AB)']^{-1} = A^{-1}B^{-1}$.

(b) If $C = X(X'X)^{-1}X'$, $C^2 = C = C'$.

(c) $(ABCD)^{-1} = D^{-1}C^{-1}B^{-1}A^{-1}$.

(d) If $A^{-1} = A'$ and $B^{-1} = B'$, $(AB)(AB)' = I$.

(e) If $A = \begin{bmatrix} I & B \\ 0 & -I \end{bmatrix}$, $A^2 = I$.

(f) If B is an $n \times n$ matrix and λ is a scalar, and if $A = \lambda B$, then
   i. A is an $n \times n$ matrix;
   ii. $|A| = \lambda^n |B|$;
   iii. $A^{-1} = \dfrac{1}{\lambda}B^{-1}$. When is iii not true?

(g) If A is a nonsingular square matrix of order n, $|\text{adjoint } (A)| = |A|^{n-1}$.

(h) The product of two matrices being a null matrix does not necessarily imply that one of them is null.

(i) If A is an orthogonal matrix its determinant has the value $+1$ or -1.

(j) If λ_1, λ_2, λ_3 and λ_4 are scalars and A is a square matrix of order n, then

$$\begin{vmatrix} \lambda_1 A & \lambda_2 A \\ \lambda_3 A & \lambda_4 A \end{vmatrix} = \left\{ \begin{vmatrix} \lambda_1 & \lambda_2 \\ \lambda_3 & \lambda_4 \end{vmatrix} \right\}^n (|A|)^2.$$

19.

By partitioning $M = \begin{bmatrix} -5 & 3 & 5 & 1 & 0 \\ -4 & 8 & 10 & 0 & -1 \\ 2 & 13 & 11 & 2 & 1 \\ 0 & 1 & 1 & 3 & 2 \\ 3 & 1 & 0 & 7 & 5 \end{bmatrix}$, find its inverse.

20. Inverting a partitioned matrix is discussed in Section 7. Show that the expressions for X, Y, Z and W as given in procedure **a** are equivalent to those of procedure **b**.

21. Show that

(a) $\begin{bmatrix} I & P \\ Q & I \end{bmatrix}^{-1} = \begin{bmatrix} (I - PQ)^{-1} & -(I - PQ)^{-1}P \\ -Q(I - PQ)^{-1} & I + Q(I - PQ)^{-1}P \end{bmatrix}$,

(b) $\begin{bmatrix} 0 & P \\ Q & I \end{bmatrix}^{-1} = \begin{bmatrix} -(PQ)^{-1} & (PQ)^{-1}P \\ Q(PQ)^{-1} & I - Q(PQ)^{-1}P \end{bmatrix}$,

(c) $\begin{bmatrix} 0 & P \\ Q & R \end{bmatrix}^{-1} = \begin{bmatrix} -(PR^{-1}Q)^{-1} & (PR^{-1}Q)^{-1}PR^{-1} \\ R^{-1}Q(PR^{-1}Q)^{-1} & R^{-1} - R^{-1}Q(PR^{-1}Q)^{-1}PR^{-1} \end{bmatrix}$,

(d) $\begin{bmatrix} S & P \\ 0 & R \end{bmatrix}^{-1} = \begin{bmatrix} S^{-1} & -S^{-1}PR^{-1} \\ 0 & R^{-1} \end{bmatrix}$.

REFERENCES

Bodewig, E. (1959). *Matrix Calculus*. North-Holland Publishing Company, Amsterdam.

Dwyer, P. (1951). *Linear Computations*. Wiley, New York.

Fadeeva, V. N. (1958). *Computational Methods of Linear Algebra*. Dover, New York.

Hildebrand, F. B. (1956). *Introduction to Numerical Analysis*. McGraw-Hill, New York.

Livingstone, J. L. (1968). Matrix algebra and cost allocation. *The Accounting Review*, **43**, 503–508.

Leontief, W. W. (1936). Quantitative input and output relations in the economic system of the United States. *Review of Economic Statistics*, **18**, 105–125.

Wilkinson, J. H. (1963). *Rounding Errors in Algebraic Processes*. Prentice-Hall, Englewood Cliffs, N.J.

CHAPTER 6

LINEAR INDEPENDENCE AND RANK

The crux of solving equations $Ax = b$ as $x = A^{-1}b$ is the existence of the inverse A^{-1}. Only if A^{-1} exists can the solution be $x = A^{-1}b$; and A^{-1} does exist only if the determinant of A is non-zero. Therefore, in any situation in which we wish to have equations of the form $Ax = b$ solved as $x = A^{-1}b$, we must first investigate the determinant of A to see if it is non-zero.

If A is a square matrix of many rows and columns, finding its determinant by the methods discussed in Chapter 4 may be almost as lengthy a procedure as solving the equations themselves, and although the determinant is an integral part of the description of A^{-1} given in Chapter 5, its explicit value is not required for the more useful (computer) methods of calculating A^{-1}. But we must know that $|A|$ is non-zero before attempting to calculate A^{-1} in order to know that A^{-1} does exist. In this chapter we present methods that indicate whether or not the determinant of a matrix is zero without having to expand the determinant in full.

1. LINEAR INDEPENDENCE[1] OF VECTORS

Suppose two vectors, v_1 and v_2, are multiplied by two scalars, k_1 and k_2, respectively. The weighted sum $k_1 v_1 + k_2 v_2$ is called a *linear combination* of the vectors v_1 and v_2. If the linear combination equals the null vector only when both k_1 and k_2 are zero, the two vectors v_1 and v_2 are said to be *linearly independent*. Consider the equation

$$k_1 v_1 + k_2 v_2 = 0. \tag{1}$$

---

[1] The reader should not confuse the concept of linear independence of vectors with that of statistical independence of random variables.

[*129*]

If v_1 and v_2 are such that this equation holds true only when $k_1 = k_2 = 0$ then v_1 and v_2 are linearly independent. Conversely, if equation (1) is true for some non-zero values of k_1 and k_2, i.e. if the weighted sum $k_1v_1 + k_2v_2$ is zero for some values of k_1 and k_2 that are not both zero, then the two vectors v_1 and v_2 are said to be *linearly dependent* (1).

Note that these definitions depend on equation (1). This is a vector equation, representing as many scalar equations as there are elements in v_1, and it is meaningful only when v_1 and v_2 are of the same order. Hence any discussion of the linear dependence or linear independence of vectors relates only to vectors of the same order.

Examples.

If
$$v_1 = \begin{bmatrix} 2 \\ 6 \\ 4 \end{bmatrix} \quad \text{and} \quad v_2 = \begin{bmatrix} 3 \\ 9 \\ 6 \end{bmatrix} \tag{2}$$

and we multiply v_1 by 3 and v_2 by -2 we find that

$$3v_1 - 2v_2 = 0. \tag{3}$$

Hence v_1 and v_2 are linearly dependent. But if

$$v_3 = \begin{bmatrix} 2 \\ 1 \\ 7 \end{bmatrix} \quad \text{and} \quad v_4 = \begin{bmatrix} 1 \\ -3 \\ 0 \end{bmatrix}$$

there are no two scalars k_3 and k_4, other than $k_3 = 0 = k_4$, such that $k_3v_3 + k_4v_4 = 0$. Hence v_3 and v_4 are linearly independent.

When the vectors are linearly dependent the constants k_1 and k_2 are not unique, for if $k_1v_1 + k_2v_2 = 0$ so does $ck_1v_1 + ck_2v_2 = 0$ where c, a scalar, is any constant. Thus if (3) is true so is, for example, $9v_1 - 6v_2 = 0$, and so too is $4.5v_1 - 3v_2 = 0$. The values of the constants can be altered so long as their ratio stays the same.

Although linear independence and linear dependence have been defined in terms of only two vectors, the definitions apply equally as well to more than two. Thus the vectors v_1, v_2, v_3 and v_4 are linearly independent if

$$k_1v_1 + k_2v_2 + k_3v_3 + k_4v_4 = 0 \tag{4}$$

is true only when $k_1 = k_2 = k_3 = k_4 = 0$. Likewise they are linearly dependent if (4) is true for some set of k's other than all of them equal to zero. Thus if λ_1 and λ_2 are two non-zero constants for which

$$\lambda_1v_1 + \lambda_2v_2 = 0,$$

the v's are said to be linearly dependent because (4) is then true for the following set of k's: $k_1 = \lambda_1$, $k_2 = \lambda_2$, $k_3 = 0$ and $k_4 = 0$, not all of which are zero.

Note that when a set of v's is a linearly dependent set there must be at least two k's that are non-zero. If only one were, k_2 say, equation (4) would reduce to $k_2 v_2 = 0$, which allows $k_2 \neq 0$ only if $v_2 = 0$, i.e. only if v_2 is null. If we exclude this possibility, i.e. exclude null vectors, then all the k's would be zero. Therefore, if (4) is satisfied by any non-zero k's, at least two of them are non-zero.

An important consequence of linear dependence of vectors is that at least one of the v's concerned can be expressed as a linear combination of the others. Let us assume four vectors v_1, v_2, v_3 and v_4 are linearly dependent. Then (4) is true for some set of k's not all zero. Suppose $k_1 \neq 0$. Then we may divide both sides of (4) by k_1 and rearrange to obtain

$$v_1 = (-k_2/k_1)v_2 + (-k_3/k_1)v_3 + (-k_4/k_1)v_4; \qquad (5)$$

since at least one of the other k's is also non-zero, equation (5) shows that v_1 is a linear combination of v_2, v_3 and v_4.

Linear dependence has been defined and illustrated in terms of only four vectors, but the general definition applies to any number of vectors. Thus in general, n vectors all of the same order, $u_1, u_2, \ldots, u_n$, are said to be linearly dependent if constants $k_1, k_2, \ldots, k_n$ not all zero exist such that

$$k_1 u_1 + k_2 u_2 + \cdots + k_n u_n = 0.$$

If zeros are the only values of the k's for which this equation is true, the vectors are said to be linearly independent; otherwise they are linearly dependent.

Example. Consider

$$u_1 = \begin{bmatrix} 0 \\ 1 \\ 0 \end{bmatrix}, \quad u_2 = \begin{bmatrix} 1 \\ 0 \\ 0 \end{bmatrix}, \quad u_3 = \begin{bmatrix} 0 \\ 0 \\ 1 \end{bmatrix}, \quad u_4 = \begin{bmatrix} 1 \\ 2 \\ 3 \end{bmatrix} \quad \text{and} \quad u_5 = \begin{bmatrix} 2 \\ 4 \\ 5 \end{bmatrix}.$$

$$(6)$$

We observe that

$$2u_1 + u_2 + 3u_3 - u_4 = 0,$$

and therefore u_1, u_2, u_3 and u_4 are linearly dependent. Similarly

$$2u_1 + u_2 + 3u_3 - u_4 + 0(u_5) = 0$$

so that u_1, u_2, u_3, u_4 and u_5 are also linearly dependent. Likewise the inclusion of further vectors u_6, u_7, ... to the set u_1, u_2, ..., u_5 would not alter the linear dependence of the set because, for example,

$$2u_1 + u_2 + 3u_3 - u_4 + 0u_5 + 0u_6 + 0u_7 = 0$$

would still be true. On the other hand, notice that no values k_1, k_2 and k_3 can be found, other than all of them zero, such that

$$k_1 u_1 + k_2 u_2 + k_3 u_3 = 0.$$

Hence the three vectors u_1, u_2 and u_3 are linearly independent.

2. LINEAR DEPENDENCE AND DETERMINANTS

Definition. For the sake of verbal simplicity we will henceforth refer to "linear independence" as just "independence" and to "linear dependence" as "dependence," dropping the description "linear" but without changing the meaning in so doing.

As shown above, a set of vectors being dependent implies that at least one of them can be expressed as a linear combination of the others. Recall, now, that a determinant is zero if multiples of columns can be subtracted from another column to give a column of zeros. But this is just the condition under which the one column is a linear combination of the others. And this in turn is the condition that the columns are dependent. Hence we see that if the columns of a square matrix are dependent the determinant of the matrix is zero and consequently the inverse of the matrix does not exist. The same is true for rows.

Example. We saw that the vectors v_1 and v_2 of (2) are linearly dependent because $3v_1 - 2v_2 = 0$. Hence any square matrix having these vectors as columns has a determinant of zero: e.g.

$$\begin{vmatrix} 2 & 3 & 1 \\ 6 & 9 & 2 \\ 4 & 6 & 5 \end{vmatrix} = 0$$

because column 2 minus 1.5 times column 1 gives a column of zeros. However, the vectors u_1, u_2 and u_3 of (6) are independent and so the determinant formed from them is non-zero:

$$\begin{vmatrix} 0 & 1 & 0 \\ 1 & 0 & 0 \\ 0 & 0 & 1 \end{vmatrix} = -1.$$

We have here a most important result: <u>if the rows (or columns) of a square matrix are dependent its determinant is zero and its inverse does not exist; conversely, if they are independent its determinant is non-zero and the inverse does exist.</u>

<u>The minimum condition for the linear dependence of rows (or columns) of a matrix is that one of them be a linear combination of the others.</u> Investigating a matrix to see whether or not this is explicitly the case is not necessarily an easy task, especially when the matrix is of large order. Fortunately, alternative procedures are available.

3. SETS OF LINEARLY INDEPENDENT VECTORS

A question that might be asked about a set of vectors is, "How many independent vectors are there?" The following theorem sets a limit on the answer.

Theorem 1. A set of linearly independent vectors of order n cannot contain more than n such vectors.[1]

Corollary. A set of r vectors of order n can be linearly independent only if $r \leq n$.

Null vectors (vectors whose every element is zero) are automatically excluded here because any set of vectors containing a null vector is a dependent set. For, if u_1 and u_2 are non-null vectors and u_3 is a null vector

$$0(u_1) + 0(u_2) + k_3 u_3 = 0$$

for any value of k_3. It is therefore unnecessary to mention non-null vectors when discussing sets of linearly independent vectors; they must be non-null in order to be linearly independent.

Note the generality of the theorem. It does not state that there is only one, unique set of n independent vectors of order n, but that in any set of vectors of order n there cannot be more than n vectors if they are to be independent. In fact, there are an infinite number of such sets, but none of them can consist of more than n vectors and still be a set of independent vectors.

Examples. The set of vectors $\begin{bmatrix} 1 \\ 3 \end{bmatrix}$ and $\begin{bmatrix} 2 \\ -1 \end{bmatrix}$ are a set of independent vectors of order 2; but include in the set any other vector of order 2, $\begin{bmatrix} 8 \\ 3 \end{bmatrix}$

---

[1] Proved in appendix, Sec. 6.12a.

say, and the set will not be independent, for the third vector can be expressed as a linear combination of the other two; for example,

$$\begin{bmatrix} 8 \\ 3 \end{bmatrix} = 2\begin{bmatrix} 1 \\ 3 \end{bmatrix} + 3\begin{bmatrix} 2 \\ -1 \end{bmatrix}.$$

The vectors $\begin{bmatrix} 0 \\ 1 \end{bmatrix}$ and $\begin{bmatrix} -1 \\ 2 \end{bmatrix}$ also form an independent set, and again any other vector (of order 2) can be expressed as a linear combination of them:

$$\begin{bmatrix} 8 \\ 3 \end{bmatrix} = 19\begin{bmatrix} 0 \\ 1 \end{bmatrix} - 8\begin{bmatrix} -1 \\ 2 \end{bmatrix}.$$

Not any two vectors of order 2 form an independent set, nor do any three vectors of order 3. In all cases the vectors must satisfy the definition of independence in order to form an independent set. But if three vectors of order 3 are independent, then any other third-order vector can be expressed as a linear combination of them. For example, the vectors $\begin{bmatrix} 1 \\ 0 \\ 0 \end{bmatrix}$, $\begin{bmatrix} 0 \\ 1 \\ 0 \end{bmatrix}$, $\begin{bmatrix} 0 \\ 0 \\ 1 \end{bmatrix}$ are independent, and any other vector of order 3, $\begin{bmatrix} a \\ b \\ c \end{bmatrix}$, can be expressed as a linear combination of them as follows:

$$\begin{bmatrix} a \\ b \\ c \end{bmatrix} = a\begin{bmatrix} 1 \\ 0 \\ 0 \end{bmatrix} + b\begin{bmatrix} 0 \\ 1 \\ 0 \end{bmatrix} + c\begin{bmatrix} 0 \\ 0 \\ 1 \end{bmatrix}. \tag{7}$$

No matter what the values of a, b and c are, this expression holds true—i.e., every third-order vector can be expressed as a linear combination of this set of three independent vectors. In general, every n-order vector can be expressed as a linear combination of any set of n independent vectors of order n. The maximum number of vectors in a set of independent vectors is therefore n, the order of the vectors, as stated in Theorem 1.

Example. The vector $\begin{bmatrix} a \\ b \end{bmatrix}$ is expressed as a linear combination of the independent vectors $\begin{bmatrix} 1 \\ 3 \end{bmatrix}$ and $\begin{bmatrix} 2 \\ -1 \end{bmatrix}$ by the equation

$$\begin{bmatrix} a \\ b \end{bmatrix} = \lambda_1\begin{bmatrix} 1 \\ 3 \end{bmatrix} + \lambda_2\begin{bmatrix} 2 \\ -1 \end{bmatrix},$$

where λ_1 and λ_2 are scalars. This can be written in the form

$$\begin{bmatrix} 1 & 2 \\ 3 & -1 \end{bmatrix}\begin{bmatrix} \lambda_1 \\ \lambda_2 \end{bmatrix} = \begin{bmatrix} a \\ b \end{bmatrix}, \tag{8}$$

and because the columns of the matrix are independent vectors its determinant is nonsingular; hence we obtain

$$\begin{bmatrix} \lambda_1 \\ \lambda_2 \end{bmatrix} = -\tfrac{1}{7}\begin{bmatrix} -1 & -2 \\ -3 & 1 \end{bmatrix}\begin{bmatrix} a \\ b \end{bmatrix} = \tfrac{1}{7}\begin{bmatrix} a + 2b \\ 3a - b \end{bmatrix}.$$

Thus
$$\begin{bmatrix} a \\ b \end{bmatrix} = \tfrac{1}{7}(a + 2b)\begin{bmatrix} 1 \\ 3 \end{bmatrix} + \tfrac{1}{7}(3a - b)\begin{bmatrix} 2 \\ -1 \end{bmatrix}.$$

Definition. For convenience we shall use the abbreviation LIN to mean "linearly independent". Thus a set of LIN vectors shall mean a set of linearly independent vectors.

4. RANK

a. Independent rows and columns of a matrix

An explanation has been given of how a determinant is zero if any of its rows (or columns) are linear combinations of other rows (or columns). In other words, a determinant is zero if its rows (or columns) do not form a set of independent vectors. It follows that a determinant cannot have its rows forming a dependent set and its columns an independent set. More generally, for any matrix the relationship between the number of independent rows of a matrix and the number of independent columns is now stated.

Theorem 2. The number of independent rows in a matrix equals the number of independent columns and vice versa.[1]

Example. In $A = \begin{bmatrix} 1 & 0 & -1 & 2 \\ 3 & 1 & 4 & 2 \\ 5 & 2 & 9 & 2 \end{bmatrix}$ the last two columns are linear combinations of the first two:

$$\begin{bmatrix} -1 \\ 4 \\ 9 \end{bmatrix} = -1\begin{bmatrix} 1 \\ 3 \\ 5 \end{bmatrix} + 7\begin{bmatrix} 0 \\ 1 \\ 2 \end{bmatrix}$$

and
$$\begin{bmatrix} 2 \\ 2 \\ 2 \end{bmatrix} = 2\begin{bmatrix} 1 \\ 3 \\ 5 \end{bmatrix} - 4\begin{bmatrix} 0 \\ 1 \\ 2 \end{bmatrix}.$$

_____

[1] Proved in appendix, Sec. 6.12a.

Therefore A has only two independent columns, and by the theorem it also has only two independent rows. This is indeed the case, since

$$[5 \quad 2 \quad 9 \quad 2] = 2[3 \quad 1 \quad 4 \quad 2] - 1[1 \quad 0 \quad -1 \quad 2].$$

b. Definition of rank
The *rank* of any matrix is the number of linearly independent rows (or columns) in the matrix.

Notation. We will use the notation $r(A)$ to mean the rank of A. Thus $r(A) = r$ means that A has rank r, that there are r linearly independent rows (columns) in A.
Certain properties and consequences of rank hold:

i. Rank is zero only for a null matrix; otherwise it is always a positive integer.
ii. The rank of a rectangular matrix $A_{p \times q}$ is equal to or less than the smaller of p and q.
iii. The rank of a square matrix is equal to or less than its order.
iv. If $r(A) = r$ there is at least one non-zero minor of order r in A, and all minors of order greater than r are zero.
v. If $r(A_n) < n$, then A^{-1} does not exist and $|A|$ equals zero.

The first three of these statements result from the definition of rank as the number of independent rows (columns) in a matrix. Statement (iv) is true because $r(A) = r$ implies that A has r independent rows and r independent columns, and therefore the determinant of the matrix formed by crossing out all but these rows and columns is non-zero. All minors of order greater than r are zero because at least one of their rows will be a linear combination of the other r rows. Result (v) is true since by (iv) $r < n$ implies $|A| = 0$ and the non-existence of A^{-1}.

c. Rank and the inverse of a matrix
Ascertaining whether or not equations $Ax = y$ can be solved as $x = A^{-1}y$ is greatly facilitated by the notion of rank. The solution $x = A^{-1}y$ exists only if A^{-1} does; this occurs only if $|A| \neq 0$, and this is true only if the rows (and the columns) are linearly independent, i.e., only if the rank of A equals its order. By introducing the concept of rank we therefore alter the problem of finding whether the value of a determinant of a square matrix is zero or non-zero to one of finding if the rank of the matrix is less than its order. If it is, the determinant is zero, but if the rank equals its order the determinant is non-zero. Because investigating rank relative to order is easier than evaluating a determinant, we use this relationship between rank and order for

If r is < Matrix order, (N×N), The Det.
is 0. If r = N×N Det is non-zero.

[6.5]   ELEMENTARY OPERATORS   137

ascertaining the existence or non-existence of an inverse; for, if $r(A_n) = r$, then

$r < n$ implies $|A| = 0$ and the non-existence of A^{-1};

and $r = n$ implies $|A| \neq 0$ and the existence of A^{-1}.

In the latter case A is nonsingular. The rank of the inverse is also n; if the rank were less than n, $|A^{-1}|$ would be zero, which we know is not so when A^{-1} exists.

The problem of ascertaining the existence or non-existence of an inverse is therefore equivalent to ascertaining whether or not the rank of a square matrix equals its order. More generally we may wish to derive the rank exactly of both square and rectangular matrices. If the rank of a matrix is r, r of its rows are linearly independent, but locating a set of r such rows is not always an easy task. On the other hand, to actually derive the value of r in any particular case is not at all difficult. To do so, extensive use is made of matrices known as elementary operators, which we consider in Section 5.

d. Full rank matrices

A square matrix whose determinant is non-zero is said to be of *full rank*. Thus a square matrix A that has full rank is also one that is nonsingular; $|A| \neq 0$, and A^{-1} exists.

Rectangular matrices having rank equal to their number of rows are described as having *full row rank*; and if rank equals the number of columns they are of *full column rank*.

Example.

$$A = \begin{bmatrix} 1 & 0 & 0 & 3 & 2 \\ 0 & 1 & 0 & 6 & 3 \\ 0 & 0 & 1 & 7 & 5 \end{bmatrix}$$

has rank 3; it has full row rank. Any matrix of the partitioned form $[I \;\; K]$ has full row rank—no row of I can be formed as a linear combination of the other rows, and hence the same is true for $[I \;\; K]$.

5. ELEMENTARY OPERATORS

a. Definitions

Three kinds of matrices known as elementary operators play an important role in matrix theory. They are all square matrices derived from the identity matrix. When used in a product with another matrix the effect of each of them is to produce in that matrix the same kind of a manipulation of rows (or columns) as is used in simplifying determinants (see Section 4.3). Furthermore, the product matrix has the same rank as the original matrix. These two properties of elementary operators provide us with methods for determining the rank of any matrix. The elementary operators are as follows:

i. E_{ij} is I with its ith and jth rows interchanged. For example,

$$E_{12} = \begin{bmatrix} 0 & 1 & 0 \\ 1 & 0 & 0 \\ 0 & 0 & 1 \end{bmatrix}.$$

Pre-multiplication of a matrix A by E_{ij} interchanges the ith and jth rows of A. Thus

$$E_{12}A = \begin{bmatrix} 0 & 1 & 0 \\ 1 & 0 & 0 \\ 0 & 0 & 1 \end{bmatrix}\begin{bmatrix} 1 & 1 & 1 \\ 2 & 2 & 2 \\ 3 & 3 & 3 \end{bmatrix} = \begin{bmatrix} 2 & 2 & 2 \\ 1 & 1 & 1 \\ 3 & 3 & 3 \end{bmatrix}.$$

ii. $R_{ii}(\lambda)$ is I with λ replacing the 1 in the ith diagonal term. Pre-multiplication of A by $R_{ii}(\lambda)$ leads to the ith row of A being multiplied by λ:

$$[R_{22}(4)]A = \begin{bmatrix} 1 & 0 & 0 \\ 0 & 4 & 0 \\ 0 & 0 & 1 \end{bmatrix}\begin{bmatrix} 1 & 1 & 1 \\ 2 & 2 & 2 \\ 3 & 3 & 3 \end{bmatrix} = \begin{bmatrix} 1 & 1 & 1 \\ 8 & 8 & 8 \\ 3 & 3 & 3 \end{bmatrix}.$$

iii. $P_{ij}(\lambda)$ is I with the scalar λ replacing the zero in the ith row and jth column for $i \neq j$. Pre-multiplication of A by $P_{ij}(\lambda)$ results in adding λ times the jth row of A to its ith row:

$$[P_{12}(2)]A = \begin{bmatrix} 1 & 2 & 0 \\ 0 & 1 & 0 \\ 0 & 0 & 1 \end{bmatrix}\begin{bmatrix} 1 & 1 & 1 \\ 2 & 2 & 2 \\ 3 & 3 & 3 \end{bmatrix} = \begin{bmatrix} 5 & 5 & 5 \\ 2 & 2 & 2 \\ 3 & 3 & 3 \end{bmatrix}.$$

b. Post-multiplication

The effects of pre-multiplying a matrix by an elementary operator have just been stated and illustrated. Post-multiplication performs similar manipulations on the columns of A as are performed on the rows of A by pre-multiplication, except that $AP_{ji}(\lambda)$ does to columns of A what $P_{ij}(\lambda)A$ does to rows of A.

c. Determinants

From the general form of the elementary operators the reader should check that

$$|E_{ij}| = -1, \qquad |R_{ii}(\lambda)| = \lambda \qquad \text{and} \qquad |P_{ij}(\lambda)| = 1.$$

Since the determinant of a product matrix equals the product of the determinants, we have the following results when A is square:

$$|E_{ij}A| = -|A|, \qquad |R_{ii}(\lambda)A| = \lambda\,|A| \qquad \text{and} \qquad |P_{ij}(\lambda)A| = |A|.$$

d. Transposes

Verification of the following results is straightforward:

$$E'_{ij} = E_{ij}, \qquad R'_{ii}(\lambda) = R_{ii}(\lambda) \qquad \text{and} \qquad P'_{ij}(\lambda) = P_{ji}(\lambda).$$

Thus an elementary operator and its transpose are of the same form, and indeed transposing has no effect on the E- and R-type operators; i.e. these two are symmetric. A P-type operator is not symmetric but its transpose is also a P-type operator.

e. Inverses

The following will be found true:

$$E_{ij}^{-1} = E_{ij}, \qquad [R_{ii}(\lambda)]^{-1} = R_{ii}(1/\lambda) \qquad \text{and} \qquad [P_{ij}(\lambda)]^{-1} = P_{ij}(-\lambda).$$

Thus an E-type operator is the same as its inverse, and the effect of inverting R- and P-type operators is simply that of changing the constants, replacing λ by $1/\lambda$ in the R-type and λ by $-\lambda$ in the P-type. From this and the preceding paragraph we see that the underlying form of any one of these elementary operators is unchanged when the operator is either inverted or transposed.

6. RANK AND THE ELEMENTARY OPERATORS

The rank of a matrix is unaffected when it is multiplied by an elementary operator; i.e., if the matrix A is multiplied by an E-, P- or R-type operator the rank of the product is the same as the rank of A. We discuss the validity of this statement for each operator in turn. If A is pre-multiplied by an E-type operator the product EA is A with two rows interchanged. Hence EA is the same as A except with the rows in a different sequence. The number of rows that are linearly independent will be unaltered. Therefore $r(EA) = r(A)$. In the product of A and an R-type operator every element of some row (or column) of A is multiplied by a constant, and the product of a P-type operator with A is the same as A except it has a multiple of one row (column) added to another. In both cases the independence of rows (columns) is unaffected and the same number will be linearly independent after making the product as before. Thus multiplication of any matrix by an elementary operator does not alter rank.

7. FINDING THE RANK OF A MATRIX

We have just seen that in multiplying a matrix by an elementary operator the rank of the product is the same as that of the original matrix; therefore the rank of a matrix is the same as the rank of its successive products with several elementary operators. Suppose in any matrix $A_{r \times c}$ we refer to the diagonal as being the elements $a_{11}, a_{22}, \ldots, a_{dd}$ with d defined as equal to the lesser of r and c. (When A is square, $d = r = c$ and these elements constitute the true diagonal of the matrix.) By the nature of the manipulations involved in multiplying A by elementary operators (the same as in simplifying determinants), it is possible to reduce various elements to zero. In particular, elementary operators can be used to reduce to zero the elements below the diagonal $a_{11}, \ldots, a_{dd}$; the number of non-zero elements remaining in this diagonal then represents the size of the largest non-zero minor in the matrix and is accordingly the rank of the matrix.

Example. For

$$A = \begin{bmatrix} 3 & 6 & 5 & 2 \\ 6 & 16 & 18 & 7 \\ 3 & 6 & 5 & 2 \end{bmatrix}, \qquad P = P_{21}(-2) = \begin{bmatrix} 1 & 0 & 0 \\ -2 & 1 & 0 \\ 0 & 0 & 1 \end{bmatrix}$$

and $\qquad P^* = P_{31}(-1) = \begin{bmatrix} 1 & 0 & 0 \\ 0 & 1 & 0 \\ -1 & 0 & 1 \end{bmatrix}$,

we have

$$PA = \begin{bmatrix} 3 & 6 & 5 & 2 \\ 0 & 4 & 8 & 3 \\ 3 & 6 & 5 & 2 \end{bmatrix} \qquad \text{and} \qquad P^*PA = \begin{bmatrix} 3 & 6 & 5 & 2 \\ 0 & 4 & 8 & 3 \\ 0 & 0 & 0 & 0 \end{bmatrix}.$$

This matrix P^*PA has two diagonal elements that are non-zero. Its largest order minor that is nonsingular is therefore a 2×2 and so the rank of the matrix is 2. Consequently the rank of A is also 2, because P^*PA is A pre-multiplied by elementary operators, a procedure that does not affect rank.

The above example demonstrates how the operators P and P^* affect A and how they lead to being able to establish the rank of A. In practice, to ascertain rank, the operators are not explicitly needed. We simply perform the various elementary operations on a matrix, recognizing that its rank is thereby unchanged. When the operators themselves are needed they are easily derived (see Section 9).

The basic objective of this procecure for finding the rank of a matrix is to add multiples of rows to other rows to reduce the sub-diagonal elements to zero, where by diagonal we continue to mean the elements $a_{11}, a_{22}, \ldots, a_{dd}$. Initially the first row is used to reduce the sub-diagonal elements of the first column to zeros, basing the calculations on the first term of the first row. Then the elements below the diagonal in the second column are reduced to zeros using the second row, based on its diagonal term. Because the sub-diagonal elements of the first column have already been made zero in the first step of this procedure, this second step does not affect their values. And so the process is continued, using each row this way in turn, until all remaining rows are zero or until the last row is reached. The number of non-zero diagonal elements is then the rank.

Matrices formed by the successive products in this procedure are said to be *equivalent*. Thus A is equivalent to B if B can be obtained from A by multiplying A by elementary operators. We write $A \cong B$ for A equivalent to B.

Example. Suppose we wish to find the rank of

$$B = \begin{bmatrix} 1 & 4 & 6 \\ 4 & -1 & 2 \\ -7 & 6 & 2 \end{bmatrix}.$$

Subtract 4 times the first row from the second, and add 7 times the first row to the third. Then

$$B \cong \begin{bmatrix} 1 & 4 & 6 \\ 0 & -17 & -22 \\ 0 & 34 & 44 \end{bmatrix}.$$

Now add twice the second row of the new matrix to its third row, and

$$B \cong \begin{bmatrix} 1 & 4 & 6 \\ 0 & -17 & -22 \\ 0 & 0 & 0 \end{bmatrix}.$$

Then, since the number of non-zero elements in the diagonal of this matrix is 2, the rank of B is 2.

As soon as all elements to the right and below the first zero diagonal element are zero, the process is complete. To achieve this, interchanging of rows and/or of columns is sometimes necessary. For example, consider

$$C = \begin{bmatrix} 1 & 4 & 6 \\ 4 & -1 & 2 \\ -7 & 6 & 2 \\ 2 & 8 & 13 \end{bmatrix}$$

which is B with one more row. If we perform on C those operations earlier performed on B, together with subtracting twice the first row from the fourth, we get

$$C \cong \begin{bmatrix} 1 & 4 & 6 \\ 0 & -17 & -22 \\ 0 & 0 & 0 \\ 0 & 0 & 1 \end{bmatrix}.$$

Interchanging rows 3 and 4 now gives

$$C \cong \begin{bmatrix} 1 & 4 & 6 \\ 0 & -17 & -22 \\ 0 & 0 & 1 \\ 0 & 0 & 0 \end{bmatrix}$$

and hence $r(C) = 3$.

Now consider the matrix D formed from B by the addition of a fourth column:

$$D = \begin{bmatrix} 1 & 4 & 6 & 0 \\ 4 & -1 & 2 & 0 \\ -7 & 6 & 2 & 1 \end{bmatrix}.$$

Performing on D the same operations as were performed on B gives

$$D \cong \begin{bmatrix} 1 & 4 & 6 & 0 \\ 0 & -17 & -22 & 0 \\ 0 & 0 & 0 & 1 \end{bmatrix}.$$

In this case the procedure is not yet complete because the $(3, 3)$ element is zero but the $(3, 4)$ element is not. However, by interchanging columns 3 and

4, the (3, 3) element can be made non-zero, giving

$$D \cong \begin{bmatrix} 1 & 4 & 0 & 6 \\ 0 & -17 & 0 & -22 \\ 0 & 0 & 1 & 0 \end{bmatrix}.$$

This has three non-zero diagonal elements and so we conclude that $r(D) = 3$.

A general description of this procedure for finding the rank of a matrix is to carry out the following steps for each element in the (i, i)th position of the matrix, for $i = 1, 2, \ldots, d$, where d is the lesser of r and c.

(1) If the element is non-zero proceed to (2) below. If it is zero make it non-zero by interchanging rows i and k (for k being any value $i + 1$, $i + 2, \ldots$, or r), and/or interchanging columns i and j (for j being any value $i + 1, i + 2, \ldots$, or c).

(2) Add multiples of the ith row to rows $i + 1, i + 2, \ldots, r$ to change to zero all elements below the diagonal in column i.

(3) Repeat steps (1) and (2) until (1) can no longer be performed.

The number of non-zero elements in the diagonal is then the rank of the original matrix.

8.   EQUIVALENCE

As has already been indicated, A is equivalent to B if B can be derived from A by multiplying A by a series of elementary operators. That is, A is equivalent to B if

$$P_1 P_2 \cdots P_s A Q_1 Q_2 \cdots Q_t = B$$

where $P_1, \ldots, P_s$ and $Q_1, \ldots, Q_t$ are any of the elementary operators, with the P's operating on rows and the Q's on columns.

We now show that this property of equivalence is a reflexive one; i.e. if A is equivalent to B, then B is equivalent to A. Suppose A is $r \times c$; then each of the P's is square with order r, each Q is square with order c and B is the same order as A, namely $r \times c$. If we write P for the product of the P's and Q for the product of the Q's, P and Q are square, their inverses exist and we have $PAQ = B$, and hence

$$A = P^{-1}BQ^{-1} \tag{9}$$

where $P^{-1} = P_s^{-1} P_{s-1}^{-1} \cdots P_2^{-1} P_1^{-1}$ and $Q^{-1} = Q_t^{-1} Q_{t-1}^{-1} \cdots Q_2^{-1} Q_1^{-1}$. Then, since the inverse of an elementary operator is itself an elementary operator, P^{-1} and Q^{-1} are products of elementary operators. Therefore equation (9) shows that multiplying B by elementary operators can lead to A. Consequently B

is equivalent to A and the reflexive nature of equivalence is established. We therefore write $A \cong B$ to mean that A and B are equivalent, namely that simultaneously A is equivalent to B and B is equivalent to A.

9. REDUCTION TO EQUIVALENT CANONICAL FORM

Any matrix A having rank r is equivalent to a matrix of the form $C = \begin{bmatrix} I_r & 0 \\ 0 & 0 \end{bmatrix}$, where C is the same order as A and I_r is the identity matrix of order r, the zeros being null matrices of appropriate order.

Example. The matrix

$$A = \begin{bmatrix} 2 & 6 & 4 & 2 \\ 4 & 15 & 14 & 7 \\ 2 & 9 & 10 & 5 \end{bmatrix},$$

is of rank 2 (its second row equals the sum of the other two). To develop C from A we carry out the following operations on A.

Operations	Effect on A
i. row 2 − 2(row 1) row 3 − row 1	$\begin{bmatrix} 2 & 6 & 4 & 2 \\ 0 & 3 & 6 & 3 \\ 0 & 3 & 6 & 3 \end{bmatrix}$
ii. row 3 − row 2	$\begin{bmatrix} 2 & 6 & 4 & 2 \\ 0 & 3 & 6 & 3 \\ 0 & 0 & 0 & 0 \end{bmatrix}$
iii. column 2 − 3(column 1) column 3 − 2(column 1) column 4 − column 1	$\begin{bmatrix} 2 & 0 & 0 & 0 \\ 0 & 3 & 6 & 3 \\ 0 & 0 & 0 & 0 \end{bmatrix}$
iv. column 3 − 2(column 2) column 4 − column 2	$\begin{bmatrix} 2 & 0 & 0 & 0 \\ 0 & 3 & 0 & 0 \\ 0 & 0 & 0 & 0 \end{bmatrix}$

At this stage A has been amended to be of the form

$$A \cong \Delta = \begin{bmatrix} D_r & 0 \\ 0 & 0 \end{bmatrix} \tag{10}$$

where D_r is a diagonal matrix with r non-zero elements. This is often referred to as the *diagonal form* of A. A can now be reduced to the C form. Suppose we operate on columns and

v. multiply column 1 by $\frac{1}{2}$ and column 2 by $\frac{1}{3}$.
This gives

$$A \cong C = \begin{bmatrix} 1 & 0 & 0 & 0 \\ 0 & 1 & 0 & 0 \\ 0 & 0 & 0 & 0 \end{bmatrix} = \begin{bmatrix} I_2 & 0 \\ 0 & 0 \end{bmatrix},$$

the expected result, since A has rank 2.

For a matrix of order $m \times n$ and rank r, operations of this nature lead to

$$A_{m \times n} \cong \begin{bmatrix} I_r & 0_{r \times n-r} \\ 0_{m-r \times r} & 0_{m-r \times n-r} \end{bmatrix},$$

the zeros being null matrices of the orders shown. If $r = m < n$ the form is $[I_m \quad 0]$; it is $\begin{bmatrix} I_n \\ 0 \end{bmatrix}$ if $r = n < m$ and I_n if $r = m = n$.

Implied in the above result is the existence of matrices P and Q such that $PAQ = C$ where P and Q are products of elementary operators. If A is $m \times n$, P is square of order m and Q is square of order n, and

$$P_m A_{m \times n} Q_n = C = \begin{bmatrix} I_r & 0 \\ 0 & 0 \end{bmatrix}.$$

The matrix C is usually known as the *equivalent canonical form* of A, or the *canonical form under equivalence*. The procedure for obtaining it is often called *reduction to canonical form under equivalence*.

When only the C matrix is required it can be found, as above, by carrying out elementary operations on A without explicitly deriving P and Q. In situations when these are needed, however, they can be obtained from noting that $PAQ = (PI)A(IQ)$. Hence performing on identity matrices the same operations performed on A to reduce it to canonical form produces the matrices P and Q, grouping row operations together to get P and column operations to get Q. Thus in the example we obtain P by applying operations (i) and (ii) to I_3. Hence

$$I_3 = \begin{bmatrix} 1 & 0 & 0 \\ 0 & 1 & 0 \\ 0 & 0 & 1 \end{bmatrix} \cong \begin{bmatrix} 1 & 0 & 0 \\ -2 & 1 & 0 \\ -1 & 0 & 1 \end{bmatrix} \cong \begin{bmatrix} 1 & 0 & 0 \\ -2 & 1 & 0 \\ 1 & -1 & 1 \end{bmatrix} = P;$$

and carrying out operations (iii), (iv) and (v) on I_4 gives Q:

$$
I_4 = \begin{bmatrix} 1 & 0 & 0 & 0 \\ 0 & 1 & 0 & 0 \\ 0 & 0 & 1 & 0 \\ 0 & 0 & 0 & 1 \end{bmatrix} \cong \begin{bmatrix} 1 & -3 & -2 & -1 \\ 0 & 1 & 0 & 0 \\ 0 & 0 & 1 & 0 \\ 0 & 0 & 0 & 1 \end{bmatrix}
$$

$$
\cong \begin{bmatrix} 1 & -3 & 4 & 2 \\ 0 & 1 & -2 & -1 \\ 0 & 0 & 1 & 0 \\ 0 & 0 & 0 & 1 \end{bmatrix} \cong \begin{bmatrix} \frac{1}{2} & -1 & 4 & 2 \\ 0 & \frac{1}{3} & -2 & -1 \\ 0 & 0 & 1 & 0 \\ 0 & 0 & 0 & 1 \end{bmatrix} = Q.
$$

To check on the results we find that

$$
PA = \begin{bmatrix} 1 & 0 & 0 \\ -2 & 1 & 0 \\ 1 & -1 & 1 \end{bmatrix} \begin{bmatrix} 2 & 6 & 4 & 2 \\ 4 & 15 & 14 & 7 \\ 2 & 9 & 10 & 5 \end{bmatrix} = \begin{bmatrix} 2 & 6 & 4 & 2 \\ 0 & 3 & 6 & 3 \\ 0 & 0 & 0 & 0 \end{bmatrix}
$$

and
$$
PAQ = \begin{bmatrix} 2 & 6 & 4 & 2 \\ 0 & 3 & 6 & 3 \\ 0 & 0 & 0 & 0 \end{bmatrix} \begin{bmatrix} \frac{1}{2} & -1 & 4 & 2 \\ 0 & \frac{1}{3} & -2 & -1 \\ 0 & 0 & 1 & 0 \\ 0 & 0 & 0 & 1 \end{bmatrix} = \begin{bmatrix} 1 & 0 & 0 & 0 \\ 0 & 1 & 0 & 0 \\ 0 & 0 & 0 & 0 \end{bmatrix}
$$

as required.

The matrices P and Q are not unique in this process. For instance, in the example we might perform an operation equivalent to (v) on rows instead of columns. PAQ would still be C although P and Q would now be

$$
P = \begin{bmatrix} \frac{1}{2} & 0 & 0 \\ -\frac{2}{3} & \frac{1}{3} & 0 \\ 1 & -1 & 1 \end{bmatrix} \quad \text{and} \quad Q = \begin{bmatrix} 1 & -3 & 4 & 2 \\ 0 & 1 & -2 & -1 \\ 0 & 0 & 1 & 0 \\ 0 & 0 & 0 & 1 \end{bmatrix}.
$$

The order in which the individual elementary operations are carried out determines the final forms of P and Q; but C will always be the same for any particular matrix. It will be the same order as A, and for all matrices of rank r it can be partitioned as $C = \begin{bmatrix} I_r & 0 \\ 0 & 0 \end{bmatrix}$ where the zeros are null matrices of appropriate order.

10.   CONGRUENT REDUCTION OF SYMMETRIC MATRICES

Derivation of the canonical form is simplified for symmetric matrices, which frequently occur in situations where the canonical form is of particular use, notably the simplification of quadratic forms.

In equation (10) the matrix has been reduced to what is called the diagonal form,

$$\Delta = \begin{bmatrix} D_r & 0 \\ 0 & 0 \end{bmatrix},$$

where D_r is a diagonal matrix of r non-zero elements. Consider the operations for deriving Δ for a symmetric matrix. Since each column is the same as the corresponding row, the same operations as are made on the rows of A to reduce the sub-diagonal elements to zero will, if performed on the columns of A, reduce the elements above the diagonal to zeros also. Therefore, when A is symmetric, performing the same operations on columns as on rows reduces A to the diagonal form Δ. This means that in PAQ the matrix Q equals the transpose of P, i.e., $Q = P'$. Hence we have $PAP' = \Delta$.

Suppose all diagonal elements of D_r are positive and from D_r we form the diagonal matrix R_r whose elements are the reciprocals of the square roots of the diagonal elements of D_r, i.e. $R_r^2 = D_r^{-1}$. If A is of order n and

$$F = \begin{bmatrix} R_r & 0 \\ 0 & I_{n-r} \end{bmatrix},$$

then $(FP)A(FP)' = FPAP'F = C = \begin{bmatrix} I_r & 0 \\ 0 & 0 \end{bmatrix}$ since $F' = F$.

This means that, provided all elements of D_r are positive, a symmetric matrix can be reduced to canonical form under equivalence by pre-multiplying by a matrix FP and post-multiplying by its transpose $P'F$. This is called the *congruent reduction* of the symmetric matrix A, and C is known as the *canonical form under congruence*.

Suppose now that not all diagonal elements of D_r are positive, but that q of them are negative. Then pre- and post-multiply PAP' by an E-type elementary operator so that the first $r - q$ diagonal elements of D_r are positive and the last q are negative. Denote the new product $P^*AP^{*'}$. Now define F as before except that R_r has as elements the reciprocals of the square roots of the elements of D_r disregarding sign. We then have

$$(FP^*)A(FP^*)' = FP^*AP^{*'}F = \begin{bmatrix} I_{r-q} & 0 & 0 \\ 0 & -I_q & 0 \\ 0 & 0 & 0 \end{bmatrix}.$$

This is the general canonical form under congruence, for the symmetric matrix A. When $q = 0$ it reduces to C. The difference between the order of I_{r-q} and that of I_q is known as the *signature* of A; i.e., signature $= r - 2q$. Retaining the negative signs in the diagonal means that the reduction to this form is entirely in terms of real numbers. If we are prepared to use imaginary numbers, involving $i = \sqrt{-1}$, then reduction to C can be made, using an F that involves imaginary numbers.

Example. For the symmetric matrix

$$A = \begin{bmatrix} 4 & 12 \\ 12 & 27 \end{bmatrix}$$

the following operations reduce A to diagonal form.

Operations	Reduction of A
i. row 2 $-$ 3(row 1)	$A \cong \begin{bmatrix} 4 & 12 \\ 0 & -9 \end{bmatrix}$
ii. column 2 $-$ 3(column 1)	$A \cong \begin{bmatrix} 4 & 0 \\ 0 & -9 \end{bmatrix}$

P is obtained by carrying out operation (i) on I_2:

$$I_2 = \begin{bmatrix} 1 & 0 \\ 0 & 1 \end{bmatrix} \cong \begin{bmatrix} 1 & 0 \\ -3 & 1 \end{bmatrix} = P$$

and

$$PAP' = \begin{bmatrix} 1 & 0 \\ -3 & 1 \end{bmatrix}\begin{bmatrix} 4 & 12 \\ 12 & 27 \end{bmatrix}\begin{bmatrix} 1 & -3 \\ 0 & 1 \end{bmatrix} = \begin{bmatrix} 4 & 0 \\ 0 & -9 \end{bmatrix}.$$

We then have

$$R = \begin{bmatrix} 1/\sqrt{4} & 0 \\ 0 & 1/\sqrt{9} \end{bmatrix} = \begin{bmatrix} \tfrac{1}{2} & 0 \\ 0 & \tfrac{1}{3} \end{bmatrix}$$

and in this case $F = R$ so that

$$FP = RP = \begin{bmatrix} \tfrac{1}{2} & 0 \\ 0 & \tfrac{1}{3} \end{bmatrix}\begin{bmatrix} 1 & 0 \\ -3 & 1 \end{bmatrix} = \begin{bmatrix} \tfrac{1}{2} & 0 \\ -1 & \tfrac{1}{3} \end{bmatrix}.$$

Then

$$FPA(FP)' = \begin{bmatrix} \frac{1}{2} & 0 \\ -1 & \frac{1}{3} \end{bmatrix} \begin{bmatrix} 4 & 12 \\ 12 & 27 \end{bmatrix} \begin{bmatrix} \frac{1}{2} & -1 \\ 0 & \frac{1}{3} \end{bmatrix} = \begin{bmatrix} 1 & 0 \\ 0 & -1 \end{bmatrix},$$ the desired form.

Example. In dealing with quadratic forms (see Section 3.3) it is often convenient to see if $x'Ax$ can be expressed as a sum of squares of linear combinations of the elements of x; if it can, $x'Ax$ is positive semi-definite. For example, suppose

$$x = \begin{bmatrix} x_1 \\ x_2 \\ x_3 \end{bmatrix} \quad \text{and} \quad A = \begin{bmatrix} 4 & 12 & 20 \\ 12 & 45 & 78 \\ 20 & 78 & 136 \end{bmatrix}.$$

Then [see equation (10), Section 3.3]

$$x'Ax = 4x_1^2 + 45x_2^2 + 136x_3^2 + 24x_1x_2 + 40x_1x_3 + 156x_2x_3.$$

Further, it can be shown that if

$$P = \begin{bmatrix} \frac{1}{2} & 0 & 0 \\ -1 & \frac{1}{3} & 0 \\ 1 & -2 & 1 \end{bmatrix} \quad \text{then} \quad PAP' = \begin{bmatrix} 1 & 0 & 0 \\ 0 & 1 & 0 \\ 0 & 0 & 0 \end{bmatrix}.$$

Suppose we now make the linear transformation $y = P'^{-1}x$ or, equivalently, $x = P'y$. Then the quadratic form $x'Ax$ becomes

$$x'Ax = y'PAP'y = \begin{bmatrix} y_1 & y_2 & y_3 \end{bmatrix} \begin{bmatrix} 1 & 0 & 0 \\ 0 & 1 & 0 \\ 0 & 0 & 0 \end{bmatrix} \begin{bmatrix} y_1 \\ y_2 \\ y_3 \end{bmatrix} = y_1^2 + y_2^2.$$

Hence, with

$$y = \begin{bmatrix} y_1 \\ y_2 \\ y_3 \end{bmatrix} = P'^{-1}x = \begin{bmatrix} 2 & 6 & 10 \\ 0 & 3 & 6 \\ 0 & 0 & 1 \end{bmatrix} \begin{bmatrix} x_1 \\ x_2 \\ x_3 \end{bmatrix} = \begin{bmatrix} 2x_1 + 6x_2 + 10x_3 \\ 3x_2 + 6x_3 \\ x_3 \end{bmatrix},$$

the quadratic form becomes

$$x'Ax = y_1^2 + y_2^2 = (2x_1 + 6x_2 + 10x_3)^2 + (3x_2 + 6x_3)^2.$$

Note that the rank of A, the number of non-zero elements in PAP', is the number of squared terms in the final form of $x'Ax$.

11. THE RANK OF A PRODUCT MATRIX

The following theorem concerns the relationship between the rank of a matrix product AB and the ranks of A and B.

Theorem 3. The rank of a product, $A_{m \times q} B_{q \times n}$, cannot exceed the rank of either A or B.

Proof. If $r(A) = r$ there exist matrices P and Q such that

$$PAQ = \begin{bmatrix} I_r & 0 \\ 0 & 0 \end{bmatrix}.$$

Therefore

$$PA = \begin{bmatrix} I_r & 0 \\ 0 & 0 \end{bmatrix} Q^{-1}$$

and

$$PAB = \begin{bmatrix} I_r & 0 \\ 0 & 0 \end{bmatrix} Q^{-1}B = \begin{bmatrix} G \\ 0 \end{bmatrix} \text{ say,}$$

where G is $r \times n$. Therefore $r(PAB)$ cannot exceed r; and because P is a product of elementary operators, $r(PAB) = r(AB)$, so that $r(AB)$ cannot exceed the rank of A. Similarly it may be shown that $r(AB)$ cannot exceed the rank of B. Hence $r(AB)$ cannot exceed the rank of either A or B.

12. APPENDIX

a. Proof of two theorems

Theorem 1. A set of linearly independent vectors of order n cannot contain more than n such vectors.

Proof. Let $u_1, u_2, \ldots, u_n$ be a set of n independent vectors of order n. We show that any other non-null vector of order n, u_{n+1} say, is not independent of them.

Consider the equations

$$[u_1 \quad u_2 \quad \cdots \quad u_n] \begin{bmatrix} q_1 \\ q_2 \\ \cdot \\ \cdot \\ \cdot \\ q_n \end{bmatrix} = -u_{n+1}. \tag{11}$$

Since the vectors $u_1, u_2, \ldots, u_n$ are independent, the determinant of the $n \times n$ matrix $[u_1 \quad u_2 \quad \cdots \quad u_n]$ is non-zero; thus for any non-null vector u_{n+1} equation (11) has a unique solution for the q's. Therefore in (11), which can be rewritten as

$$q_1 u_1 + q_2 u_2 + \cdots + q_n u_n + u_{n+1} = 0, \tag{12}$$

not all the q's are zero. Then, since (12) is a special case of the general equation

$$k_1 u_1 + k_2 u_2 + \cdots + k_n u_n + k_{n+1} u_{n+1} = 0 \tag{13}$$

with not all the k's zero, we have shown that the vectors $u_1, u_2, \ldots, u_n, u_{n+1}$ are linearly dependent.

Theorem 2. The number of independent rows in a matrix equals the number of independent columns.

Proof. Consider A as a matrix of order $p \times q$ having k independent rows and m independent columns. The rows are then vectors of order q, so that by the corollary of Theorem 1 $k \leq q$, and likewise $m \leq p$. Note also that the property of independence of rows (or columns) of a matrix relates only to the rows (or columns) themselves, and not to their sequence in the matrix. Assuming the first k rows to be independent (and the first m columns) therefore has no effect on arguments based on consequences of linear independence. We find this convenient, and accordingly denote by X the leading $k \times m$ sub-matrix of A formed by the intersection of the k independent rows and m independent columns. A is then partitioned as

$$A = \begin{bmatrix} X_{k \times m} & Y_{k \times (q-m)} \\ Z_{(p-k) \times m} & W_{(p-k) \times (q-m)} \end{bmatrix}.$$

Further, let
$$X_{k \times m} = [u_1 \quad u_2 \quad \cdots \quad u_m]$$
and
$$Z_{(p-k) \times m} = [v_1 \quad v_2 \quad \cdots \quad v_m]$$

where the u's are vectors of order k and the v's are vectors of order $p - k$.

We seek to show that $k = m$. For the moment let us assume that the columns of X are linearly dependent, i.e., for some non-zero scalars $\lambda_1, \lambda_2, \ldots, \lambda_m$,

$$\lambda_1 u_1 + \lambda_2 u_2 + \cdots + \lambda_m u_m = 0. \tag{14}$$

Now because the first k rows of A (those that contain X) are LIN, the rows of Z are linear combinations of those of X; i.e., for some matrix K

$$Z = KX.$$

Therefore $[v_1 \quad v_2 \quad \cdots \quad v_m] = K[u_1 \quad u_2 \quad \cdots \quad u_m]$

and so $v_j = K u_j \qquad$ for $j = 1, 2, \ldots, m$.

Hence
$$\lambda_j v_j = \lambda_j K u_j$$
$$= K \lambda_j u_j$$

because λ_j is a scalar. Thus

$$\lambda_1 v_1 + \lambda_2 v_2 + \cdots + \lambda_m v_m = K[\lambda_1 u_1 + \lambda_2 u_2 + \cdots + \lambda_m u_m] = K(0) = 0$$

because of (14). Therefore

$$\lambda_1 \begin{bmatrix} u_1 \\ v_1 \end{bmatrix} + \lambda_2 \begin{bmatrix} u_2 \\ v_2 \end{bmatrix} + \cdots + \lambda_m \begin{bmatrix} u_m \\ v_m \end{bmatrix} = 0.$$

But this means the m columns of A containing X and Z are dependent, which is contrary to what is given. Therefore the assumption that the columns of X are dependent is false; i.e., the m u's are independent. But the u's are of order k, and therefore by Theorem 1 they can be independent only if $m \le k$. By the same form of argument we could also show that the k rows of X, having order m, are also independent, and hence $k \le m$. Therefore $k = m$ and the number of independent rows of A is the same as the number of independent columns.

b. Factorization of matrices

If the rows of a matrix Q are linear combinations of the rows of P then, for some matrix X, $Q = XP$. This is a consequence of the definition of matrix multiplication. Likewise, if the columns of R are linear combinations of the columns of S, then for some matrix Y, $R = SY$.

Suppose that the columns and rows of a matrix $M_{p \times q}$ of rank r are arranged such that the first r columns and the first r rows are independent. Then M can be partitioned as

$$M = \begin{bmatrix} A_{r \times r} & B_{r \times (q-r)} \\ C_{(p-r) \times r} & D_{(p-r) \times (q-r)} \end{bmatrix}$$

and we can write $[C \quad D] = K_{(p-r) \times r}[A \quad B] = [KA \quad KB]$

so that
$$M = \begin{bmatrix} A & B \\ KA & KB \end{bmatrix}.$$

Similarly we can write

$$\begin{bmatrix} B \\ KB \end{bmatrix} = \begin{bmatrix} A \\ KA \end{bmatrix} L_{r \times (q-r)} = \begin{bmatrix} AL \\ KAL \end{bmatrix}$$

so that
$$M = \begin{bmatrix} A & AL \\ KA & KAL \end{bmatrix} = \begin{bmatrix} I \\ K \end{bmatrix} A[I \quad L], \tag{15}$$

where I has order r, K is $(p - r) \times r$ and L is $r \times (q - r)$, with A being $r \times r$. The rank of M is r; so is that of A. (If it were not, some of its columns would be linear combinations of others and hence $\begin{bmatrix} A \\ KA \end{bmatrix}$ would contain more columns of M than just the independent ones.) Therefore A^{-1} exists.

The preceding discussion is based on assuming it is the first r rows of M that are independent, and likewise the first r columns. This may not always be so, but by using E-type elementary operators, rows (and columns) of M can be interchanged to bring the independent rows (and columns) to being the first. If N is the resulting matrix, and E_1 and E_2 are the products of the necessary E-type operators, $N = E_1 M E_2$. Then N can be factorized as above,

$$N = \begin{bmatrix} I \\ K \end{bmatrix} A[I \quad L] \quad \text{with} \quad M = E_1^{-1} N E_2^{-1}.$$

A corollary of this factorization is that if $M_{r \times q}$ has full row rank, r, then

$$M = A[I \quad L] \tag{16}$$

with A^{-1} existing. Furthermore, $(MM')^{-1}$ exists. For, using (16)

$$MM' = A(I + LL')A'.$$

Assuming that $(I + LL')$ is not of full rank, some linear combination of its columns is null: i.e.,

$$(I + LL')\lambda = 0 \quad \text{for some non-null vector } \lambda. \tag{17}$$

Pre-multiplication by λ' gives

$$\lambda'\lambda + (L'\lambda)'L'\lambda = 0. \tag{18}$$

Each term on the left-hand side of (18) is a sum of squares and therefore (18) is true only for $\lambda = 0$, i.e. λ a null vector. This contradicts the assumption in (17) of λ being non-null. Hence $(I + LL')$ has full rank. Therefore MM' has full rank (because $|MM'| = |A| |I + LL'| |A'| \neq 0$, being a product of non-zero determinants). Thus, we have proved that if M has full row rank then MM' has an inverse; similarly, if some matrix N has full column rank then $N'N$ has an inverse.

These results are useful in many matrix proofs. For example, if A has full row rank, then $ABA = A$ implies that $AB = I$. *Proof*: If $ABA = A$ then $ABAA' = AA'$ and since A has full row rank $(AA')^{-1}$ exists; therefore $AB = I$.

For a symmetric matrix the factorization (15) simplifies. Since, if $M = M'$, the relationship between the independent rows and the other rows is the same as that between the corresponding columns, (15) becomes

$$M = \begin{bmatrix} I \\ K \end{bmatrix} A[I \quad K'],$$

where A^{-1} exists.

Canonical forms also provide factorizations; in particular, they show that a full rank square matrix can be expressed as the product of two triangular matrices. Thus if

$$PAQ = C = \begin{bmatrix} I_r & 0 \\ 0 & 0 \end{bmatrix},$$

$$A = P^{-1} \begin{bmatrix} I_r & 0 \\ 0 & 0 \end{bmatrix} Q^{-1} = P^{-1} \begin{bmatrix} I_r & 0 \\ 0 & 0 \end{bmatrix} \begin{bmatrix} I_r & 0 \\ 0 & 0 \end{bmatrix} Q^{-1} = RS \qquad (19)$$

where $\qquad R = P^{-1} \begin{bmatrix} I_r & 0 \\ 0 & 0 \end{bmatrix} \quad$ and $\quad S = \begin{bmatrix} I_r & 0 \\ 0 & 0 \end{bmatrix} Q^{-1}.$

When A is square of full rank $r = n$, then $R = P^{-1}$ and $S = Q^{-1}$; and when P and Q are lower triangular (elements above the diagonal zero, see Section 1.6), so are their inverses. Then $A = P^{-1}Q^{-1}$ is a product of triangular matrices, as is $A^{-1} = QP$.

When A is symmetric, $Q = P'$ and $S = R'$, (19) becomes $A = RR'$, and when A has full rank, $A = P^{-1}(P^{-1})'$, the product of a lower triangular matrix and its transpose.

c. Left and right inverses

Brief mention has already been made of left and right inverses of rectangular matrices in Section 5.9. It was shown there that for $A_{r \times c}$ no matrix $B_{c \times r}$ exists for which both $AB = I_r$ and $BA = I_c$. We now consider the conditions under which one or other of these equations can be true. From Theorem 3 the equation $A_{r \times c} B_{c \times r} = I_r$ can be true only if both A and B have rank at least equal to r. Likewise $D_{c \times r}$ for $D_{c \times r} A_{r \times c} = I_c$ exists only if the rank of A is at least equal to c. But for $r \neq c$, A has rank equal or less than the smaller of r and c so that it cannot be equal or greater than both of them. Therefore $B_{c \times r}$ and $D_{c \times r}$ for which $AB = I_r$ and $DA = I_c$ cannot both exist when $r \neq c$. Only one of them can: if A has full row rank then B, a right inverse, exists; or if A has full column rank then D, a left inverse, exists.

[6.13] EXERCISES 155

13. EXERCISES

1. For $A = \begin{bmatrix} 1 & 2 & 3 \\ 4 & 5 & 7 \\ 9 & 8 & 6 \end{bmatrix}$

and $E = \begin{bmatrix} 0 & 1 & 0 \\ 1 & 0 & 0 \\ 0 & 0 & 1 \end{bmatrix}$, $R = \begin{bmatrix} 1 & 0 & 0 \\ 0 & 1 & 0 \\ 0 & 0 & 2 \end{bmatrix}$ and $P = \begin{bmatrix} 1 & 3 & 0 \\ 0 & 1 & 0 \\ 0 & 0 & 1 \end{bmatrix}$,

show that

(a) $EA = \begin{bmatrix} 4 & 5 & 7 \\ 1 & 2 & 3 \\ 9 & 8 & 6 \end{bmatrix}$, $AE = \begin{bmatrix} 2 & 1 & 3 \\ 5 & 4 & 7 \\ 8 & 9 & 6 \end{bmatrix}$, $EAE = \begin{bmatrix} 5 & 4 & 7 \\ 2 & 1 & 3 \\ 8 & 9 & 6 \end{bmatrix}$;

(b) $RA = \begin{bmatrix} 1 & 2 & 3 \\ 4 & 5 & 7 \\ 18 & 16 & 12 \end{bmatrix}$, $AR = \begin{bmatrix} 1 & 2 & 6 \\ 4 & 5 & 14 \\ 9 & 8 & 12 \end{bmatrix}$, $RAR = \begin{bmatrix} 1 & 2 & 6 \\ 4 & 5 & 14 \\ 18 & 16 & 24 \end{bmatrix}$;

(c) $PA = \begin{bmatrix} 13 & 17 & 24 \\ 4 & 5 & 7 \\ 9 & 8 & 6 \end{bmatrix}$, $AP = \begin{bmatrix} 1 & 5 & 3 \\ 4 & 17 & 7 \\ 9 & 35 & 6 \end{bmatrix}$, $PAP = \begin{bmatrix} 13 & 56 & 24 \\ 4 & 17 & 7 \\ 9 & 35 & 6 \end{bmatrix}$;

(d) $REA = \begin{bmatrix} 4 & 5 & 7 \\ 1 & 2 & 3 \\ 18 & 16 & 12 \end{bmatrix}$, $REAP = \begin{bmatrix} 4 & 17 & 7 \\ 1 & 5 & 3 \\ 18 & 70 & 12 \end{bmatrix}$;

(e) $-|E| = 1 = |P| = \frac{1}{2}|R|$;
(f) $E = E^{-1} = E'$; and $R = R'$

(g) $R^{-1} = \begin{bmatrix} 1 & 0 & 0 \\ 0 & 1 & 0 \\ 0 & 0 & \frac{1}{2} \end{bmatrix}$ and $P^{-1} = \begin{bmatrix} 1 & -3 & 0 \\ 0 & 1 & 0 \\ 0 & 0 & 1 \end{bmatrix}$.

2. Show that $\begin{bmatrix} 1 \\ 2 \\ 1 \end{bmatrix}$, $\begin{bmatrix} -1 \\ 3 \\ 2 \end{bmatrix}$ and $\begin{bmatrix} -13 \\ -1 \\ 2 \end{bmatrix}$ is not a set of LIN vectors and find a

linear relationship existing among the three vectors.

3. Show that $\begin{bmatrix} 1 \\ 2 \\ 1 \end{bmatrix}$, $\begin{bmatrix} -1 \\ 3 \\ 2 \end{bmatrix}$ and $\begin{bmatrix} 1 \\ 1 \\ 0 \end{bmatrix}$ is a set of LIN vectors and find the linear combination of them that equals the vector $\begin{bmatrix} a \\ b \\ c \end{bmatrix}$.

4. Find the rank of

$$A = \begin{bmatrix} 1 & 6 \\ 2 & 9 \\ 4 & 3 \end{bmatrix}, \quad B = \begin{bmatrix} 6 & 4 & -1 & 2 & 5 \\ 3 & 0 & -1 & 2 & 7 \\ 18 & 3 & 4 & -2 & 0 \\ 6 & 8 & 0 & 0 & -4 \end{bmatrix},$$

$$C = \begin{bmatrix} 3 & 7 & 6 \\ 2 & 1 & 7 \\ 4 & 6 & 3 \\ -1 & -1 & 0 \\ 6 & 8 & 3 \end{bmatrix}, \quad D = \begin{bmatrix} 1 & 1 & 0 & 2 \\ -1 & 3 & 6 & 0 \\ 1 & 5 & 6 & 4 \\ 6 & 4 & -3 & 11 \end{bmatrix},$$

$$E = \begin{bmatrix} 1 & -1 & 0 \\ 3 & 2 & -4 \\ 5 & 0 & -4 \\ 1 & -6 & 4 \end{bmatrix}, \quad \text{and} \quad F = \begin{bmatrix} 21 & 6 & 3 \\ 12 & 16 & 36 \\ -63 & 13 & 18 \\ 0 & 93 & 81 \end{bmatrix}.$$

5. Express the matrices A, D and E of Exercise 4 as a product XY, where the number of rows in Y is the rank of the matrix.

6. For $A = \begin{bmatrix} 1 & 2 & 4 & 0 \\ -2 & -3 & -1 & 1 \\ 0 & 1 & 7 & 1 \\ -2 & -2 & 6 & 2 \\ -3 & -6 & -12 & 0 \end{bmatrix}$

show the following:
(a) All minors of A of order 3 or more are zero.
(b) The rank of A is 2.
(c) For a pair of linearly independent rows each of the other three rows is a linear function thereof.

(d) A can be expressed as XY, where X is 5×2 and Y is 2×4, in the following forms:

$$A = XY = \begin{bmatrix} 1 & 0 \\ 0 & 1 \\ 2 & 1 \\ 2 & 2 \\ -3 & 0 \end{bmatrix} \begin{bmatrix} 1 & 2 & 4 & 0 \\ -2 & -3 & -1 & 1 \end{bmatrix}$$

$$= \begin{bmatrix} -1 & \frac{1}{2} \\ 1 & 0 \\ -1 & 1 \\ 0 & 1 \\ 3 & -\frac{3}{2} \end{bmatrix} \begin{bmatrix} -2 & -3 & -1 & 1 \\ -2 & -2 & 6 & 2 \end{bmatrix}.$$

7. Reduce the following matrices to diagonal form $PAQ = \begin{bmatrix} D_r & . \\ . & . \end{bmatrix}$ and thence to canonical form $PAQ^* = \begin{bmatrix} I_r & . \\ . & . \end{bmatrix}$ where $Q^* = Q \begin{bmatrix} D_r^{-1} & . \\ . & I \end{bmatrix}$:

$$A_1 = \begin{bmatrix} 7 & 13 \\ 2 & 9 \end{bmatrix} \qquad A_2 = \begin{bmatrix} 3 & 6 & 48 \\ 1 & 9 & 2 \\ 4 & 1 & 3 \end{bmatrix}$$

$$A_3 = \begin{bmatrix} 1 & 0 & -1 \\ 3 & -4 & 2 \\ 5 & -4 & 0 \\ 1 & 4 & -6 \end{bmatrix} \qquad A_4 = \begin{bmatrix} 3 & 6 & 2 & 4 \\ 9 & 1 & 3 & 2 \\ 6 & -5 & 1 & -2 \end{bmatrix}.$$

8. Reduce the following symmetric matrices to diagonal form and thence to canonical form:

$$B_1 = \begin{bmatrix} 4 & 4 & 10 \\ 4 & 20 & 18 \\ 10 & 18 & 29 \end{bmatrix} \qquad B_2 = \begin{bmatrix} 4 & 6 & 12 \\ 6 & 8 & 1 \\ 12 & 1 & 5 \end{bmatrix}$$

$$B_3 = \begin{bmatrix} 4 & -2 & 0 \\ -2 & 3 & -2 \\ 0 & -2 & 2 \end{bmatrix} \qquad B_4 = \begin{bmatrix} 1 & 8 & 6 & 7 \\ 8 & 65 & 99 & 40 \\ 6 & 99 & 81 & 78 \\ 7 & 40 & 78 & 21 \end{bmatrix}.$$

9. Find the rank of

$$\begin{bmatrix} 1 & 4 & 6 & 0 & 2 & 1 \\ 4 & -1 & 2 & 0 & 7 & 6 \\ 7 & 6 & 2 & 1 & -9 & -12 \\ -7 & 6 & 2 & 0 & -6 & -12 \end{bmatrix}.$$

10. A chain of four ski shops must determine a set of uniform prices for four of its products: skis, poles, boots and boot laces. The volume of expected sales is shown in the following matrix, where rows represent the products in the order just listed and columns represent the four shops.

$$A = \begin{bmatrix} 1 & 2 & 3 & 2 \\ 3 & 1 & 4 & 0 \\ 1 & 3 & 4 & 2 \\ 1 & 3 & 4 & 2 \end{bmatrix}.$$

It is assumed that sales volumes are unaffected by price, and the four shops must have total revenues from these products of [200 340 540 280] respectively.
   (a) Can the prices which the shops should charge be derived from the above information? Justify your answer.
   (b) If the answer to (a) is no, change the problem in such a way that you can solve it and do so. What question is left unanswered by your method?

11. Show that if the operations used on rows to reduce the lower triangular elements

of

$$A = \begin{bmatrix} 1 & 2 & 3 \\ 2 & 1 & 2 \\ 3 & 2 & 4 \end{bmatrix}$$

to zero are then performed on columns of the resultant equivalent matrix, a diagonal matrix results. What are P and Q such that $PAQ = \Delta$ in this case?

12. Given a symmetric matrix A_n of rank r and its diagonal equivalent

$$\Delta = \begin{bmatrix} D_r & 0 \\ 0 & 0 \end{bmatrix},$$

show that if all diagonal elements of D_r are positive, then

$$(FP)A(FP)' = \begin{bmatrix} I_r & 0 \\ 0 & 0 \end{bmatrix},$$

where

$$F = \begin{bmatrix} R_r & 0 \\ 0 & I_{n-r} \end{bmatrix} \quad \text{and} \quad R_r^2 = D_r^{-1}.$$

13. Samuelson (1939) considers the interaction of the multiplier and accelerator in relation to business cycle theory. He presents the following model, where the symbols are scalars:

$$Y_t = g_t + C_t + I_t,$$
$$C_t = \alpha Y_{t-1},$$
$$I_t = \beta[C_t - C_{t-1}] = \alpha\beta Y_{t-1} - \alpha\beta Y_{t-2},$$

and
$$g_t = 1$$

where, in period t, Y_t is national income, g_t is government expenditures, C_t is consumption expenditures and I_t is induced private investment.

(a) Find A such that

$$Ax_t = y_t$$

where
$$x_t = \begin{bmatrix} Y_t \\ g_t \\ C_t \\ I_t \end{bmatrix} \quad \text{and} \quad y_t = \begin{bmatrix} 0 \\ \alpha Y_{t-1} \\ \alpha\beta Y_{t-1} - \alpha\beta Y_{t-2} \\ 1 \end{bmatrix}.$$

(b) What is the rank of A? Is it of full column rank or full row rank? Can the system be solved?

(c) Solve the above equation for x_t if it is possible.

REFERENCE

Samuelson, P.A. (1939). Interactions between the multiplier analysis and the principle of acceleration. *Review of Economic Statistics*, **21**, 75–78.

CHAPTER 7

LINEAR EQUATIONS AND GENERALIZED INVERSES

We have seen that linear equations in several unknowns can be represented in matrix form as $Ax = y$ where x is the vector of unknowns, y is a vector of known values and A is the matrix of coefficients. So far, the only form of $Ax = y$ considered has been where there are as many equations as unknowns. The matrix A is then square and, if it has full rank, A^{-1} exists and the solution is $x = A^{-1}y$. However, the case of A being square but not of full rank has not been considered. Neither has the case where the number of equations differs from the number of unknowns and A is rectangular. These three situations are all included in the general case of p equations in q unknowns, which can be written as

$$A_{p \times q} x_{q \times 1} = y_{p \times 1}$$

with the rank of A being r. If $p = q = r$ we have A square and of full rank; if $p = q$ and $r < p$, A is square and not of full rank; and if $p \neq q$, A is rectangular (with r equal to or less than the lesser of p and q).

In this chapter we consider solutions to this general equation $A_{p \times q} x_{q \times 1} = y_{p \times 1}$. When A is square and of full rank the solution is $x = A^{-1}y$; but if A is square and not of full rank, or if A is rectangular, then A^{-1} does not exist. Nevertheless, if certain conditions are satisfied, solutions can be found. We here discuss these conditions and show how they lead to solving the equations. An improved method of solving the general equation $Ax = y$ is then developed, based on a generalized inverse of the matrix A, a technique that is a relatively recent innovation in matrix theory.

Throughout the chapter we deal with linear equations only (e.g. of the type $3a + b - 2c = 17$). Thus by "equations" we always mean "linear equations." The unknown scalars in $Ax = y$, the elements of x, will be denoted by the letters $a, b, c, \ldots$, and not by subscripted x's, which are used to refer to different x vectors. Thus notation will be of the form

$$x_1' = [a_1 \quad b_1 \quad c_1] \quad \text{and} \quad x_2' = [a_2 \quad b_2 \quad c_2].$$

1. EQUATIONS HAVING MANY SOLUTIONS

Equations $Ax = y$ may have either one solution, many solutions or no solution at all, depending on the form of A and of y. Thus the equations

$$a + b = 7$$

and

$$2a + b = 10$$

have just one solution: $a = 3$ and $b = 4$. But

$$a + b = 5$$

and

$$2a + 2b = 10$$

have an infinite number of solutions, in each of which b equals $5 - a$. It might be said that the equation $2a + 2b = 10$ is exactly equivalent to $a + b = 5$, and therefore there is really only one equation, not two. For another example consider the equations

$$2a + 3b + c = 14$$
$$a + b + c = 6 \tag{1}$$
$$3a + 5b + c = 22.$$

Substituting into the left-hand sides of (1) the values $a = 4$, $b = 2$ and $c = 0$ shows that these values satisfy the equations. So do the values $a = 6$, $b = 1$ and $c = -1$; and likewise $a = 1.2$, $b = 3.4$ and $c = 1.4$. Hence these are three solutions to the equations; and it can be shown that there are many more. Thus equations (1) are a second example of equations having not just one solution, but many solutions. This situation may be found novel at first encounter, but it is one that occurs quite frequently in quantitative analyses of business and economic problems.

Before proceeding, notice a characteristic of equations (1): twice the first equation minus the second gives the third. Thus the third equation is a linear combination of the other two, and there are only two independent equations with three unknowns.

2. CONSISTENT EQUATIONS

a. Definition

Consider the two equations

$$a + b = 5$$

and
$$2a + 2b = 11.$$

If one of them is true the other cannot be, for the second is incompatible, or inconsistent, with the first. Considered as a set of linear equations they are said to be *inconsistent*. Similarly, consider

$$2a + 3b + c = 14$$
$$a + b + c = 6 \qquad (2)$$
$$3a + 5b + c = 19.$$

Twice the first equation minus the second gives

$$2(2a + 3b + c) - (a + b + c) = 2(14) - 6$$

i.e.
$$3a + 5b + c = 22.$$

This is a direct outcome of the first two equations in (2), and yet it and the third equation of (2) cannot be true together. This inconsistency within the equations of (2) leads to their being described as inconsistent equations.

With equations (1) we saw that twice the first equation minus the second equals the third. Writing the equations in matrix form $Ax = y$,

$$\begin{bmatrix} 2 & 3 & 1 \\ 1 & 1 & 1 \\ 3 & 5 & 1 \end{bmatrix} \begin{bmatrix} a \\ b \\ c \end{bmatrix} = \begin{bmatrix} 14 \\ 6 \\ 22 \end{bmatrix},$$

we see that this same relationship applies to both the rows of A and the corresponding elements of y. The last row of A (the last element of y) is twice the first row (the first element) minus the second row (element). Equations of this nature are said to be *consistent*; i.e. linear equations $Ax = y$ are defined as consistent if any linear relationships existing among the rows of A also exist among the corresponding elements of y.

The definition of consistent equations does not require that linear relationships exist among the rows of A; but if they do, the same relationships must also exist among the corresponding elements of y for the equations to be consistent. When A is square and of full rank the equations $Ax = y$ are always consistent, for there are no linear relationships among the rows of A and therefore none that the elements of y must satisfy. But whenever the rank of A is less than its number of rows, whether A is square or rectangular, then we must ask whether the linear relationships existing among the rows of A also exist among the corresponding elements of y. If they do exist, then the equations $Ax = y$ are consistent; if they do not exist, then the equations are inconsistent.

b. Existence of solutions

The importance of consistency is that consistent equations can be solved and inconsistent equations cannot. Before giving two useful tests for ascertaining whether equations are consistent or not, we show that consistent equations do always have at least one solution.

Consider again the set of equations

$$A_{p \times q} x_{q \times 1} = y_{p \times 1},$$

with the rank of A being r. Then r of the p rows in A are linearly independent (LIN). Suppose we rearrange the equations so that the first r rows of A are LIN and denote them by the matrix A^*. The remaining rows of A are therefore linear combinations of the first r LIN rows now denoted by A^*, and can be written as CA^* where C represents the matrix of coefficients of these linear combinations. Hence A can be expressed in partitioned form as $A = \begin{bmatrix} A^* \\ CA^* \end{bmatrix}$. Assuming the equations are consistent the vector of right-hand sides, y, can be similarly partitioned as $y = \begin{bmatrix} y_1 \\ Cy_1 \end{bmatrix}$, where y_1 has r elements and Cy_1 has $p - r$. The equations $Ax = y$ then become

$$\begin{bmatrix} A^* \\ CA^* \end{bmatrix} x = \begin{bmatrix} y_1 \\ Cy_1 \end{bmatrix}, \tag{3}$$

or

$$A^*x = y_1 \tag{4}$$

and

$$CA^*x = Cy_1. \tag{5}$$

Since any solution to (4) also satisfies (5), the problem reduces to solving (4).

In (4) the matrix A^* has r LIN rows. Let us interchange columns of A^* and corresponding elements of x so that the first r columns of A^* are LIN. Then A^* can be partitioned as $A^* = [A_1 \quad A_2]$ where A_1 is $r \times r$ and nonsingular and A_2 is $r \times (q - r)$. Conformable with partitioning A^*, the vector x can also be partitioned as $\begin{bmatrix} x_1 \\ x_2 \end{bmatrix}$ where x_1 has r elements and x_2 has $q - r$.

Equation (4) then becomes

$$[A_1 \quad A_2]\begin{bmatrix} x_1 \\ x_2 \end{bmatrix} = y_1 \tag{6}$$

or

$$A_1 x_1 + A_2 x_2 = y_1 \tag{7}$$

and since A_1 is nonsingular we can pre-multiply by A_1^{-1} and solve for x_1:

$$x_1 = A_1^{-1} y_1 - A_1^{-1} A_2 x_2. \tag{8}$$

This gives x_1 in terms of x_2. Hence, if for any vector x_2, equation (8) is used for calculating x_1 then the vector

$$x = \begin{bmatrix} x_1 \\ x_2 \end{bmatrix} = \begin{bmatrix} A_1^{-1} y_1 - A_1^{-1} A_2 x_2 \\ x_2 \end{bmatrix} \tag{9}$$

is a solution to $Ax = y$. Note that this result relies on the assumption made in (3) that the equations $Ax = y$ are consistent. In this way we have shown that consistent equations have a solution; in the appendix (Sec. 7.6a) we show the converse, that any equations having solutions are consistent.

Example. To solve the system of equations (1) by expression (9) we start from the matrix form given in Section 2a. Partitioned as (3) this is

$$\begin{bmatrix} A^* \\ CA^* \end{bmatrix} x = \begin{bmatrix} 2 & 3 & 1 \\ 1 & 1 & 1 \\ \hline 3 & 5 & 1 \end{bmatrix}\begin{bmatrix} a \\ b \\ c \end{bmatrix} = \begin{bmatrix} 14 \\ 6 \\ \hline 22 \end{bmatrix} = \begin{bmatrix} y_1 \\ Cy_1 \end{bmatrix}$$

so that equation (6) is then

$$[A_1 \quad A_2]\begin{bmatrix} x_1 \\ x_2 \end{bmatrix} = \begin{bmatrix} 2 & 3 & 1 \\ 1 & 1 & 1 \end{bmatrix}\begin{bmatrix} a \\ b \\ c \end{bmatrix} = \begin{bmatrix} 14 \\ 6 \end{bmatrix} = y_1.$$

Consequently (8) is

$$x_1 = \begin{bmatrix} a \\ b \end{bmatrix} = \begin{bmatrix} 2 & 3 \\ 1 & 1 \end{bmatrix}^{-1}\begin{bmatrix} 14 \\ 6 \end{bmatrix} - \begin{bmatrix} 2 & 3 \\ 1 & 1 \end{bmatrix}^{-1}\begin{bmatrix} 1 \\ 1 \end{bmatrix} c$$

which reduces to

$$x_1 = \begin{bmatrix} a \\ b \end{bmatrix} = \begin{bmatrix} 4 - 2c \\ 2 + c \end{bmatrix}. \tag{10}$$

In this example x_2 is simply the scalar c. Applying (10) to expression (9) now gives the solution[1] as

$$\tilde{x} = \begin{bmatrix} 4 - 2c \\ 2 + c \\ c \end{bmatrix}. \tag{11}$$

This demonstrates that equations (1) have many solutions because, for any value of c, expression (11) satisfies (1). For example, with $c = 0$ the

solution is $\tilde{x} = \begin{bmatrix} 4 \\ 2 \\ 0 \end{bmatrix}$; with $c = -1$ it is $\tilde{x} = \begin{bmatrix} 6 \\ 1 \\ -1 \end{bmatrix}$; and when $c = 2$ it is

$\tilde{x} = \begin{bmatrix} 0 \\ 4 \\ 2 \end{bmatrix}$. That (11) satisfies (1) for any value of c is seen by direct substitu-

tion. Thus we find that

$$\begin{bmatrix} 2 & 3 & 1 \\ 1 & 1 & 1 \\ 3 & 5 & 1 \end{bmatrix} \begin{bmatrix} 4 - 2c \\ 2 + c \\ c \end{bmatrix} = \begin{bmatrix} 14 \\ 6 \\ 22 \end{bmatrix}$$

no matter what the value of c is. This means that there is an infinite number of solutions.

Note that although the result (9) is derived on the basis of r being less than both p and q it holds true even when r equals either or both these values. In these latter cases one of the partitions will not exist, and either or both CA^* or A_2 will contain no elements. If A_2 is non-existent, so is x_2.

In general, consistent equations $Ax = y$ have the solution $\tilde{x} = \begin{bmatrix} \tilde{x}_1 \\ \tilde{x}_2 \end{bmatrix}$ as in

(9), with $\tilde{x}_1$ constructed from (8) for any $\tilde{x}_2$. Thus (9) not only represents a method of obtaining solutions to equations $Ax = y$, but it also demonstrates the possibility of there being many solutions. Depending on the non-existence or existence of x_2 there will be either a single unique solution or an infinite number of solutions. When the rank r equals q, the number of unknowns, there is no vector x_2; A^* and A_1 are then the same matrix and solution (9) reduces to $x = A_1^{-1}y_1$, which is the sole, unique solution. If, in addition, $p = r = q$, then A, A^* and A_1 are all the same and the unique solution is the familiar $x = A^{-1}y$. When $r < q$, x_2 exists and there is an infinite number of solutions.

---

[1] The notation $\tilde{x}$ is used for a solution, to distinguish it from x, the vector of unknowns.

Examples. The only solution to

$$\begin{bmatrix} 2 & 3 \\ 1 & 1 \\ 3 & 5 \\ 4 & 1 \\ 1 & -2 \end{bmatrix} \begin{bmatrix} a \\ b \end{bmatrix} = \begin{bmatrix} 14 \\ 6 \\ 22 \\ 18 \\ 0 \end{bmatrix}$$

is

$$\begin{bmatrix} a \\ b \end{bmatrix} = \begin{bmatrix} 2 & 3 \\ 1 & 1 \end{bmatrix}^{-1} \begin{bmatrix} 14 \\ 6 \end{bmatrix} = \begin{bmatrix} 4 \\ 2 \end{bmatrix}.$$

The rank of the matrix equals the number of unknowns ($r = q = 2$) and there is only one solution. But as shown above,

$$\begin{bmatrix} 2 & 3 & 1 \\ 1 & 1 & 1 \\ 3 & 5 & 1 \end{bmatrix} \begin{bmatrix} a \\ b \\ c \end{bmatrix} = \begin{bmatrix} 14 \\ 6 \\ 22 \end{bmatrix}$$

has an infinite number of solutions ($r < q$):

$$\begin{bmatrix} a \\ b \\ c \end{bmatrix} = \begin{bmatrix} 4 - 2c \\ 2 + c \\ c \end{bmatrix}.$$

Consistent equations $Ax = y$ are usually described as being equations *of full rank* if the A matrix is square and $r = p = q$; the solution is then $x = A^{-1}y$. If A is square and not of full rank the equations are said to be "equations not of full rank," and they have infinitely many solutions. These descriptions could be broadened to include rectangular matrices $A_{p \times q}$, by referring to all equations for which $r = q$ as being equations of full rank; they have a unique solution. And all equations for which $r < q$ could be described as equations not of full rank; they have an infinite number of solutions. Whatever description is used, the important fact remains: when the rank of A equals the number of unknowns (q) in consistent equations $Ax = y$ they have a unique solution; otherwise they have infinitely many solutions. If the equations are inconsistent they have no solution.

c. Tests for consistency

Before discussing further procedures for solving consistent equations we consider methods of testing whether or not equations are consistent. The most common test for consistency is one involving the partitioned matrix

$[A \quad y]$ formed by adding y as an extra column to A. Such a matrix is referred to as an *augmented matrix*, A augmented by y. With it we find that the equations $Ax = y$ are consistent if, and only if, the rank of the augmented matrix $[A \quad y]$ equals the rank of A.

This can be shown as follows. If $Ax = y$ is consistent, the augmented matrix can, by equation (3), be written as

$$[A \quad y] = \begin{bmatrix} A^* & y_1 \\ CA^* & Cy_1 \end{bmatrix}.$$

This has the same number of LIN rows as A, namely r, and hence the same rank as A. Conversely, if $[A \quad y]$ and A have the same rank, then so do $\begin{bmatrix} A^* & y_1 \\ CA^* & y_2 \end{bmatrix}$ and $\begin{bmatrix} A^* \\ CA^* \end{bmatrix}$. This can only be true if $y_2 = Cy_1$, i.e. if $Ax = y$ is consistent.

This test for consistency requires ascertaining the rank of $[A \quad y]$ and the rank of A. If the ranks are equal the equations are consistent; otherwise they are inconsistent.

Example. For equations $Ax = y$ with

$$A = \begin{bmatrix} 2 & 6 & 4 & 2 \\ 4 & 15 & 14 & 7 \\ 2 & 9 & 10 & 5 \end{bmatrix} \quad \text{and} \quad y = \begin{bmatrix} 1 \\ 6 \\ 5 \end{bmatrix}$$

the rank of A could be determined by row operations, as in Section 6.7, to get

$$A \cong \begin{bmatrix} 2 & 6 & 4 & 2 \\ 0 & 3 & 6 & 3 \\ 0 & 0 & 0 & 0 \end{bmatrix}.$$

Hence the rank of A is 2; i.e., $r(A) = 2$. The augmented matrix is

$$[A \quad y] = \begin{bmatrix} 2 & 6 & 4 & 2 & 1 \\ 4 & 15 & 14 & 7 & 6 \\ 2 & 9 & 10 & 5 & 5 \end{bmatrix}$$

and by the same row operations

$$[A \quad y] \cong \begin{bmatrix} 2 & 6 & 4 & 2 & 1 \\ 0 & 3 & 6 & 3 & 4 \\ 0 & 0 & 0 & 0 & 0 \end{bmatrix}.$$

Hence $r[A \quad y] = 2 = r(A)$ and so the equations $Ax = y$ are consistent.

Since in later sections of this chapter we derive methods for solving equations that utilize the diagonal form of A, we give an alternative test for consistency that can be used in conjunction with those methods: If A is a matrix of p rows and rank r, and if PAQ is a diagonal form of A, then the equations $Ax = y$ are consistent if and only if the last $p - r$ elements of Py are zero.[1]

This provides us with a method of ascertaining whether or not the equations are consistent. In calculating the diagonal matrix PAQ the rank of A, namely r, is obtained as the number of non-zero elements in PAQ. Then, knowing p and r, we investigate the last $p - r$ elements of Py: if they are zero the equations are consistent, but if any of those elements are non-zero the equations are not consistent.

Example. For equations (2) the diagonal form PAQ is

$$\begin{bmatrix} 1 & 0 & 0 \\ -\frac{1}{2} & 1 & 0 \\ -2 & 1 & 1 \end{bmatrix} \begin{bmatrix} 2 & 3 & 1 \\ 1 & 1 & 1 \\ 3 & 5 & 1 \end{bmatrix} \begin{bmatrix} 1 & -\frac{3}{2} & -2 \\ 0 & 1 & 1 \\ 0 & 0 & 1 \end{bmatrix} = \begin{bmatrix} 2 & 0 & 0 \\ 0 & -\frac{1}{2} & 0 \\ 0 & 0 & 0 \end{bmatrix}.$$

The rank of A is 2, and for the equations to be consistent the last $p - r = 3 - 2 = 1$ elements of Py must be zero. Since

$$Py = \begin{bmatrix} 1 & 0 & 0 \\ -\frac{1}{2} & 1 & 0 \\ -2 & 1 & 1 \end{bmatrix} \begin{bmatrix} 14 \\ 6 \\ 19 \end{bmatrix} = \begin{bmatrix} 14 \\ -1 \\ -3 \end{bmatrix},$$

the equations $Ax = y$ are not consistent.

d.   Summary

We have just seen how consistent equations $Ax = y$ can be solved using expressions (8) and (9). These require selecting r LIN rows of A and partitioning them as $[A_1 \quad A_2]$ where A_1 is square, of order r and nonsingular. A_1^{-1} then exists and so (8) and (9) can be used. The ability to proceed in this way

---

[1] With PAQ being a diagonal form of A, PA can be written as $PA = \begin{bmatrix} A_r \\ 0 \end{bmatrix}$ where A_r is a matrix of r LIN rows, P being a product of elementary operators. Therefore, if the equations $Ax = y$ are consistent, so are $PAx = Py$; i.e. so are $\begin{bmatrix} A_r x \\ 0 \end{bmatrix} = Py$. Hence the last $p - r$ elements of Py must be zero. Conversely, if the last $p - r$ elements of Py are zero, the equations $\begin{bmatrix} A_r \\ 0 \end{bmatrix} x = Py$ are consistent, and since P^{-1} exists the equations $P^{-1} \begin{bmatrix} A_r \\ 0 \end{bmatrix} x = y$ are also consistent; i.e. $Ax = y$ is a set of consistent equations.

relies on first finding a suitable A_1, which involves (i) ascertaining the rank r of A, (ii) locating r LIN rows of A and r LIN columns among those rows and (iii) interchanging, if need be, rows and columns of A and elements of x and y to recreate A in a form suited to partitioning.

There is usually little difficulty in carrying out these steps when only a few equations are involved, but when the number of equations is large any one of the steps may be quite tedious. Fortunately, however, they can be avoided by using a method that requires neither ascertaining the rank of A nor searching for a nonsingular submatrix. Furthermore, all cases of either more or fewer equations than unknowns (rectangular A) are handled similarly, and in a manner only slightly different from that used when the number of equations is the same as the number of unknowns (square A). The method involves matrices known as generalized inverses.

3. GENERALIZED INVERSE MATRICES

a. Definitions

Penrose (1955) shows that for any matrix A there is a unique matrix K which satisfies the following four conditions:

$$\text{i. } AKA = A \qquad \text{ii. } KAK = K$$
$$\text{iii. } (KA)' = KA \qquad \text{iv. } (AK)' = AK$$

He called K the generalized inverse of A and showed that it exists and is unique no matter what the form of A is, be it square (singular or non-singular) or rectangular.

Example. Conditions (i) through (iv) are satisfied for

$$A = \begin{bmatrix} 1 & 0 & 2 \\ 0 & -1 & 1 \\ -1 & 0 & -2 \\ 1 & 2 & 0 \end{bmatrix} \quad \text{and} \quad K = \tfrac{1}{66} \begin{bmatrix} 6 & -2 & -6 & 10 \\ 0 & -11 & 0 & 22 \\ 12 & 7 & -12 & -2 \end{bmatrix}.$$

Not only did Penrose show the existence and uniqueness of a matrix[1] satisfying conditions (i) through (iv), but he also demonstrated its use in solving linear equations. However, K is not easily computed when A is at all large in size, and variations on K exist that are just as useful in solving equations but not so difficult to compute. Several different names have been

---

[1] Also considered by Moore (1935) and Bjehammer (1951); see Malik (1968) for historical development.

used for these variants. Rao (1955) and Greville (1957), for example, use the term "pseudo inverse," Wilkinson (1958) considers a matrix which he calls an "effective inverse" and Goldman and Zelen (1964) refer to any matrix satisfying conditions (i), (ii) and (iii) as a "weak generalized inverse." The term "generalized inverse" has also been used by Rao (1962) for any matrix T for which $x = Ty$ is a solution to consistent equations $Ax = y$. This, as shown in Theorem 1, below, implies that $ATA = A$, namely condition (i). For this reason, and also because most variants of Penrose's generalized inverse that have been proposed satisfy condition (i), we choose here to define a *generalized inverse* as being any matrix G for which $AGA = A$, that is to say, it is any matrix satisfying just the first of Penrose's four conditions.

The theorem already alluded to ensures that Rao's (1962) definition is included in the one just given, and most of the other definitions that have been used are also included because they incorporate condition (i). Such others of the conditions (ii), (iii) and (iv) as may be included in any definition do not negate the definition just given; they only restrict it, the greatest restriction being when all of them are included. In this case we may refer to the generalized inverse so defined, namely that considered by Penrose, as the "unique generalized inverse." [1]

b. The product H = GA

Two simple properties of the product $H = GA$ arise from the definition of G as any matrix for which $AGA = A$. First, H and A have the same rank for, since $H = GA$, $r(H) \leq r(A)$, and, because $A = AGA = AH$, $r(A) \leq r(H)$. Therefore $r(H) = r(A)$. Secondly, $H^2 = H$: this is so because $H^2 = GAGA = G(AGA) = GA = H$. Matrices having this property are often described as *idempotent: H* is idempotent because $H^2 = H$.

c. Deriving a generalized inverse

A procedure for obtaining from A a generalized inverse G is now presented. Although the method to be described applies equally well to rectangular matrices as to square ones, it will be discussed initially in terms of square matrices. The minor modifications needed for rectangular matrices are given at the end of the chapter.

Section 6.9 deals with the method of reducing any matrix A to the following diagonal form

$$PAQ = \Delta = \begin{bmatrix} D_r & 0 \\ 0 & 0 \end{bmatrix},$$

---

[1] The left and right inverses discussed in Section 5.9 each satisfy three of the conditions (i) through (iv). Thus if L is a left inverse of A, $LA = I$, in which case conditions (i), (ii) and (iii) are satisfied for $K = L$. Similarly, right inverses satisfy (i), (ii) and (iv).

where Δ is the same order as A, r is the rank of A, D_r is a diagonal matrix having r non-zero elements and the 0's are null matrices of appropriate order. The matrices P and Q are products of elementary operators.

Let us now define a new matrix Δ^- as

$$\Delta^- = \begin{bmatrix} D_r^{-1} & 0 \\ 0 & 0 \end{bmatrix},$$

where the symbol Δ^- is read as "delta minus" and where D_r^{-1} is the inverse of D_r. Then if the matrix G is defined as

$$G = Q\,\Delta^-P, \tag{12}$$

it will be shown that G is a generalized inverse of A, i.e. $AGA = A$.

Example. For the matrix of equations (1) considered earlier,

$$A = \begin{bmatrix} 2 & 3 & 1 \\ 1 & 1 & 1 \\ 3 & 5 & 1 \end{bmatrix},$$

a diagonal form is obtained using

$$P = \begin{bmatrix} 1 & 0 & 0 \\ -\tfrac{1}{2} & 1 & 0 \\ -2 & 1 & 1 \end{bmatrix} \quad \text{and} \quad Q = \begin{bmatrix} 1 & -\tfrac{3}{2} & -2 \\ 0 & 1 & 1 \\ 0 & 0 & 1 \end{bmatrix}.$$

Thus

$$PAQ = \Delta = \begin{bmatrix} 2 & 0 & 0 \\ 0 & -\tfrac{1}{2} & 0 \\ 0 & 0 & 0 \end{bmatrix} = \begin{bmatrix} D_2 & 0 \\ 0 & 0 \end{bmatrix} \quad \text{where } D_2 = \begin{bmatrix} 2 & 0 \\ 0 & -\tfrac{1}{2} \end{bmatrix}.$$

Then

$$\Delta^- = \begin{bmatrix} \tfrac{1}{2} & 0 & 0 \\ 0 & -2 & 0 \\ 0 & 0 & 0 \end{bmatrix}$$

and

$$G = Q\,\Delta^-P$$

$$= \begin{bmatrix} 1 & -\tfrac{3}{2} & -2 \\ 0 & 1 & 1 \\ 0 & 0 & 1 \end{bmatrix}\begin{bmatrix} \tfrac{1}{2} & 0 & 0 \\ 0 & -2 & 0 \\ 0 & 0 & 0 \end{bmatrix}\begin{bmatrix} 1 & 0 & 0 \\ -\tfrac{1}{2} & 1 & 0 \\ -2 & 1 & 1 \end{bmatrix}$$

$$= \begin{bmatrix} -1 & 3 & 0 \\ 1 & -2 & 0 \\ 0 & 0 & 0 \end{bmatrix}.$$

Carrying out the multiplication shows that $AGA = A$.

The equality $AGA = A$ is based upon properties of Δ and Δ^-, namely that

$$\Delta\Delta^- = \begin{bmatrix} D_r & 0 \\ 0 & 0 \end{bmatrix}\begin{bmatrix} D_r^{-1} & 0 \\ 0 & 0 \end{bmatrix} = \begin{bmatrix} I_r & 0 \\ 0 & 0 \end{bmatrix} = \begin{bmatrix} D^{-1} & 0 \\ 0 & 0 \end{bmatrix}\begin{bmatrix} D_r & 0 \\ 0 & 0 \end{bmatrix} = \Delta^-\Delta \quad (13)$$

and that consequently $\Delta^-\Delta\Delta^- = \Delta^-$ and $\Delta\Delta^-\Delta = \Delta$. In addition, because P and Q are products of elementary operators their inverses exist; hence $PAQ = \Delta$ implies $A = P^{-1}\Delta Q^{-1}$. Substituting this and (12) into AGA and then using (13) gives

$$AGA = P^{-1}\Delta Q^{-1}Q\,\Delta^-PP^{-1}\Delta Q^{-1} = P^{-1}\Delta\Delta^-\Delta Q^{-1} = P^{-1}\Delta Q^{-1} = A.$$

Thus, having shown that $AGA = A$ for G given in (12), we conclude that G is a generalized inverse of A. Notice that G is called "a" generalized inverse and not "the" generalized inverse of A because, with one exception, it is not unique[1] (the matrices P and Q are in general not unique and therefore neither is $G = Q\,\Delta^-P$).

Two further properties of G derived in this manner are of interest. First, the expression GAG reduces to G (Penrose's second condition)[2].

$$GAG = (Q\,\Delta^-P)(P^{-1}\Delta Q^{-1})(Q\,\Delta^-P) = Q\,\Delta^-P = G.$$

Second, if $H = GA$, the rank of $H - I$ is $q - r$ where q is the number of columns in A.[3] These two results are additional to those proved before for any G for which $AGA = A$, namely that $r(H) = r(A)$ and $H^2 = H$ for $H = GA$. All four results play a part in the application of G to the problem of solving the equations $Ax = y$ for x.

---

[1] The exception is when A is square and of full rank; for then $\Delta = D$, and $G = Q\,\Delta^-P = Q(Q^{-1}A^{-1}P^{-1})P = A^{-1}$. Thus G is unique, the (regular) inverse of A, when A is square and of full rank.

[2] Although the method of deriving G just described leads to the equality $GAG = G$, this equality is not necessary for solving $Ax = y$. Any matrix G satisfying just $AGA = A$ will suffice.

[3] Substituting $G = Q\,\Delta^-P$ and $A = P^{-1}\Delta Q^{-1}$ into $H = GA$ gives

$$H = GA = Q\,\Delta^-\Delta Q^{-1} = Q\begin{bmatrix} I_r & 0 \\ 0 & 0 \end{bmatrix}Q^{-1}$$

so that

$$H - I_q = Q\left\{\begin{bmatrix} I_r & 0 \\ 0 & 0 \end{bmatrix} - I_q\right\}Q^{-1}.$$

Therefore, because Q is a product of elementary operators,

$$\text{rank } [H - I_q] = \text{rank }\left\{\begin{bmatrix} I_r & 0 \\ 0 & 0 \end{bmatrix} - I_q\right\} = q - r.$$

4. SOLVING LINEAR EQUATIONS USING GENERALIZED INVERSES

a. Obtaining a solution

The relationship between a generalized inverse of A and the consistent equations $Ax = y$ is set out in the following theorem adapted from Rao (1962).

Theorem 1. The consistent equations $Ax = y$ have the solution $x = Gy$ if, and only if, $AGA = A$.

Proof. Let a_j be the jth column of A and consider the equations $Ax = a_j$: they are consistent, because a solution is the vector x_0 defined as a null vector except for unity as its jth element. Now if the equations $Ax = y$ have the solution $x = Gy$ then equations $Ax = a_j$ have the solution $x = Ga_j$. Substituting this solution back into its equation gives $AGa_j = a_j$. This is true for all j, i.e. for all columns of A; therefore $AGA = A$. Thus we have shown that if $Ax = y$ has solution $x = Gy$ then $AGA = A$.

Conversely, if $AGA = A$, $AGAx = Ax$, and if $Ax = y$ this becomes $AGy = y$, i.e., $A(Gy) = y$. Therefore $x = Gy$ is a solution of $Ax = y$, and so the theorem is proved.

Example. Equations (1) continue to be used as an example. Thus as shown above,

$$A = \begin{bmatrix} 2 & 3 & 1 \\ 1 & 1 & 1 \\ 3 & 5 & 1 \end{bmatrix} \quad \text{has} \quad G = \begin{bmatrix} -1 & 3 & 0 \\ 1 & -2 & 0 \\ 0 & 0 & 0 \end{bmatrix}$$

as a generalized inverse and hence

$$\tilde{x} = Gy = \begin{bmatrix} -1 & 3 & 0 \\ 1 & -2 & 0 \\ 0 & 0 & 0 \end{bmatrix} \begin{bmatrix} 14 \\ 6 \\ 22 \end{bmatrix} = \begin{bmatrix} 4 \\ 2 \\ 0 \end{bmatrix}$$

is a solution, as may be verified by direct substitution in equations (1).

Theorem 1 indicates how a solution of equations $Ax = y$ may be obtained: find G for which $AGA = A$ and Gy is a solution. But, as shown earlier, when the rank of A is less than the number of unknowns there are many solutions. The following theorem provides a method of deriving all solutions from the one solution given by Theorem 1.

Theorem 2. If A is a matrix of q columns, and if $AGA = A$ and $H = GA$, then $\tilde{x} = Gy + (H - I)z$ is a solution of consistent equations $Ax = y$ for z being any arbitrary vector of order q.

Proof. If
$$\tilde{x} = Gy + (H - I)z,$$
$$A\tilde{x} = AGy + A(H - I)z$$
$$= AGy + (AGA - A)z$$
$$= AGy, \text{ because } AGA = A;$$

but by Theorem 1 the equations $Ax = y$ have a solution $x = Gy$. Therefore

$$A\tilde{x} = AGy = Ax = y;$$

i.e. $A\tilde{x} = y$, showing that $\tilde{x} = Gy + (H - I)z$ is a solution to consistent equations $Ax = y$ for any q-order vector z.

Example (*continued*). For equations (1)

$$Gy = \begin{bmatrix} 4 \\ 2 \\ 0 \end{bmatrix} \quad \text{and} \quad H = GA = \begin{bmatrix} 1 & 0 & 2 \\ 0 & 1 & -1 \\ 0 & 0 & 0 \end{bmatrix}.$$

Therefore $\quad \tilde{x} = Gy + (H - I)z = \begin{bmatrix} 4 \\ 2 \\ 0 \end{bmatrix} + \begin{bmatrix} 0 & 0 & 2 \\ 0 & 0 & -1 \\ 0 & 0 & -1 \end{bmatrix} z$

is a solution. Taking $z = \begin{bmatrix} z_1 \\ z_2 \\ z_3 \end{bmatrix}$ as a vector of arbitrary elements z_1, z_2, z_3, the

solution becomes

$$\tilde{x} = \begin{bmatrix} 4 \\ 2 \\ 0 \end{bmatrix} + \begin{bmatrix} 2z_3 \\ -z_3 \\ -z_3 \end{bmatrix} = \begin{bmatrix} 4 + 2z_3 \\ 2 - z_3 \\ -z_3 \end{bmatrix}$$

which is exactly the same form (with $z_3 = -c$) as solution (11) obtained earlier by partitioning. With $z_3 = 0$, 1 and -2 we get the same explicit solutions as before:

$$\tilde{x}_1 = \begin{bmatrix} 4 \\ 2 \\ 0 \end{bmatrix}, \quad \tilde{x}_2 = \begin{bmatrix} 6 \\ 1 \\ -1 \end{bmatrix} \quad \text{and} \quad \tilde{x}_3 = \begin{bmatrix} 0 \\ 4 \\ 2 \end{bmatrix}.$$

The solution $\tilde{x} = Gy + (H - I)z$ given by Theorem 2 is essentially the same form as that obtained in equation (9). Both forms will generate, through allocating different sets of values to the arbitrary vectors z and x_2, the same series of vectors that satisfy $Ax = y$. Since the rank of $H - I$ is $q - r$ the vector $(H - I)z$ contains only $q - r$ independent elements, the same number as x_2. Hence both forms of solution generate the same number of LIN vectors that are solutions of $Ax = y$. Vectors of this nature are referred to as *LIN solutions*.

An important difference between the solutions given by Theorem 2 [generalized inverse] and equation (9) [partitioning] lies in the relative effort involved in calculating them. The earlier one, (9), requires choosing a nonsingular matrix A_1 from among the rows and columns of A whereas Theorem 2 only demands finding a generalized inverse. This is a somewhat easier task computationally than selecting the necessary LIN rows and columns required for choosing A_1, especially when handling large numbers of equations; moreover, the calculations for obtaining a generalized inverse are more easily programmed for a high-speed computer than those needed for choosing A_1.

b. Independent solutions

Having established a method for solving linear equations and having shown in doing so that it is possible for them to have an infinite number of solutions, we ask two questions: (i) To what extent are the solutions linearly independent? (ii) What relationships exist among the solutions? Since each solution is a vector of order q there can of course, be no more than q LIN solutions. In fact there are fewer, as the following theorem shows.

Theorem 3. If A is a matrix of q columns and rank r, and if y is a non-null vector, the number of LIN solutions to the consistent equations $Ax = y$ is $q - r + 1$.

As the proof[1] of this theorem shows, $q - r$ LIN solutions are obtained from $\tilde{x} = Gy + (H - I)z$ by using $q - r$ LIN vectors z, and another LIN solution is obtained by setting $z = 0$, i.e., $\tilde{x} = Gy$. All other solutions will be linear combinations of those forming a set of LIN solutions. Note that the theorem does not apply if y is a null vector; that case is covered in Section 4c below.

Example (*continued*).

$$\tilde{x}_1 = \begin{bmatrix} 4 \\ 2 \\ 0 \end{bmatrix}, \quad \tilde{x}_2 = \begin{bmatrix} 6 \\ 1 \\ -1 \end{bmatrix} \quad \text{and} \quad \tilde{x}_3 = \begin{bmatrix} 0 \\ 4 \\ 2 \end{bmatrix}$$

---

[1] See appendix, Sec. 7.6b.

are solutions to equations (1) in which $q = 3$ and $r = 2$. There are therefore $3 - 2 + 1 = 2$ LIN solutions. It will be found that any two of the above solutions are LIN and therefore the third is a linear combination of them; e.g. $\tilde{x}_3 = 3\tilde{x}_1 - 2\tilde{x}_2$.

A means of constructing solutions as linear combinations of other solutions is contained in the following theorem.

Theorem 4. If $\tilde{x}_1, \tilde{x}_2, \ldots, \tilde{x}_s$ are any s solutions of consistent equations $Ax = y$ for which $y \neq 0$, then any linear combination of these solutions,

$$x^* = \sum_{i=1}^{s} \lambda_i \tilde{x}_i,$$ is also a solution of the equations if, and only if,

$$\sum_{i=1}^{s} \lambda_i = 1.$$

Proof. (To simplify notation, the limits of summation are omitted.) Because

$$x^* = \Sigma \lambda_i \tilde{x}_i,$$
$$Ax^* = A \Sigma \lambda_i \tilde{x}_i = \Sigma \lambda_i A\tilde{x}_i.$$

And because $\tilde{x}_i$ is a solution, $A\tilde{x}_i = y$ for all i, so giving

$$Ax^* = \Sigma \lambda_i y \tag{14}$$
$$= y(\Sigma \lambda_i). \tag{15}$$

Now if x^* is a solution of $Ax = y$, $Ax^* = y$, and by comparison with (15) this means, y being non-null, that $\Sigma \lambda_i = 1$. Conversely, if $\Sigma \lambda_i = 1$, equation (15) implies that $Ax^* = y$, namely that x^* is a solution. So the theorem is proved.

Notice that the theorem is in terms of *any* s solutions. Hence, for any number of solutions, whether LIN or not, any linear combination of them is itself a solution provided the coefficients in that combination sum to unity.

Example (*continued*). We have already seen that $\tilde{x}_3 = 3\tilde{x}_1 - 2\tilde{x}_2$ is a solution of equations (1) and the sum of the coefficients in $\tilde{x}_3$ is $3 - 2 = 1$. The same is true of

$$\tilde{x}_4 = 0.73\tilde{x}_2 + 0.27\tilde{x}_1,$$

and of
$$\tilde{x}_5 = 0.23\tilde{x}_1 + 0.45\tilde{x}_2 + 0.32\tilde{x}_3.$$

The reader should calculate these vectors explicitly and satisfy himself that they are solutions of equations (1).

Example. For equations

$$Ax = \begin{bmatrix} 1 & 2 & -1 & 9 \\ 2 & 4 & 3 & 3 \\ -1 & -2 & 6 & -24 \\ 1 & 2 & 4 & -6 \end{bmatrix} x = \begin{bmatrix} 4 \\ 13 \\ 1 \\ 9 \end{bmatrix} = y \tag{16}$$

a diagonal form of A comes from using

$$P = \begin{bmatrix} 1 & 0 & 0 & 0 \\ -2 & 1 & 0 & 0 \\ 3 & -1 & 1 & 0 \\ 1 & -1 & 0 & 1 \end{bmatrix} \quad \text{and} \quad Q = \begin{bmatrix} 1 & 1 & -2 & -6 \\ 0 & 0 & 1 & 0 \\ 0 & 1 & 0 & 3 \\ 0 & 0 & 0 & 1 \end{bmatrix},$$

giving

$$PAQ = \Delta = \begin{bmatrix} 1 & 0 & 0 & 0 \\ 0 & 5 & 0 & 0 \\ 0 & 0 & 0 & 0 \\ 0 & 0 & 0 & 0 \end{bmatrix}.$$

The rank of A is therefore 2, and because

$$Py = \begin{bmatrix} 4 \\ 5 \\ 0 \\ 0 \end{bmatrix}$$

the equations are consistent (the last $q - r$ elements of Py are zero). The generalized inverse $G = Q \, \Delta^- P$ is found to be

$$G = \begin{bmatrix} 0.6 & 0.2 & 0 & 0 \\ 0 & 0 & 0 & 0 \\ -0.4 & 0.2 & 0 & 0 \\ 0 & 0 & 0 & 0 \end{bmatrix}, \quad \text{with} \quad H = GA = \begin{bmatrix} 1 & 2 & 0 & 6 \\ 0 & 0 & 0 & 0 \\ 0 & 0 & 1 & -3 \\ 0 & 0 & 0 & 0 \end{bmatrix}.$$

Hence solutions to the equations are obtained as

$$\tilde{x} = Gy + (H - I)z$$

$$= \begin{bmatrix} 5 \\ 0 \\ 1 \\ 0 \end{bmatrix} + \begin{bmatrix} 0 & 2 & 0 & 6 \\ 0 & -1 & 0 & 0 \\ 0 & 0 & 0 & -3 \\ 0 & 0 & 0 & -1 \end{bmatrix} \begin{bmatrix} z_1 \\ z_2 \\ z_3 \\ z_4 \end{bmatrix} \tag{17}$$

$$
= \begin{bmatrix} 5 + 2z_2 + 6z_4 \\ -z_2 \\ 1 - 3z_4 \\ -z_4 \end{bmatrix}, \tag{18}
$$

where z_2 and z_4 are arbitrary scalars. For example, with $z_2 = 1$ and $z_4 = 0$

the explicit solution is $\tilde{x}_1 = \begin{bmatrix} 7 \\ -1 \\ 1 \\ 0 \end{bmatrix}$; and for $z_2 = 0$ and $z_4 = 1$ the solution

is $\tilde{x}_2 = \begin{bmatrix} 11 \\ 0 \\ -2 \\ -1 \end{bmatrix}$. Since $q = 4$ and $r(A) = 2$ there are $q - r + 1 = 3$ LIN

solutions. Because $\tilde{x}_1$ and $\tilde{x}_2$ have been derived from $\tilde{x} = Gy + (H - I)z$

using two values of z that are independent, namely $\begin{bmatrix} 0 \\ 1 \\ 0 \\ 0 \end{bmatrix}$ and $\begin{bmatrix} 0 \\ 0 \\ 0 \\ 1 \end{bmatrix}$, the

solutions $\tilde{x}_1$ and $\tilde{x}_2$ are independent; and a third solution, independent of

these two, is $\tilde{x}_3 = Gy = \begin{bmatrix} 5 \\ 0 \\ 1 \\ 0 \end{bmatrix}$. All other solutions will be linear combinations

of $\tilde{x}_1$, $\tilde{x}_2$ and $\tilde{x}_3$.

c. The equations $Ax = 0$

The two preceding theorems apply only to cases where y is a non-null vector. Consideration is now given to equations $Ax = 0$, namely $Ax = y$ when y is null.

First note that in the scalar counterpart of $Ax = 0$, namely $ax = 0$ where a and x are scalars, the conclusion is that a and/or x are zero. But this is not so for $Ax = 0$, for with A being non-null there are often non-null vectors x for which $Ax = 0$.

Example. Consider equations (16) with $y = 0$:

$$a + 2b - c + 9d = 0$$
$$2a + 4b + 3c + 3d = 0$$
$$-a - 2b + 6c - 24d = 0 \qquad (19)$$
$$a + 2b + 4c - 6d = 0.$$

An obvious solution is $a = 0 = b = c = d$, corresponding to the null vector $x = 0$ in $Ax = 0$. But non-null solutions can also be found. For example, $a = 8, b = -1, c = -3$ and $d = -1$ is a solution and so is $a = 2, b = -4, c = 3$ and $d = 1$.

By the definition of consistency the equations $Ax = 0$ are always consistent, and with $y = 0$ in Theorem 2 their general solution is $\tilde{x} = (H - I)z$ for arbitrary z of order q. Now the rank of $H - I$ is $q - r$, so that when $r = q$, $H - I$ is null and so is $\tilde{x}$; i.e. the only solution is $x = 0$. Otherwise, for $r < q$, the solution is $\tilde{x} = (H - I)z$ which can be non-null. This means that the only condition under which $Ax = 0$ has non-null solutions is when the rank of A is less than its number of columns, and under this condition there will always be non-null solutions. Furthermore, because $r(H - I) = q - r$ there are only $q - r$ LIN vectors $(H - I)z$ and so only $q - r$ LIN solutions to $Ax = 0$, in contrast to the $q - r + 1$ LIN solutions of $Ax = y$ when y is non-null (Theorem 3).

In addition, in considering Theorem 4, (14) reduces to $Ax^* = 0$ when $y = 0$, indicating that for solutions $\tilde{x}_i$, $x^* = \Sigma \, \lambda_i \tilde{x}_i$ is also a solution no matter what values the λ_i's take. Hence any linear combination of any solutions of $Ax = 0$ is itself a solution.

Example (*continued*). Equations (19) can be written as

$$\begin{bmatrix} 1 & 2 & -1 & 9 \\ 2 & 4 & 3 & 3 \\ -1 & -2 & 6 & -24 \\ 1 & 2 & 4 & -6 \end{bmatrix} x = 0.$$

Their non-null solutions $\tilde{x} = (H - I)z$, where $H - I$ is the same as in (17), are

$$\begin{bmatrix} 2z_2 + 6z_4 \\ -z_2 \\ -3z_4 \\ -z_4 \end{bmatrix},$$

and there are $q - r = 4 - 2 = 2$ LIN solutions to (19). Thus for $z_2 = 1$ and $z_4 = 1$ and for $z_2 = 4$ and $z_4 = -1$ the respective solutions are $\tilde{x}_1 =$

$$\begin{bmatrix} 8 \\ -1 \\ -3 \\ -1 \end{bmatrix} \text{ and } \tilde{x}_2 = \begin{bmatrix} 2 \\ -4 \\ 3 \\ 1 \end{bmatrix} ; \text{ and any linear combination of these is also a}$$

solution, for example,

$$\begin{bmatrix} 62.24 \\ -19.48 \\ -11.64 \\ -3.88 \end{bmatrix} = 7\tilde{x}_1 + 3.12\tilde{x}_2$$

is a solution.

d. Comprehensive example

We here present a comprehensive example. Suppose the equations $Ax = y$ are

$$\begin{bmatrix} 5 & 2 & -1 & 2 \\ 2 & 2 & 3 & 1 \\ 1 & 1 & 4 & -1 \\ 2 & -1 & -3 & -1 \end{bmatrix} x = \begin{bmatrix} 7 \\ 9 \\ 5 \\ -6 \end{bmatrix}.$$

For row and column operations represented respectively by

$$P = \begin{bmatrix} 0 & 0 & 1 & 0 \\ 1 & 0 & -5 & 0 \\ 0 & 1 & -2 & 0 \\ -1 & 2 & -1 & 1 \end{bmatrix} \text{ and } Q = \begin{bmatrix} 1 & -1 & 3 & 7/15 \\ 0 & 1 & -7 & -28/15 \\ 0 & 0 & 1 & 9/15 \\ 0 & 0 & 0 & 1 \end{bmatrix},$$

we have

$$PAQ = \Delta = \begin{bmatrix} 1 & 0 & 0 & 0 \\ 0 & -3 & 0 & 0 \\ 0 & 0 & -5 & 0 \\ 0 & 0 & 0 & 0 \end{bmatrix},$$

$$
\text{and} \qquad G = Q\,\Delta^- P = \tfrac{1}{15}
\begin{bmatrix}
5 & -9 & 8 & 0 \\
-5 & 21 & -17 & 0 \\
0 & -3 & 6 & 0 \\
0 & 0 & 0 & 0
\end{bmatrix}.
$$

From Δ we see that $r(A) = 3$ and in Py we find that the last element is zero. Hence the equations are consistent and can be solved. We find

$$
H = GA =
\begin{bmatrix}
1 & 0 & 0 & -7/15 \\
0 & 1 & 0 & 28/15 \\
0 & 0 & 1 & -9/15 \\
0 & 0 & 0 & 0
\end{bmatrix}.
\tag{20}
$$

Then solutions to the equations are given by $\tilde{x} = Gy + (H - I)z$ where the vector z is arbitrary, and if z is taken as $z' = [z_1 \quad z_2 \quad z_3 \quad z_4]$ where the subscripted z's are scalars, the solution is

$$
\tilde{x} = \tfrac{1}{15}
\begin{bmatrix}
-6 - 7z_4 \\
69 + 28z_4 \\
3 - 9z_4 \\
-15z_4
\end{bmatrix},
\tag{21}
$$

z_4 being any arbitrary scalar value.

Because $q = 4$ and $r(A) = 3 = r(H)$, we can find $q - r + 1 = 4 - 3 + 1 = 2$ LIN solutions. For example, putting $z_4 = 0$ in (21) gives

$$
\tilde{x}_1 =
\begin{bmatrix}
-0.4 \\
4.6 \\
0.2 \\
0
\end{bmatrix},
\quad \text{and} \quad
z_4 = -3 \text{ gives } \tilde{x}_2 =
\begin{bmatrix}
1 \\
-1 \\
2 \\
3
\end{bmatrix}.
$$

Any solution other than $\tilde{x}_1$ and $\tilde{x}_2$ will be a linear combination of these two, with coefficients summing to unity. For example,

$$
\tilde{x}_3 =
\begin{bmatrix}
-1.8 \\
10.2 \\
-1.6 \\
-3.0
\end{bmatrix}
$$

is a solution, and $\tilde{x}_3 = 2\tilde{x}_1 - \tilde{x}_2$.

This procedure may appear somewhat lengthy for such a small example, but its advantages are real if there are a large number of equations, say 80 to 100. Each step in the procedure involves matrix operations that are easily carried out no matter what the size or rank of the matrix is, and each step is readily amenable to computer programming. As a result, the problem of solving linear equations can be reduced to a straightforward computing procedure, available either as a desk-calculator operation for cases involving relatively few variables or as a computer operation where large numbers of variables are involved.

5.   RECTANGULAR MATRICES

A method of solving a set of equations which has a different number of equations from unknowns ($p \neq q$) is illustrated in the following example.

Example. Suppose the equations $Ax = y$ are

$$\begin{bmatrix} 1 & 4 \\ 2 & 7 \\ 3 & 10 \end{bmatrix} x = \begin{bmatrix} 7 \\ 13 \\ 19 \end{bmatrix}.$$

Let us redefine A by adding a column of zeros to it to make it square. A generalized inverse of the redefined A can then be obtained. Thus if

$$A = \begin{bmatrix} 1 & 4 & 0 \\ 2 & 7 & 0 \\ 3 & 10 & 0 \end{bmatrix}, \qquad P = \begin{bmatrix} 1 & 0 & 0 \\ -2 & 1 & 0 \\ 1 & -2 & 1 \end{bmatrix}$$

and

$$Q = \begin{bmatrix} 1 & -4 & 0 \\ 0 & 1 & 0 \\ 0 & 0 & 1 \end{bmatrix},$$

it will be found that

$$PAQ = \Delta = \begin{bmatrix} 1 & 0 & 0 \\ 0 & -1 & 0 \\ 0 & 0 & 0 \end{bmatrix} \quad \text{and} \quad G = Q\,\Delta^- P = \begin{bmatrix} -7 & 4 & 0 \\ 2 & -1 & 0 \\ 0 & 0 & 0 \end{bmatrix}.$$

Let us now redefine G by dropping off the same number of rows of zeros as columns were added to A, namely one, at the same time reverting to the

original form of A. Then

$$G = \begin{bmatrix} -7 & 4 & 0 \\ 2 & -1 & 0 \end{bmatrix}$$

is a generalized inverse of

$$A = \begin{bmatrix} 1 & 4 \\ 2 & 7 \\ 3 & 10 \end{bmatrix}, \quad \text{with } H = GA = \begin{bmatrix} 1 & 0 \\ 0 & 1 \end{bmatrix}.$$

Before solving the equations they must first be tested for consistency. The redefining of A does not affect this, and since $r(A) = 2$ and $p = 3$, we find that the last $p - r = 1$ term of Py is zero, and so the equations are consistent. They can therefore be solved. The solution takes the same form as previously,

$$\tilde{x} = Gy + (H - I)z = \begin{bmatrix} 3 \\ 1 \end{bmatrix}$$

in this case, there being only one solution in this example because $H - I = 0$, and $q - r + 1 = 2 - 2 + 1 = 1$.

The key to this method is at once apparent: if A is rectangular it is redefined by adding sufficient rows (columns) of zeros to make it square. A generalized inverse is then derived in the usual fashion, and is redefined by dropping off the same number of columns (rows) of zeros as had rows (columns) been added to A. If A as originally given has order $p \times q$, the order of G is $q \times p$ and the same properties of G and procedures for solving equations $Ax = $ hold as previously. There is nothing more to the procedure than this; jus add rows (columns) of zeros to A, obtain G and drop off columns (rows) o zeros. Note the uniformity of this procedure: it is the same whether there ar more or fewer equations than unknowns.

6.   APPENDIX

a.   Equations having solutions

In Section 7.2 it was shown that consistent equations have solutions. Th converse is now demonstrated: that equations which do have solutions mus be consistent.

Equation (3) in Section 2b was derived as a result of assuming co sistency. Without this assumption equation (3) would be

$$\begin{bmatrix} A^* \\ CA^* \end{bmatrix} x = \begin{bmatrix} y_1 \\ y_2 \end{bmatrix}. \tag{2}$$

Now suppose $\tilde{x}$ is a solution to this equation. Then

$$A^*\tilde{x} = y_1 \tag{23}$$

and
$$CA^*\tilde{x} = y_2. \tag{24}$$

Pre-multiplying (23) by C and substituting in (24) gives $y_2 = Cy_1$. Therefore

(22) becomes $\begin{bmatrix} A^* \\ CA^* \end{bmatrix} x = \begin{bmatrix} y_1 \\ Cy_1 \end{bmatrix}$; i.e. the equations are consistent.

b.   Proof of Theorem 3

Theorem 3. If A is a matrix of q columns and rank r, and if y is a non-null vector, the number of LIN solutions to the consistent equations $Ax = y$ is $q - r + 1$.

Proof.   For $PAQ = \Delta$, $G = Q\,\Delta^- P$ and $H = GA$, a solution to $Ax = y$ is $\tilde{x} = Gy + (H - I)z$. Also, because $r(H - I) = q - r$, there are only $q - r$ arbitrary elements in $(H - I)z$ for arbitrary z; the other r elements are linear combinations of those $q - r$. Therefore there are only $q - r$ LIN vectors $(H - I)z$, and using these in $\tilde{x}$ gives $q - r$ LIN solutions. For $i = 1, 2, \ldots, q - r$, let $\tilde{x}_i = Gy + (H - I)z_i$ be these solutions. $\tilde{x} = Gy$ is also a solution. Assume it is linearly dependent on the x_i. Then, for scalars λ_i, $i = 1, 2, \ldots, q - r$, not all of which are zero,

$$Gy = \Sigma\, \lambda_i \tilde{x}_i = \Sigma\, \lambda_i [Gy + (H - I)z_i], \tag{25}$$

i.e.
$$Gy = Gy\, \Sigma\, \lambda_i + (H - I)\, \Sigma\, \lambda_i z_i. \tag{26}$$

Now the left-hand side of (26) contains no z's. Therefore, on the right-hand side the second summation must be null, i.e. $\Sigma\, \lambda_i z_i = 0$. But since the z_i are LIN this can be true only if every λ_i is zero. This means (25) is no longer true for some λ_i non-zero. Therefore Gy is independent of the $\tilde{x}_i$ and so Gy and $\tilde{x}_i$ for $i = 1, 2, \ldots, q - r$ form a set of $q - r + 1$ LIN solutions.

c.   Linear combinations of elements of a solution

Linear combinations of elements of a solution that are invariant to whatever solution is used have important applications in certain statistical analyses of quantitative data [regression on $(0, 1)$ variables, and their equivalent form, linear statistical models; see Chapter 11]. We therefore have a final theorem, relating to this topic. It is due to Rao (1962).

Theorem 5. For consistent equations $Ax = y$ for which $AGA = A$ and $H = GA$, the linear combination $k'\tilde{x}$ of elements of a solution $\tilde{x}$ is invariant to whatever solution is used for $\tilde{x}$ if, and only if, $k'H = k'$.

Proof.   For a solution $\tilde{x}$ given by Theorem 2,

$$k'\tilde{x} = k'Gy + k'(H - I)z.$$

This is independent of the arbitrary z if $k'H = k'$; and since *any* solution can be put in the form $\tilde{x}$ by appropriate choice of z, the value of $k'\tilde{x}$ for any $\tilde{x}$ is $k'Gy$ provided that $k'H = k'$. Also, when $k'H = k'$ the value of $k'\tilde{x} = k'Gy$ is unique no matter which of the many generalized inverses is used for the matrix G. We now prove this point. First, by Theorem 3 there are $q - r + 1$ LIN solutions of the form $\tilde{x} = Gy + (H - I)z$. Let these solutions be $\tilde{x}_i$ for $i = 1, 2, \ldots, q - r + 1$. Suppose for some other generalized inverse G^* say, we have a solution

$$x^* = G^*y + (H^* - I)z.$$

Then, since the $\tilde{x}_i$'s are a LIN set of $q - r + 1$ solutions, x^* must be a linear combination of them; that is, there is a set of scalars λ_i, for $i = 1, 2, \ldots, q - r + 1$, such that

$$x^* = \sum_{i=1}^{q-r+1} \lambda_i \tilde{x}_i$$

where not all the λ_i's are zero, and, further, by Theorem 4, $\Sigma \lambda_i = 1$. Further, if $k'H = k'$, $k'\tilde{x}_i = k'Gy$ for all i, so that the same linear combination of the elements of x^* is

$$k'x^* = k' \Sigma \lambda_i \tilde{x}_i = \Sigma \lambda_i k'\tilde{x}_i = \Sigma \lambda_i k'Gy$$
$$= k'Gy(\Sigma \lambda_i) = k'Gy.$$

Hence $k'x^*$ for *any* solution x^* equals $k'Gy$ if $k'H = k'$. This proves the theorem.

Example (*continued*). In the solution of equations (16) derived earlier, it will be found that $k'H = k'$ for

$$k' = [1 \quad 2 \quad 1 \quad 3].$$

The value of $k'\tilde{x}$ for the solutions $\tilde{x}_1$ and $\tilde{x}_2$ derived below equation (18) is

$$k'\tilde{x}_1 = [1 \quad 2 \quad 1 \quad 3] \begin{bmatrix} 7 \\ -1 \\ 1 \\ 0 \end{bmatrix} = 6$$

and

$$k'\tilde{x}_2 = [1 \quad 2 \quad 1 \quad 3] \begin{bmatrix} 11 \\ 0 \\ -2 \\ -1 \end{bmatrix} = 6.$$

In fact, for this vector k' the value of $k'\tilde{x}$ is 6 for any solution $\tilde{x}$, as can be seen from the product of k' with the general form of $\tilde{x}$ given in (18):

$$k'\tilde{x} = \begin{bmatrix} 1 & 2 & 1 & 3 \end{bmatrix} \begin{bmatrix} 5 + 2z_2 + 6z_4 \\ -z_2 \\ 1 - 3z_4 \\ -z_4 \end{bmatrix} = 6, \tag{27}$$

regardless of the values of z_2 and z_4.

The values of k' for which $k'\tilde{x}$ is invariant for all solutions $\tilde{x}$ are those given by the equation $k'H = k'$. By the idempotency property of $H(H^2 = H)$ it can be seen that $k' = w'H$ is a solution of $k'H = k'$, no matter what the elements of w' are. Hence, for any vector w', $k' = w'H$ gives $k'\tilde{x}$ as

$$k'\tilde{x} = w'H\tilde{x} = w'HGy + w'H(H - I)z = w'HGy$$
$$= w'GAGy.$$

Since Gy is a solution to $Ax = y$, $AGy = y$, and so therefore

$$k'\tilde{x} = w'Gy.$$

Thus for any vector w' and for $k' = w'H$, we have $k'\tilde{x} = w'Gy$, the same for all solutions $\tilde{x}$. Since $r(H) = r$, $k' = w'H$ provides a set of r LIN vectors k' that have this property; for example, for two such vectors k'_1 and k'_2 say, $k'_1\tilde{x}$ and $k'_2\tilde{x}$ are different but each has a value that is the same for all solutions $\tilde{x}$.

Example (*continued*). Using $w' = \begin{bmatrix} w_1 & w_2 & w_3 & w_4 \end{bmatrix}$ and the H used in (17) gives

$$k' = w'H = \begin{bmatrix} w_1 & w_2 & w_3 & w_4 \end{bmatrix} \begin{bmatrix} 1 & 2 & 0 & 6 \\ 0 & 0 & 0 & 0 \\ 0 & 0 & 1 & -3 \\ 0 & 0 & 0 & 0 \end{bmatrix}$$

$$= \begin{bmatrix} w_1 & 2w_1 & w_3 & (6w_1 - 3w_3) \end{bmatrix}$$

and with Gy as in (17)

$$k'\tilde{x} = w'Gy = \begin{bmatrix} w_1 & w_2 & w_3 & w_4 \end{bmatrix} \begin{bmatrix} 5 \\ 0 \\ 1 \\ 0 \end{bmatrix}$$

$$= 5w_1 + w_3. \tag{28}$$

Thus we see that for any values of w_1 and w_3,

$$k' = [w_1 \quad 2w_1 \quad w_3 \quad (6w_1 - 3w_3)]$$

satisfies $k' = k'H$; hence any expression $k'\tilde{x}$ is invariant no matter what solution $\tilde{x}$ is used, its value being $k'\tilde{x} = 5w_1 + w_3$. For example, the $k' = [1 \quad 2 \quad 1 \quad 3]$ used in (27) is derived by putting $w_1 = 1$ and $w_3 = 1$; and from (28), $k'\tilde{x}$ then equals 6, as obtained previously.

d. An alternative procedure for rectangular matrices

A procedure alternative to adding null rows or columns to A before deriving G involves a minor modification to the definition of Δ^-, but no redefining of A and G. It is based on the fact that reduction to canonical form under equivalence can be carried out for rectangular matrices as readily as it can for square ones. The matrices P and Q are then of different orders and the process of finding G is as exemplified below.

Example. For $A = \begin{bmatrix} 2 & 4 \\ 2 & 7 \\ 4 & 2 \end{bmatrix}$,

and for $\quad P = \begin{bmatrix} 1 & 0 & 0 \\ -1 & 1 & 0 \\ -4 & 2 & 1 \end{bmatrix} \quad$ and $\quad Q = \begin{bmatrix} 1 & -2 \\ 0 & 1 \end{bmatrix}$,

$$PAQ = \Delta = \begin{bmatrix} 2 & 0 \\ 0 & 3 \\ 0 & 0 \end{bmatrix}.$$

From this, Δ^- is defined as

$$\Delta^- = \begin{bmatrix} \frac{1}{2} & 0 & 0 \\ 0 & \frac{1}{3} & 0 \end{bmatrix}$$

and the method proceeds as before.

The only difference between this method and the one used for square matrices is that Δ^- is defined as the transpose of Δ with its non-zero elements replaced by their reciprocals; the reciprocals arise from taking the inverse of the diagonal sub-matrix in Δ. If A is $p \times q$ so is Δ, but Δ^- will be $q \times p$.

7. EXERCISES

1. Find a generalized inverse of each of the following matrices.

(a) $\begin{bmatrix} 2 & 1 & 4 \\ 6 & 9 & 3 \\ 4 & 4 & 5 \end{bmatrix}$

(b) $\begin{bmatrix} 1 & 0 & -1 & 2 \\ 3 & 1 & 2 & 1 \\ 4 & 3 & -2 & 1 \\ 13 & 11 & 3 & -5 \end{bmatrix}$

(c) $\begin{bmatrix} 1 & 2 & 1 & 2 \\ 1 & 3 & 2 & 1 \\ 0 & 1 & 1 & 1 \\ -1 & 2 & 3 & 1 \end{bmatrix}$ (d) $\begin{bmatrix} 4 & 3 \\ 1 & 2 \end{bmatrix}$

2. Find a generalized inverse of each of the matrices given in Exercise 1 of Chapter 6.

3. Find a set of LIN solutions for each of the following sets of equations and show that any other solution is a linear combination of them.

(a) $\begin{bmatrix} 6 & 2 & 0 \\ -1 & 0 & 3 \\ 3 & 2 & 9 \end{bmatrix} x = \begin{bmatrix} 8 \\ 2 \\ 14 \end{bmatrix}$ (b) $\begin{bmatrix} 4 & -9 & -1 & 2 \\ 3 & 1 & 0 & 1 \\ 10 & -7 & -1 & 4 \\ 25 & -2 & -1 & 9 \end{bmatrix} x = \begin{bmatrix} 7 \\ 5 \\ 17 \\ 42 \end{bmatrix}$

(c) $\begin{bmatrix} 1 & 1 & 0 & 1 \\ -1 & -1 & 1 & 1 \\ 1 & 0 & 0 & 1 \\ -1 & 0 & 1 & 1 \end{bmatrix} x = \begin{bmatrix} 8 \\ -1 \\ 6 \\ 1 \end{bmatrix}$

(d) $\begin{bmatrix} 1 & 2 & 3 & 4 \\ 5 & 6 & 7 & 8 \end{bmatrix} x = \begin{bmatrix} 10 \\ 26 \end{bmatrix}$ (e) $\begin{bmatrix} 1 & 0 & 2 \\ 4 & 1 & 7 \\ 3 & 2 & 3 \end{bmatrix} x = \begin{bmatrix} 12 \\ 46 \\ 27 \end{bmatrix}$

(f) $\begin{bmatrix} 6 & 1 & 4 & 2 & 1 \\ 3 & 0 & 1 & 4 & 2 \\ -3 & -2 & -5 & 8 & 4 \end{bmatrix} x = \begin{bmatrix} 11 \\ 6 \\ -4 \end{bmatrix}$

4. Using both methods given in Section 2, show that the equations of Exercise 3 are consistent.

5. An industrialist has three different machines that he wants to keep fully utilized, 8 hours a day, in the production of four different products. The numbers of hours that each machine must be used in the production of one unit of each product are shown in the following matrix, in which rows represent machines and columns represent products:

$$\begin{bmatrix} 1 & 2 & 1 & 2 \\ 7 & 0 & 2 & 0 \\ 1 & 0 & 0 & 4 \end{bmatrix}.$$

Let n_1, n_2, n_3 and n_4 be the number of units of each product produced.

(a) In terms of the n's, write down equations that represent the full utilization of the machines.

(b) Use the generalized inverse to find an algebraic solution to the equations in (a).

(c) The solution in (b) will include the possibility of negative values which, the industrialist says, are nonsense. What condition applied to the algebraic solution ensures that its elements are non-negative (fractional solutions are permissible)?

(d) What will the solution be if the industrialist finds that product 1 is the best seller and he wants to maximize its production while still fully utilizing his three machines?

(e) What will be the solution if product 4 is all-important but full utilization is again required?

6. (Exercise 5 continued). Repeat Exercise 5 for keeping the machines fully utilized for a 40-hour week instead of an 8-hour day, with solutions required to be in integers for this exercise.

7. (Exercise 5 continued). Repeat Exercise 5 for keeping the machines fully utilized for a 40-hour week, except that machine number 3 is unavailable 8 hours a week, when it is being adjusted and repaired.

8. Suppose the total revenue function of a firm can be adequately represented over the relevant range by the quadratic equation

$$y_i = a + bq_i + cq_i^2$$

where y_i is the total revenue and q_i is the quantity of output in period i. Suppose also that there are only two observations on quantities sold and their corresponding revenues generated:

Period i	q_i	y_i
1	10	100
2	20	150

(a) Express the set of simultaneous equations implied by the two observations and the governing total revenue equation in matrix form as $A \begin{bmatrix} a \\ b \\ c \end{bmatrix} = y$.

(b) Find a general set of revenue functions that satisfies the system of equations in (a).

(c) How many LIN solutions are there? Relate your answer to (b).

9. Consider Exercise 10 of Chapter 4, in which goods are shipped from production plants to warehouses under the constraints that available plant capacity is not exceeded and that all warehouse requirements are met. Use the generalized

REFERENCES   191

inverse method to solve the equations $Ax = y$ in that exercise. Derive a set of LIN solutions. How many LIN solutions are there?

10. If G is a generalized inverse of A show that $\tilde{x} = (G + GA - I)y$ is a solution of $Ax = y$, and that $\tilde{x} = [G(I + \lambda A) - \lambda I]y$ is also a solution, where λ is any scalar.

11. For the generalized inverse based on P, Q and Δ as used in this chapter, show that (a) $GAQP = PQAG$; and (b) $H = Q \Delta\Delta^- Q^{-1}$.

12. If $A = A'$, $U'U = I$ and $U'AU = \Delta$, the diagonal form used in this chapter, show that for $R = U \Delta^- U'$;

(a) $RAR = R$ (b) $ARA = A$

(c) $(RA)' = RA$ (d) $(AR)' = AR$

(e) $(RA)^2 = RA$ (f) $RA = AR$

(g) $A^2 = U\Delta^2 U'$ (h) $I - RA$ is idempotent.

13. If G is any generalized inverse of $X'X$, prove that (a) GX' is a generalized inverse of X, and (b) XGX' is unique. *Hint*: If $(A - B)'(A - B) = 0$, then $A - B = 0$.

REFERENCES

Bjehammer, A. (1951). Rectangular reciprocal matrices with special reference to Geodetic calculations. *Bull. Géodésique*, 188–220.

Goldman, A. J., and M. Zelen (1964). Weak generalized inverses and minimum variance linear unbiased estimators. *J. Res. Nat. Bur. of Standards*, **68B,** 151–172.

Greville, T. N. E. (1957). The pseudo inverse of a rectangular or singular matrix and its application to the solution of systems of linear equations. *Soc. Industrial and Applied Math. Newsletter*, **5,** 3–6.

Malik, H. J. (1968). A note on generalized inverses. *Naval Research Logistics Quarterly*, **15,** 605–612.

Moore, E. H. (1935). General analysis. *Memoirs of the American Philosophical Society*, **1.**

Penrose, R. A. (1955) A generalized inverse for matrices. *Proceedings of the Cambridge Philosophical Society*, **51,** 406–413.

Rao, C. R. (1955). Analysis of dispersion for multiple classified data with unequal numbers of cells. *Sankhya*, **15,** 253–280.

Rao, C. R. (1962). A note on a generalized inverse of a matrix with applications to problems in mathematical statistics. *Journal of the Royal Statistical Society (B)*, **24,** 152–158.

Wilkinson, G. N. (1958). Estimation of missing values for the analysis of incomplete data. *Biometrics*, **14,** 257–286.

CHAPTER 8

MARKOV CHAINS

---

1. INTRODUCTION

Markov chains are a particular class of probability models that are often applicable to decision-making problems in business and economics. In a Markov chain the system of variables being considered is divided into several clearly defined, mutually exclusive states, and at any point in time the system is in one, and only one, of those states. Whatever state the system is in currently, the probability of being in any one of the states at the next point in time depends only on the state the system is in now; i.e., this probability is independent of how the system reached its present state.

Illustration. Consider a situation in which a machine may either be "operating properly" or "in need of adjustment."[1] If the machine is operating properly today, the probability that it will be operating properly tomorrow is .7 and the probability that it will need adjustment tomorrow is .3. Assume the machine has a self-adjusting mechanism which functions imperfectly, so that if the machine needs adjustment today the probability that it will be operating properly tomorrow is .6 and the probability that it will still need adjustment tomorrow is .4. If State 1 represents the situation in which the machine is "operating properly" and State 2 represents "in need of adjustment," then the probabilities of change are given in Table 1 (overleaf). Note that the sum of the probabilities in any row is equal to 1.

The transition probability matrix P derived from Table 1 is:

$$P = \begin{bmatrix} .7 & .3 \\ .6 & .4 \end{bmatrix}.$$

---

[1] The major illustration of this chapter is adapted from Bierman, Bonini and Hausman (1969), but the basic source for much of the material in the chapter is Howard (1960).

TABLE 1. PROBABILITIES OF CHANGE

To:	Operating Properly (State 1)	Needing Adjustment (State 2)
From:		
Operating Properly (State 1)	.7	.3
Needing Adjustment (State 2)	.6	.4

Suppose the machine starts out in State 1 (operating properly) initially, which we shall refer to as time period 0. If the elements of the *state probability vector* x_0' represent the probabilities of the machine being in the various states in period 0, then

$$x_0' = [1 \quad 0].$$

Here the subscript "0" on x' indicates the time period referred to, and is not a subscript referring to an element of any matrix.

With this notation the state vector[1] for the machine in period 1, x_1', is calculated by post-multiplying the state vector of period 0 by the transition probability matrix P:

$$x_1' = x_0'P = [1 \quad 0]\begin{bmatrix} .7 & .3 \\ .6 & .4 \end{bmatrix} = [.7 \quad .3].$$

The state vector of the machine in period 2 may be calculated as follows:

$$x_2' = x_1'P = x_0'P^2 = [.67 \quad .33].$$

Similarly,

$$x_3' = x_0'P^3 = [.667 \quad .333]$$

and in general for period n,

$$x_n' = x_0'P^n.$$

That is, the state vector for the machine in period n is equal to the state vector in period 0 multiplied by the nth power of the transition probability matrix, P^n (see also Section 2.5d). Thus if the machine started out in State 1 in period 0, the probability that it would be in State 1 in period 2 is given by the first element of x_2', namely, .67.

The behavior of the machine over time as described by the transition probability matrix P and the unending sequence of state vectors $x_0', x_1', \ldots,$ $x_n', \ldots$ is called a *Markov chain*. The unique property of a Markov chain is that the probability of going from any state to any other state is independent

---

[1] Hereafter the term "state vector" will be used for "state probability vector."

of how the chain reached its current state. For example, if the machine is in State 2 in period 2, the probability of going to State 1 in the next time period is $p_{21} = .6$, regardless of how the machine reached State 2. This property is often called the Markovian property or the property of "no memory." There is no need to remember how the system reached a particular state at a particular period; the state of the system at a particular period of time and the transition probability matrix contain all the necessary information about the system and the probabilities of future changes.[1]

The Markovian property can exist for continuous state descriptions as well as for discrete state descriptions (our example has two discrete states). If the property is present in a continuous-state system the system is called a *Markov process;* if the states are discrete it is called a *Markov chain.* This chapter deals only with Markov chains, although most of the results can, with some modification, be carried over to Markov processes.

2.   STEADY-STATE PROBABILITIES

In applications of Markov chains to business and economic problems an important aspect is the long-run behavior of the system, after the effects of the initial conditions have worn off. Suppose after a very large number of periods, the state vector x'_n at period n is the same as the state vector x'_{n+1} at period $n + 1$ and is independent of the initial state vector x'_0. Dropping the subscript, we call such a vector x' the *steady-state vector* for the Markov chain described by P; and the elements of x' are called the *steady-state probabilities.* Thus the steady-state vector x' (if it exists) is the state vector which satisfies the equations[2]

$$x' = x'P. \tag{1}$$

Illustration. Consider the machine illustration above. If the system had started in State 2 in period 0 the state vectors would be

$$x'_0 = [0 \quad 1],$$

$$x'_1 = [0 \quad 1]\begin{bmatrix} .7 & .3 \\ .6 & .4 \end{bmatrix} = [.6 \quad .4],$$

and

$$x'_2 = [.6 \quad .4]\begin{bmatrix} .7 & .3 \\ .6 & .4 \end{bmatrix} = [.66 \quad .34].$$

---

[1] An example of a process which violates the Markovian property is Exercise 16.

[2] The conditions under which steady-state probabilities exist for a Markov chain are discussed in Section 3d below.

Comparing these results with those given above for x_1' and x_2' when the machine started in State 1, it appears that the state vector will eventually converge to $[\tfrac{2}{3} \quad \tfrac{1}{3}]$, regardless of the initial state. Equations (1) will be solved for the steady-state vector x' to confirm this result.

Given the matrix P, equations (1) are

$$x' = x' \begin{bmatrix} .7 & .3 \\ .6 & .4 \end{bmatrix}. \tag{2}$$

Writing $x' = [\pi_1 \quad \pi_2]$ and expanding (2) gives

$$\pi_1 = .7\pi_1 + .6\pi_2$$

and
$$\pi_2 = .3\pi_1 + .4\pi_2. \tag{3}$$

One of the two equations in (3) is redundant[1] because they both simplify to

$$\pi_1 = (.6/.3)\pi_2 = 2\pi_2. \tag{4}$$

However, another relationship which holds among the steady-state probabilities is that they must sum to 1, since the system has to be in one of the allowable states; hence

$$\pi_1 + \pi_2 = 1. \tag{5}$$

Equations (4) and (5) can now be written as

$$\pi_1 - 2\pi_2 = 0 \qquad \text{and} \qquad \pi_1 + \pi_2 = 1,$$

equivalent to
$$\begin{bmatrix} 1 & -2 \\ 1 & 1 \end{bmatrix}\begin{bmatrix} \pi_1 \\ \pi_2 \end{bmatrix} = \begin{bmatrix} 0 \\ 1 \end{bmatrix} \tag{6}$$

which has the solution

$$\begin{bmatrix} \pi_1 \\ \pi_2 \end{bmatrix} = \begin{bmatrix} 1 & -2 \\ 1 & 1 \end{bmatrix}^{-1}\begin{bmatrix} 0 \\ 1 \end{bmatrix} = \begin{bmatrix} \tfrac{1}{3} & \tfrac{2}{3} \\ -\tfrac{1}{3} & \tfrac{1}{3} \end{bmatrix}\begin{bmatrix} 0 \\ 1 \end{bmatrix} = \begin{bmatrix} \tfrac{2}{3} \\ \tfrac{1}{3} \end{bmatrix} \tag{7}$$

Hence the long-term probability of its being in State 1 (operating properly) is $\tfrac{2}{3}$ and the long-term probability of its being in State 2 (needing adjustment) is $\tfrac{1}{3}$. These results are independent of where the system starts, i.e. of whether the machine starts out working properly or in need of adjustment.

3.   TRANSIENT, PERIODIC AND ERGODIC BEHAVIOR

In the machine example above, the steady-state probabilities existed, each of the states had a non-zero steady-state probability and these probabilities

---

[1] The redundancy of one equation in (3) is true in general.

fully described all that was known about the long-run behavior of the system. Some situations which differ from the machine example are now considered.

a. Transient states and trapping states

$Lim \; P \to 0$
$n \to \infty$

A *transient state* in a Markov chain is a state with a steady-state probability of zero. An example will demonstrate this situation.

Example. Consider a Markov chain that has the transition probability matrix

$$P = \begin{bmatrix} .2 & .8 \\ 0 & 1 \end{bmatrix}.$$

Once the system reaches State 2 it remains there forever, since $p_{22} = 1$ (and $p_{21} = 0$).

In this system State 1 is a transient state, and the steady-state probability vector is $x' = [0 \quad 1]$. State 2 is often called a *trapping state* or an *absorbing state*.

Illustration. Suppose a machine can be in one of three states: (1) broken beyond repair, (2) in need of adjustment or (3) working properly, with the following transition probability matrix:

$$P = \begin{bmatrix} 1 & 0 & 0 \\ \frac{1}{4} & \frac{1}{2} & \frac{1}{4} \\ \frac{1}{18} & \frac{8}{18} & \frac{9}{18} \end{bmatrix}.$$

Under these conditions it might be asked: "What is the probability that the machine will be in State i after n time periods?" We may also wish to know the probability of being in State i when n becomes infinitely large (i.e. the steady-state probability). If x_0' is the state vector at period 0, then the state vector at period n is $x_n' = x_0'P^n$.

If the machine is initially in need of adjustment, then

$$x_0' = [0 \quad 1 \quad 0];$$

the probabilities two time periods later are

$$x_2' = x_0'P^2 \doteq [0 \quad 1 \quad 0] \begin{bmatrix} 1 & 0 & 0 \\ .39 & .36 & .25 \\ .20 & .44 & .36 \end{bmatrix} = [.39 \quad .36 \quad .25].$$

Suppose a used machine is to be purchased at auction and it cannot be inspected beforehand. The auction consists of hundreds of these machines, to be auctioned individually, where $\frac{1}{2}$ of them are broken beyond repair, $\frac{1}{3}$ of them are in need of adjustment and $\frac{1}{6}$ of them are in working order. Then, so far as the potential purchase is concerned,

$$x_0' = [\tfrac{1}{2} \quad \tfrac{1}{3} \quad \tfrac{1}{6}],$$

and if the used machine's behavior is characterized by the P matrix above,

$$x_2' = x_0' P^2 = [\tfrac{1}{2} \quad \tfrac{1}{3} \quad \tfrac{1}{6}] \begin{bmatrix} 1 & 0 & 0 \\ .39 & .36 & .25 \\ .20 & .44 & .36 \end{bmatrix} = [.66 \quad .19 \quad .15].$$

Hence if a machine is purchased at the auction the probability is .66 that it will be broken beyond repair within two periods of time. Note that State 1 is a trapping state; the steady-state vector is [1 0 0].

b. Periodic (cycling) behavior

Example. The transition probability matrix

$$P = \begin{bmatrix} 0 & 1 & 0 \\ 0 & 0 & 1 \\ 1 & 0 & 0 \end{bmatrix}$$

generates a Markov chain exhibiting periodic or cyclic behavior. Steady-state probabilities for such a chain do not exist because the state probabilities for large n do not tend to specific values independent of the starting state. To see this, suppose the starting state is State 1 at period 0. Then

$$\begin{aligned}
x_0' &\quad\quad\quad = [1 \quad 0 \quad 0], \\
x_1' &= x_0' P = [0 \quad 1 \quad 0], \\
x_2' &= x_1' P = [0 \quad 0 \quad 1], \\
x_3' &= x_2' P = [1 \quad 0 \quad 0] = x_0', \\
x_4' &= x_3' P = [0 \quad 1 \quad 0] = x_1',
\end{aligned}$$

and it will be found that $x_5' = x_2'$, $x_6' = x_3'$, $x_7' = x_4'$ and so on. Thus even though the probability vector $x' = [\tfrac{1}{3} \quad \tfrac{1}{3} \quad \tfrac{1}{3}]$ satisfies the equation $x' = x'P$, it cannot be said that the steady-state probabilities are each $\tfrac{1}{3}$, since the state of the system in any future period is perfectly predictable once the starting state is known. Starting at State 1 in period 0, the system is then in State 1 in periods 3, 6, 9, ... , in State 2 in periods 1, 4, 7, ... and in State 3 in periods 2, 5, 8, ... Such a system, exhibiting cyclic or periodic behavior, is said to be *periodic*. The only relevance the probability vector x' satisfying (1) may have in a periodic system is when the starting state is unknown.

In that case, if the probabilities of the system starting in each state are equal, then the likelihood of the system being in any particular state in any future period is given by the elements of x'.[1]

c. Recurrent sets and ergodic systems

Even though a Markov chain has one or more transient states it need not have a trapping state (or states). But suppose it has a *set* of states such that once the system enters this set it always makes transitions within the set and never leaves it. Such a set is called a *recurrent set*. For example, if a system has the matrix

$$P = \begin{bmatrix} .3 & .7 & 0 & 0 \\ .6 & .4 & 0 & 0 \\ 0 & .2 & .3 & .5 \\ .1 & .2 & .4 & .3 \end{bmatrix}$$

then once the system reaches either of States 1 or 2 it remains in one or the other of those states. The probabilities of going from either of these states to any of the others are all zero. Thus States 1 and 2 form a recurrent set.

A recurrent set that has just one state is a trapping state. This is the simplest form of recurrent set. Thus in the system having

$$P = \begin{bmatrix} .2 & .8 \\ 0 & 1 \end{bmatrix}$$

considered earlier, State 2 is a recurrent set (comprised of a single state). A recurrent set containing more than one state is sometimes called a *generalized trapping state*.

A whole system can be a recurrent set. Any system in which it is possible to go from any state to any other state in a finite number of transitions is said to be *ergodic*. Thus if a Markov chain is ergodic the entire set of states is a recurrent set, while if it is not ergodic then there may be more than one recurrent set in the system. It is important to recognize systems containing more than one recurrent set because in such systems the steady-state probabilities do not exist, since the long-run behavior of the system is dependent on the starting state.

---

[1] The probability vector x' in the example has an alternative interpretation: it is the long-run expected frequency of being in each state. The system will, in the long run, be in each state one-third of the time. Hence if the initial state is known but the number of periods is unknown, the elements of x' are the appropriate probabilities to assign to the various states.

Example. Consider a system whose matrix is

$$P = \begin{bmatrix} 0 & 1 & 0 & 0 \\ 1 & 0 & 0 & 0 \\ 0 & .3 & 0 & .7 \\ 0 & 0 & 0 & 1 \end{bmatrix}.$$

States 1 and 2 form one recurrent set, State 3 is a transient state and State 4 is a second recurrent set (or alternatively, a trapping state). When n is large the state probabilities of this system for period n are not independent of the starting state. For example, if this system started in State 1 or State 2 it would forever remain cycling between those states; if it started in State 4 it would forever remain in that state; and if it started in State 3 it would have a 30% chance of going to the first recurrent set and a 70% chance of going to the second recurrent set. The elements of the probability vector x' satisfying $x' = x'P$ are

$$x' = [\tfrac{1}{3} \quad \tfrac{1}{3} \quad 0 \quad \tfrac{1}{3}],$$

but here they do not have any meaningful interpretation.

d. Existence of steady-state probabilities

As stated above, if a system contains more than one recurrent set then the steady-state probabilities do not exist. However, even if a system is ergodic the steady-state probabilities will not always exist, as demonstrated by the periodic system in Section 3b above. That system was clearly ergodic, yet the long-run state probabilities were not independent of the initial state.

A Markov chain is *regular* if there exists some positive integer n such that every element of the matrix P^n except those relating to transient states is greater than zero. If a chain is regular then it can be shown[1] that as n grows large the matrix P^n approaches a limiting form (call it P^*) in which all rows are identical and equal to the steady-state probability vector x'. The steady-state probabilities for a system exist if, and only if, the system is regular.

In order to analyze whether a system is regular or not, the transient states of the system are not considered. For example, for the system with

$$P = \begin{bmatrix} .6 & 0 & .4 \\ .3 & .5 & .2 \\ .7 & 0 & .3 \end{bmatrix},$$

State 2 is a transient state whose limiting probability is zero. Then P (with the probabilities relating to State 2 eliminated from consideration) satisfies

---

[1] See Exercise 12.15.

the definition of a regular system; and the steady-state probabilities for States 1 and 3 do indeed exist. The reader may verify that as n grows large the matrix P^n approaches

$$P^* = \begin{bmatrix} \frac{7}{11} & 0 & \frac{4}{11} \\ \frac{7}{11} & 0 & \frac{4}{11} \\ \frac{7}{11} & 0 & \frac{4}{11} \end{bmatrix},$$

a matrix with all rows identical and equal to the steady-state probability vector x' satisfying $x' = x'P$.

Example. Consider the system with matrix

$$P = \begin{bmatrix} .7 & .3 & 0 \\ 0 & .8 & .2 \\ .4 & .3 & .3 \end{bmatrix}.$$

There are no transient states. To check whether the system is regular we compute P^n for increasing values of n. For $n = 2$,

$$P^2 = \begin{bmatrix} .49 & .45 & .06 \\ .08 & .70 & .22 \\ .40 & .45 & .15 \end{bmatrix}$$

with every element greater than zero. Thus the steady-state probabilities for this system will exist.

4.   MARKOV CHAINS WITH REWARDS

Up to this point we have considered only the probability of being in various states in different time periods. Now information will be added concerning the reward or payoff that may be obtained from the transitions. Let r_{ij} be the reward associated with a transition from State i to State j; it may be interpreted as the reward from the transition itself or the reward for being in State i (or State j) during one time period. The first interpretation would be appropriate if, for example, the states were locations in a city and the transitions were taxi rides; r_{ij} would be the profit from the fare collected in going from location i to location j. The second interpretation would be appropriate if, for example, the states were the alternative conditions of a machine; r_{ij} might be the profit earned from being in State i for the period of time before the transition.

Now consider the total reward that might be expected after n transitions, from period 0 to period n. To do this suppose there are N states in the system and let R be the matrix of rewards

$$
R = \begin{bmatrix}
r_{11} & r_{12} & \cdots & r_{1N} \\
r_{21} & \cdots & & r_{2N} \\
\cdot & & & \cdot \\
\cdot & & & \cdot \\
\cdot & & & \cdot \\
r_{N1} & \cdots & & r_{NN}
\end{bmatrix}.
$$

Suppose the system starts in State i; let $v_i(n)$ be the total expected reward after n transitions, starting from State i. Then $[v(n)]' = [v_1(n) \quad v_2(n) \quad \cdots \quad v_N(n)]$ is the vector of total expected rewards over n transitions for each of the N possible starting states of the system. Now suppose that at the first transition the system moves to State j. The reward for this transition is r_{ij}. Moreover, when the system has reached State j the expected reward over all n transitions can be expressed as $r_{ij} + v_j(n-1)$, with the term $v_j(n-1)$ representing the expected reward over the remaining $n-1$ transitions when the system starts in State j. But the probability of going to State j from State i is p_{ij}. Therefore the total expected reward over the next n transitions, starting in State i, may be written as

$$
v_i(n) = \sum_{j=1}^{N} {}_i p_j [r_{ij} + v_j(n-1)]
$$

$$
= \sum_{j=1}^{N} p_{ij} r_{ij} + \sum_{j=1}^{N} p_{ij} v_j(n-1). \tag{8}
$$

Now write

$$
q_i = \sum_{j=1}^{N} p_{ij} r_{ij}, \tag{9}
$$

so that q_i represents the expected reward from the next transition when i is the current state. Then

$$
v_i(n) = q_i + \sum_j p_{ij} v_j(n-1). \tag{10}
$$

Now, since equations (9) and (10) are true for $i = 1, 2, \ldots, N$, (10) can be written in vector form as

$$
v(n) = q + Pv(n-1) \tag{11}
$$

where $q' = [q_1 \quad q_2 \quad \cdots \quad q_N]$. It may be noted from (9) that q_i is the ith diagonal term of PR' and hence q' is a row vector composed of the diagonal elements of PR'.

If the Markov chain is regular, then the steady-state probabilities exist. The steady-state expected reward per period, g, can then be written as the weighted sum of the expected rewards q_i from transitions from State i, weighted by the steady-state probabilities, π_i, of being in State i:

$$g = \sum_i \pi_i q_i$$

or in vector form

$$g = x'q \tag{12}$$

where the vector $x' = [\pi_1 \quad \pi_2 \quad \cdots \quad \pi_N]$ is the steady-state vector.

Illustration. Suppose the reward matrix (in dollars) for the machine example at the beginning of the chapter is

$$R = \begin{bmatrix} 2 & 1 \\ 1 & -1 \end{bmatrix}.$$

This means that when the machine operates properly both before and after a transition the profit is \$2; when it starts off operating properly but then requires adjustment after a transition (or vice versa) the profit is \$1; and when it remains out of adjustment both before and after a transition the loss is \$1. Now, as given earlier,

$$P = \begin{bmatrix} .7 & .3 \\ .6 & .4 \end{bmatrix}$$

and

$$PR' = \begin{bmatrix} .7 & .3 \\ .6 & .4 \end{bmatrix} \begin{bmatrix} 2 & 1 \\ 1 & -1 \end{bmatrix} = \begin{bmatrix} 1.7 & .4 \\ 1.6 & .2 \end{bmatrix}$$

so from (9)

$$q' = [1.7 \quad .2].$$

Therefore in (11)

$$v(n) = \begin{bmatrix} 1.7 \\ .2 \end{bmatrix} + \begin{bmatrix} .7 & .3 \\ .6 & .4 \end{bmatrix} v(n-1),$$

where we define $v(0) = 0$. Then

$$v(1) = \begin{bmatrix} v_1(1) \\ v_2(1) \end{bmatrix} = \begin{bmatrix} 1.7 \\ .2 \end{bmatrix};$$

i.e. if the machine starts out in working condition the expected profit for one time period is \$1.70; if it starts out in need of repair the expected profit is 20 cents.

The expected profit for two periods is likewise obtained from (11) with $n = 2$:

$$v(2) = q + Pv(1)$$

$$= \begin{bmatrix} 1.7 \\ 0.2 \end{bmatrix} + \begin{bmatrix} .7 & .3 \\ .6 & .4 \end{bmatrix} \begin{bmatrix} 1.7 \\ 0.2 \end{bmatrix} = \begin{bmatrix} 2.95 \\ 1.30 \end{bmatrix}.$$

Note that using equation (11) for $n = 2$ requires knowing the value of $v(1)$; and equation (11) for $n = 3$ makes use of $v(2)$. In this manner the vector $v(n)$ may be calculated for any n by starting with $v(1)$ and using equation (11) recursively.

The steady-state expected reward g is calculated from equation (12):

$$g = x'q = \begin{bmatrix} \frac{2}{3} & \frac{1}{3} \end{bmatrix} \begin{bmatrix} 1.7 \\ .2 \end{bmatrix} = 1.20.$$

Thus, if the system has been operating for a large number of periods and the state of the system is unknown, the expected return per period in the next and every subsequent period is $1.20.

5. OPTIMAL POLICIES IN MARKOV CHAINS

In the previous section a reward matrix R has been combined with a transition probability matrix P to create a Markov chain with rewards, and we have shown how to calculate vectors $v(n)$ whose elements describe the expected reward from the system for n periods as a function of the initial state of the system. Up to this point no decisions have been involved; the model has simply described a system which presented no alternatives. Now suppose some alternative decisions relating to the system are introduced. For example, in our machine illustration consider the following decision rule: "Always repair the machine when it needs adjustment." If the same reward matrix presented above holds, and if it costs $.90 to repair the machine, an important question is: "Should this decision rule be followed, as opposed to a decision rule of doing nothing?" (Recall that the machine adjusts itself with .6 probability.)

This problem is approached in a general manner by considering K different decision rules. In general, alternative decisions may affect both the transition probability matrix and the reward matrix. Accordingly, let $p_{ij}^{(k)}$, $P^{(k)}$, $r_{ij}^{(k)}$, $R^{(k)}$, $q_i^{(k)}$ and $q^{(k)}$ represent the corresponding symbols when the kth decision rule is chosen. [Here the superscript (k) indicates that the kth decision is chosen; it does *not* represent raising to a power.] Also, $v_i(n)$ is redefined as the *maximum* expected reward over the next n transitions when the system is currently in State i *and optimal decisions are made in each of the remaining periods.* Now $v_i(n)$ is written as follows:

$$v_i(n) = \underset{\{k\}}{\text{Max}} \sum_{j=1}^{N} p_{ij}^{(k)}[r_{ij}^{(k)} + v_j(n-1)] , \qquad i = 1, 2, \ldots, N.$$

The symbol Max indicates that the maximization is performed over the set
$\{k\}$
of possible decisions, $k = 1, 2, \ldots, K$. In matrix notation,

$$v(n) = \underset{\{k\}}{\text{Max}} \{q^{(k)} + P^{(k)}v(n-1)\} \tag{13}$$

where there are N separate maximizations in equations (13), one for each element of $v(n)$, i.e. one for each possible starting state.

The solution to the decision problem proceeds as follows. First, equations (13) for $n = 1$ are solved by finding, for each initial State i, the optimal decision, i.e., the value of k which maximizes the right-hand side of (13).[1] Let the *vector of optimal decisions* when n periods remain be denoted by

$$d(n) = [d_1(n) \quad d_2(n) \quad \cdots \quad d_N(n)]'.$$

Then $d_i(n)$ is the optimal decision when the system is in State i with n periods remaining. Solving equations (13) for $n = 1$ is straightforward because $v(0) = 0$ and the equations become

$$v(1) = \underset{\{k\}}{\text{Max}} \{q^{(k)}\}. \tag{14}$$

Once equations (13) are solved for each element of $v(1)$, the optimal decisions are recorded in the vector $d(1)$, where the ith element $d_i(1)$ is the integer in the range 1 through K which maximizes the expected reward over one transition when the system starts in State i. Then equations (13) are solved for $n = 2$, recording the optimal decisions in the vector $d(2)$, and proceeding to $n = 3, 4, \ldots$ until the desired number of periods has been reached. Then the *set* of optimal decision vectors $\{d(n), d(n-1), \ldots, d(1)\}$ constitutes an *optimal policy*. The optimal policy is used as follows: with n periods remaining, if the system is in State i, then decision $k = d_i(n)$ is used. Then after a transition, the system will be in some new state; call it j. There are now $n - 1$ periods remaining and the optimal decision is $k = d_j(n-1)$. In this manner the optimal decision vectors $d(n), d(n-1), \ldots, d(1)$ enable us to make optimal decisions no matter what states are reached as the system makes transitions. As stated above, the maximum expected reward over n periods if the system starts in State i is given by $v_i(n)$.

Illustration. In our machine illustration, let decision rule $k = 1$ be the rule "Do nothing," and decision rule $k = 2$ be "Always repair the machine when it needs adjustment," the repair cost being \$.90. Under rule 1 (do

---

[1] The value of k which maximizes may be found by complete enumeration.

nothing) the P and R matrices remain as above:

$$P^{(1)} = \begin{bmatrix} .7 & .3 \\ .6 & .4 \end{bmatrix}, \qquad R^{(1)} = \begin{bmatrix} 2 & 1 \\ 1 & -1 \end{bmatrix}.$$

Under rule 2 (always repair when required) the P and R matrices are

$$P^{(2)} = \begin{bmatrix} .7 & .3 \\ 1 & 0 \end{bmatrix}, \qquad R^{(2)} = \begin{bmatrix} 2 & 1 \\ 1-.9 & -1 \end{bmatrix},$$

reflecting the certain transition from State 2 to State 1 and the repair cost of \$.90 associated with this transition.

From $P^{(k)}$ and $R^{(k)}$ we may obtain $q^{(k)}$ by equations (9):

$$q^{(1)} = \begin{bmatrix} 1.7 \\ .2 \end{bmatrix} \qquad \text{and} \qquad q^{(2)} = \begin{bmatrix} 1.7 \\ .1 \end{bmatrix}.$$

Now we write equations (13) for $n = 1$:

$$v(1) = \underset{\{k\}}{\text{Max}}\{q^{(k)}\} = \text{Max}\{q^{(1)}, q^{(2)}\}$$

$$= \text{Max}\left\{ \begin{bmatrix} 1.7 \\ .2 \end{bmatrix}, \begin{bmatrix} 1.7 \\ .1 \end{bmatrix} \right\} \tag{15}$$

where the maximization is performed separately for the two elements of $v(1)$. For the element $v_1(1)$ a tie exists, and we arbitrarily choose rule 2 as the optimal decision. For the element $v_2(1)$ rule 1 (do nothing) maximizes the expression. The optimal decision vector $d(1)$ is therefore

$$d(1) = [d_1(1) \quad d_2(1)]' = [2 \quad 1]'. \tag{16}$$

If the optimal decision vector is followed, then equation (15) becomes:

$$v(1) = \begin{bmatrix} 1.7 \\ .2 \end{bmatrix}. \tag{17}$$

Now consider equations (13) for $n = 2$:

$$v(2) = \underset{\{k\}}{\text{Max}}\{q^{(k)} + P^{(k)}v(1)\}$$

$$= \text{Max}\{q^{(1)} + P^{(1)}v(1), q^{(2)} + P^{(2)}v(1)\}$$

$$= \text{Max}\left\{ \begin{bmatrix} 1.7 \\ .2 \end{bmatrix} + \begin{bmatrix} .7 & .3 \\ .6 & .4 \end{bmatrix}\begin{bmatrix} 1.7 \\ .2 \end{bmatrix}, \begin{bmatrix} 1.7 \\ .1 \end{bmatrix} + \begin{bmatrix} .7 & .3 \\ 1 & 0 \end{bmatrix}\begin{bmatrix} 1.7 \\ .2 \end{bmatrix} \right\}$$

$$= \text{Max}\left\{ \begin{bmatrix} 2.95 \\ 1.30 \end{bmatrix}, \begin{bmatrix} 2.95 \\ 1.80 \end{bmatrix} \right\}. \tag{18}$$

For element $v_1(2)$ there is another tie, and we arbitrarily choose rule 2 as the optimal decision.[1] For element $v_2(2)$ the optimal decision is rule 2 (always repair when required). The optimal decision vector at period 2 is thus:

$$d(2) = [d_1(2) \quad d_2(2)]' = [2 \quad 2]', \tag{19}$$

and if optimal decisions are followed for two periods equation (18) becomes

$$v(2) = \begin{bmatrix} 2.95 \\ 1.80 \end{bmatrix}. \tag{20}$$

This procedure may be repeated for $n = 3, 4, \ldots$ until the desired number of periods is reached. For $n = 3$,

$$v(3) = \underset{\{k\}}{\text{Max}}\{q^{(k)} + P^{(k)}v(2)\}$$

$$= \text{Max}\left\{ \begin{bmatrix} 1.7 \\ .2 \end{bmatrix} + \begin{bmatrix} .7 & .3 \\ .6 & .4 \end{bmatrix}\begin{bmatrix} 2.95 \\ 1.80 \end{bmatrix}, \begin{bmatrix} 1.7 \\ .1 \end{bmatrix} + \begin{bmatrix} .7 & .3 \\ 1 & 0 \end{bmatrix}\begin{bmatrix} 2.95 \\ 1.80 \end{bmatrix} \right\}$$

$$= \text{Max}\left\{ \begin{bmatrix} 4.305 \\ 2.69 \end{bmatrix}, \begin{bmatrix} 4.305 \\ 3.05 \end{bmatrix} \right\} = \begin{bmatrix} 4.305 \\ 3.05 \end{bmatrix} \tag{21}$$

with
$$d(3) = [2 \quad 2]'.$$

Note that with one period remaining, it is not optimal to repair the machine, whereas with two or three periods remaining, it *is* optimal to repair the machine if it is in State 2 (needing adjustment).[2]

a. Optimal steady-state policies

The solution method just presented allows us to obtain the optimal decision rules when n transitions are remaining. Suppose we are actually interested in the long-run profitability of the system, and desire to maximize the expected reward per period, g. This problem can be solved by repeated use of the methods of Section 4; i.e., we may take each possible decision rule in turn, calculating the relevant P and R matrices, the steady-state probability vector, x', the vector of expected reward over the next transition, q, and the expected steady-state reward $g = x'q$. Then a simple comparison of the values of g for the various decision rules will indicate the *optimal steady-state policy*, i.e., that decision rule which will maximize the steady-state expected reward per period[3] (g). For example, the steady-state policy

---

[1] There will always be a tie for the first element of $v(n)$ since the two decision rules are identical for State 1 (i.e. do nothing).

[2] The general use of the recursive optimization we have been describing is called Dynamic Programming. For an introductory discussion see Bierman, Bonini and Hausman (1969) and for a more comprehensive treatment see Bellman (1957) or Bellman and Dreyfus (1962).

[3] Alternative methods exist for systematically finding the optimal steady-state policy without examining all possibilities; see Howard (1960).

in the example of the last section which maximizes expected reward is "repair the machine when in need of adjustment," and the expected reward is $1.33 per period (see Exercise 13).

b. Optimal policies with discounting

In many business and economic problems it is appropriate to take into account the time value of money when considering cash flows at different points in time. For this reason it is desirable to be able to discount future returns; also, the addition of discounting to our decision problem will cause it to converge more rapidly on a decision policy which is optimal when the number of periods, n, tends toward infinity.

Adding discounting to the framework already established is straightforward; let $\alpha = 1/(1 + r)$ be a discount factor reflecting an interest rate of r per period. Then equations (13) become

$$v(n) = \text{Max}_{\{k\}}\{q^{(k)} + \alpha P^{(k)}v(n - 1)\} \tag{22}$$

and we proceed as before.

6. EXERCISES

1. Two machines, A and B, are candidates for leasing at the same cost. The two machines have differing transition probability matrices for changing from an "operating properly" state (State 1) to a "requires adjustment" state (State 2) as follows:

$$\begin{array}{cc} \text{Machine A} & \text{Machine B} \\ P_A = \begin{bmatrix} .9 & .1 \\ .6 & .4 \end{bmatrix} & P_B = \begin{bmatrix} .8 & .2 \\ .7 & .3 \end{bmatrix}. \end{array}$$

Derive the steady-state probabilities for each machine. Which machine would be most desirable to lease?

2. The ABC Ski Resort has found that after a clear day the probability of stormy weather is .3, while after a stormy day the probability of clear weather is .8. Write down the transition probability matrix. What are the steady-state probabilities for clear days and stormy days?

3. The ACME Car Rental Company rents cars from three airports: A, B and C. Customers return cars to each of the airports according to the probabilities on the next page:

   (a) Calculate the vector x' which satisfies $x' = x'P$ and $\sum_{i=1}^{3} x_i = 1$. Does this vector represent the steady-state probabilities? Justify your answer.

   (b) The ACME Company is planning to build a maintenance facility at one of the three airports. Which airport would you recommend for this purpose? Why?

TABLE OF TRANSITION PROBABILITIES

To:	A	B	C
From:			
A	.8	.2	0
B	.2	0	.8
C	.2	.2	.6

4. Consider the following general two-state transition probability matrix:

$$P = \begin{bmatrix} p_{11} & p_{12} \\ p_{21} & p_{22} \end{bmatrix}.$$

Derive the elements of the steady-state probability vector $x' = [\pi_1 \quad \pi_2]$ in terms of the elements of P. (Recall that the elements in each row of P sum to 1.)

5. The behavior of stock market prices exhibits a systematic tendency for trans-actions in which the price change is in one direction to be followed by trans-actions in which the price change is in the opposite direction. Suppose the conditional probability of a price increase, given that the previous change was a decrease, is .75, as is the conditional probability of a decrease given a previous increase. Define two relevant states and write down a transition probability matrix for this system. What are the steady-state probabilities?

6. Consider a Markov chain with the following transition matrix:

$$P = \begin{bmatrix} .7 & .3 & 0 \\ .8 & .2 & 0 \\ 0 & .2 & .8 \end{bmatrix}$$

(a) If the process starts in State 1 and a very large number of transitions occur, what fraction of these transitions are from State 1 to State 2? *Hint:* First calculate the steady-state probability of being in State 1.

(b) Repeat part (a) assuming the process starts in State 3.

7. A cab driver has found that when he is in Town 1 there is a .8 probability that his next fare will take him to Town 2; otherwise he will stay in Town 1. When he is in Town 2 there is a .4 probability that his next fare will take him to Town 1; otherwise his next fare will keep him in Town 2. The average profit for each type of trip is as follows: Between Towns: $2 (either direction). Within either Town: $1.

(a) Write down the transition probability matrix P for two states and the reward matrix R, and compute the steady-state probabilities of being in the two towns.

(b) Calculate q, the vector of expected rewards from the next transition.

(c) Calculate g, the steady-state expected reward per period.

8. Cyert, Davidson and Thompson (1962) have used a Markov chain to describe the behavior of credit customers moving from a paid-up state to states such as 1 month late, 2 months late, 3 months late or more than 3 months late. In a simplified version of their model, consider the following three states and the associated transition probabilities and rewards:

State 1: Paid up

State 2: 1 month late

State 3: 2 or more months late

$$P = \begin{bmatrix} .8 & .2 & 0 \\ .5 & 0 & .5 \\ .3 & 0 & .7 \end{bmatrix}, \quad R = \begin{bmatrix} 10 & 5 & 0 \\ 10 & 5 & 0 \\ 10 & 5 & 0 \end{bmatrix}.$$

Suppose there are 1,000 credit customers for a certain store. Calculate the steady-state expected reward per period.

9. A coffee manufacturer is considering an advertising campaign designed to attract consumers to his brand of coffee. From panel data obtained through market research he has been able to estimate the current probabilities of consumers changing from "our brand" to "any other" and vice-versa, with both probabilities being .2. Market researchers have also estimated that after the advertising campaign has taken place, assuming no change in the advertising of competitors, the only change in the probabilities will be that the probability of changing from "any other" to "our brand" will increase from .2 to .3.

(a) Suppose the advertising campaign will cost $70 million (the present value of all future expenditures) and that there are 50 million coffee purchasers in the market. For each customer the average annual profit before taxes is $2.00. If the appropriate annual discount rate is 20%, should the manufacturer undertake the advertising campaign? (Assume the steady-state is reached very quickly.) *Hint:* A reward matrix is not necessary. Work with the steady-state probabilities directly.

(b) Now suppose the manufacturer is faced with an alternative advertising campaign for the same cost but one which alters the initial transition probabilities only by increasing the probability of retaining his own customers from .8 to .9. Should this campaign be selected?

10. Consider the machine illustration in Section 5 of this chapter. Suppose it costs $1.40 rather than $.90 to repair the machine when it is in need of adjustment. If three periods are remaining before the machine is to be disassembled and rebuilt, what is the optimal repair strategy to follow for each of the three periods?

11. Assume a machine moves from State 1 (operating properly) to State 2 (requiring

adjustment) and vice-versa with the following probabilities of change:

$$P = \begin{bmatrix} .8 & .2 \\ .4 & .6 \end{bmatrix}.$$

Assume that when the machine is in State 1 for a day a profit of $100 is gained and when the machine is in State 2 for a day a loss of $50 is incurred.

(a) Compute the expected daily profit rate.

(b) Now suppose the machine may be repaired at a cost of $120 and, when needed, the repair can be performed quickly, so a full day's operation in State 1 occurs after repairing. Assuming the state transitions occur at the end of the working day, calculate the steady-state expected reward per period under the following decision rule: always repair when required. Which decision rule (no repair vs. repair) offers the highest steady-state reward?

(c) If there were only a few days remaining before the machine was to be replaced by a new model, should the optimal decision rule obtained in (b) be followed? Justify your answer.

12. A manager wishes to establish a stocking policy for an infrequently used part. Each week there may be demands for zero, one or two units, with the following probabilities:

No. of units demanded:	0	1	2
Probability:	.8	.1	.1

The manager is considering the following decision alternatives: (1) Never order any units; (2) order one unit when inventory on hand reaches 0. Assume that the carrying cost per unit per week is $1 and that the profit on a single sale is $5. Any demands that cannot be met from stock are lost. Which decision should be followed to maximize the steady-state expected reward? Assume units ordered at the end of one week arrive at the beginning of the next week.

13. Verify that the steady-state expected reward of the machine illustration in Section 5 is $1.33 per period if the machine is repaired when needed.

14. For any transition probability matrix P, show for any positive integer n
(a) that the row sums of P^n are 1.0;
(b) that if the elements of x sum to unity then so do the elements of $x'P^n$.
    Hint: Utilize a vector **1** whose every element is unity.

15. Verify that the matrix $(I - P)$, for P being any transition probability matrix, is singular.

16. Consider the situation of Exercise 5 but suppose the state description was defined as the current price of the stock. Would this state description violate the Markovian property of "no memory"? Give your answer in terms of the formulation described in Exercise 5.

REFERENCES

Bellman, R. E. (1957). *Dynamic Programming*. Princeton University Press.

Bellman, R. E., and S. E. Dreyfus (1962). *Applied Dynamic Programming*. Princeton University Press.

Bierman, H., Jr., C. P. Bonini, and W. H. Hausman (1969). *Quantitative Analysis for Business Decisions*. Third Edition; Irwin, Homewood, Ill.

Cyert, R. M., H. J. Davidson, and G. L. Thompson (1962). Estimation of the allowance for doubtful accounts by Markov chains. *Management Science*, **8,** 287–303.

Howard, R. A. (1960). *Dynamic Programming and Markov Processes*. M.I.T. Press. Cambridge, Mass.

CHAPTER 9

LINEAR PROGRAMMING

Linear programming is an analytical technique [originated by Dantzig (1951)] that has received much attention in business and economics in the past twenty years. It is a method for maximizing (or minimizing) a linear function subject to a set of linear restrictions. The most common application is that of allocating scarce resources among various activities to maximize or minimize some objective such as profit or cost. The technique of linear programming provides solutions for a large class of problems of this nature in business and economics.

1. THE MAXIMIZATION PROBLEM

a. A graphical solution

Illustration 1. A small machine shop makes two products, each requiring time on two different machines. In particular, each unit of product one requires 3 hours on machine I and 2 hours on machine II; each unit of product two requires 2 hours on machine I and 3 hours on machine II. Machine I is available only 8 hours a day and machine II only 7 hours a day. Suppose that the profit obtained from selling one unit of each product is 20 and there is no limit to the amount of each product that can be sold. The problem then is to find how much of each product should be produced each day in order to maximize total profit, subject to the restrictions imposed by available machine hours.

The problem is first stated in algebraic form: letting x_1 denote the number of units of product one produced, x_2 the number of units of product two

[*213*]

produced and z the total profit, the problem is to

$$\begin{aligned}
\text{maximize} &\quad z = 20x_1 + 20x_2 \\
\text{subject to the restrictions} &\quad 3x_1 + 2x_2 \leq 8, \\
&\quad 2x_1 + 3x_2 \leq 7, \qquad\qquad (1)\\
&\quad x_1 \qquad\; \geq 0, \\
\text{and} &\quad x_2 \geq 0.
\end{aligned}$$

The inequality $3x_1 + 2x_2 \leq 8$, called a *constraint*, summarizes the capacity restriction associated with machine I; and the constraint $2x_1 + 3x_2 \leq 7$ does the same for machine II. The constraints x_1, $x_2 \geq 0$ specify that negative production is not feasible. The function $z = 20x_1 + 20x_2$ to be maximized is called the *objective function*.

The problem described in (1) is a *linear programming problem* or *L.P. problem*. There are three essential properties to any such problem:

(i) all relationships are linear, including the objective function;
(ii) constraints may be equalities or inequalities in either direction;
(iii) solutions must be non-negative.

Properties (ii) and (iii) usually imply that there are a large number of solutions which will satisfy the constraints. For example, if problem (1) is illustrated graphically, as in Figure 1, the shaded area satisfies all four constraints;

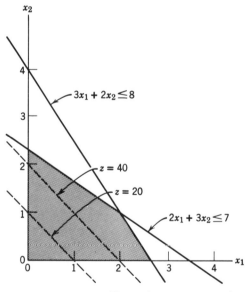

Figure 1

i.e., any (x_1, x_2) in that area satisfies $3x_1 + 2x_2 \leq 8$, $2x_1 + 3x_2 \leq 7$ and $x_1, x_2 \geq 0$. This area is called the *feasible region*, and the problem is to find the point (x_1, x_2) in the feasible region which maximizes the value of the objective function.

Having graphed the feasible region, obtaining the optimal solution for this problem is a relatively simple task. Consider the parallel lines $z = 20x_1 + 20x_2 = 20$ and $z = 20x_1 + 20x_2 = 40$ shown as dashed lines in Figure 1. Now place a pencil on either one and proceed to push it away from the origin, keeping it parallel to the two dashed lines. The pencil represents the line $20x_1 + 20x_2 = z$, where z is increasing as the pencil proceeds away from the origin. Since the goal is to obtain the largest possible value of z, the pencil will continue to be pushed upward until it touches the feasible region only at one point; that point is the optimal solution to the linear programming problem. The optimal point is $(2, 1)$ and at that point $z = 20x_1 + 20x_2 = 60$. No other point in the feasible region yields a value of the objective function equal to 60, since each of these points lies on a line $20x_1 + 20x_2 = z$ with $z < 60$, as we have seen. The dotted lines are sometimes called *iso-profit lines* (from the Greek word *isos* meaning equal). Thus any point on the line $20x_1 + 20x_2 = 60$ will produce a profit of 60; however, only one point on that line is feasible, given the constraints of the problem.

b.   The corner-point property

The graphical method of solving linear programming problems works well when the feasible region is two-dimensional, but cannot be readily used when the feasible region is described by three or more dimensions. In this case a different solution technique must be used to obtain optimal solutions. The particular method of solution presented here is the "revised simplex" method, and the crucial property it employs is that an optimal solution to any linear programming problem is a corner-point of the feasible region.[1]

Definition. A *corner-point* of any n-dimensional region in a linear programming problem is the n-element vector $x' = [x_1 \quad x_2 \quad \cdots \quad x_n]$ that is the unique solution to the set of consistent equations formed by selecting n of the constraints defining the feasible region and requiring them to hold as equalities. If a corner-point also satisfies the remaining constraints it is called a *corner-point of the feasible region* or a *feasible corner-point*.

For example, in the illustration above there are four feasible corner-points: $(0, 0)$, $(\frac{8}{3}, 0)$, $(0, \frac{7}{3})$ and $(2, 1)$. They are solutions to the following sets of

---

[1] The only exceptions occur as follows: (i) when the feasible region is unbounded, in which case the optimal solution may be undefined; (ii) when the constraints are such that there is no feasible region. These exceptions are avoided in this chapter.

equations (formed from the constraints):

(i) $x_1 = 0,\ x_2 = 0,$

(ii) $x_2 = 0,\ 3x_1 + 2x_2 = 8,$

(iii) $x_1 = 0,\ 2x_1 + 3x_2 = 7,$ and

(iv) $3x_1 + 2x_2 = 8,\ 2x_1 + 3x_2 = 7.$

Two other corner-points can be obtained (from $3x_1 + 2x_2 = 8$, $x_1 = 0$; and from $2x_1 + 3x_2 = 7$, $x_2 = 0$), but they are infeasible, i.e. not in the feasible region.

Given that an optimal solution is always a corner-point of the feasible region, the problem is considerably simplified. If there are n unknowns $x_1, \ldots, x_n$ and $m + n$ constraints, taking n of the constraints at a time to form a corner-point means there are $\dfrac{(m + n)!}{m!\,n!}$ corner-points. As we shall see, the revised simplex method systematically considers only the feasible corner-points, and does so in a fashion such that the values of the objective function at each successive feasible corner-point considered form a non-decreasing monotonic series.[1] Thus in general not all feasible corner-points will be evaluated; this is the major computational advantage of the revised simplex method as compared with, for example, a process of complete enumeration of all feasible corner-points.

c. Matrix formulation of linear programming problems

Linear programming problems take one of the following two forms:

I. Maximize $z = c'x$ II. Minimize $z = c'x$

   subject to $Ax \lesseqgtr b,$ subject to $Ax \lesseqgtr b,$

                $x \geq 0$ $x \geq 0$

where c is an $n \times 1$ vector of objective function coefficients (usually unit profits or costs),

x is an $n \times 1$ vector of decision variables or unknowns,

A is an $m \times n$ matrix of "technological coefficients,"

b is an $m \times 1$ vector of scarce resources with $b_i \geq 0$, $i = 1, \ldots, m$,

z is the value of the objective function, and

$\lesseqgtr$ denotes "greater than," "equal to" or "less than."

---

[1] The series $a_1, a_2, \ldots, a_n, \ldots$ is non-decreasing monotonic if $a_n \leq a_{n+1}$ for all n.

Example. The illustrative linear programming problem in (1) takes form I above with

$$A = \begin{bmatrix} 3 & 2 \\ 2 & 3 \end{bmatrix}, \quad c = \begin{bmatrix} 20 \\ 20 \end{bmatrix}, \quad b = \begin{bmatrix} 8 \\ 7 \end{bmatrix} \quad \text{and} \quad x = \begin{bmatrix} x_1 \\ x_2 \end{bmatrix}.$$

Both technological constraints are of the "less than" variety.

This section considers problems taking form I (maximization) with "less than" constraints. Section 2 considers form II (minimization) and Section 3 treats the problem of mixed constraints (both "less than" and "greater than") and equality constraints for both forms. Thus, the problem to be considered in this section is:

$$\text{Maximize} \quad z = c'x$$

$$\text{subject to} \quad Ax \leq b, \tag{2}$$

$$x \geq 0.$$

d. Slack variables

Let us rewrite (2) in a slightly different form:

$$\text{Maximize} \quad z = c^{*\prime}x^*$$

$$\text{subject to} \quad A^*x^* = b, \tag{3}$$

$$x^* \geq 0,$$

where $x^* = [x' \quad x_S']'$, a column vector of order $n + m$,

     x_S is a column vector of order m of "slack" variables,

     $c^* = [c' \quad 0']'$, a column vector of order $n + m$,

     $A^* = [A \quad I_m]$, a matrix of order $m \times (n + m)$

and where c, x, b and A are as defined earlier. This new form is identical to the original form except that a set of m "slack" variables have been added, one to each inequality of $Ax \leq b$, to take up the slack and thus convert the inequalities to equalities. By requiring these slack variables to be non-negative and to have zero profit associated with them we insure that if x^* is such that $A^*x^* = b$ then $Ax \leq b$, and $z = c^{*\prime}x^* = [c' \quad 0'][x' \quad x_S']' = c'x$. Thus the vector x in x^* satisfying the constraints of (3) also satisfies the constraints of (2) and produces the same value for the objective function. Furthermore, if (3) can be solved for its optimal solution $x_0^* = [x_0', x_{S0}']'$ then the optimal solution to (2) is x_0.

Example. Problem (1) can be written in the form of (3):

$$\text{Maximize} \quad z = \begin{bmatrix} 20 & 20 & 0 & 0 \end{bmatrix} \begin{bmatrix} x_1 \\ x_2 \\ x_{S1} \\ x_{S2} \end{bmatrix}$$

$$\text{subject to} \quad \begin{bmatrix} 3 & 2 & 1 & 0 \\ 2 & 3 & 0 & 1 \end{bmatrix} \begin{bmatrix} x_1 \\ x_2 \\ x_{S1} \\ x_{S2} \end{bmatrix} = \begin{bmatrix} 8 \\ 7 \end{bmatrix}$$

and $\quad x_1 \geq 0, \; x_2 \geq 0, \; x_{S1} \geq 0 \text{ and } x_{S2} \geq 0.$

The revised simplex method is directly applicable to a linear programming (L.P.) problem when it is written in the form of (3). As mentioned above, this method of solving L.P. problems examines feasible corner-points, and while an infinite number of vectors x^* exist[1] such that $A^*x^* = b$ and $x^* \geq 0$, only a finite number of corner-points exist that satisfy $Ax \leq b$, $x \geq 0$, as in (2).

e. A basis

A *basis* is an $m \times m$ matrix with columns consisting of m linearly independent columns of A^*. Since $A^* = \begin{bmatrix} A & I_m \end{bmatrix}$ there is always at least one set of m linearly independent columns in A^*, obtained from the identity matrix. Now since A^* has $m + n$ columns there are at most $\dfrac{(m + n)!}{m! \, n!}$ bases.

Each basis is uniquely associated with a corner-point, although all such corner-points need not be feasible. For example, consider the basis B consisting of k columns (from A) associated with the x's and $m - k$ columns (from I) associated with the x_S's. Recall from Chapter 7 that for the consistent equations $\begin{bmatrix} A_1 & A_2 \end{bmatrix} \begin{bmatrix} x_1 \\ x_2 \end{bmatrix} = y_1$ the solution by partitioning yields

$$x = \begin{bmatrix} x_1 \\ x_2 \end{bmatrix} = \begin{bmatrix} A_1^{-1}y_1 - A_1^{-1}A_2 x_2 \\ x_2 \end{bmatrix}. \tag{4}$$

In this case x is x^*, A_1 is B, y_1 is b and x_1 represents the $m \times 1$ solution vector for the variables associated with the basis B. We shall call this vector x_B,

---

[1] The generalized inverse of Chapter 7 could be used to find all x^* satisfying $A^*x^* = b$; however, in this case $\tilde{x}^* = Gb + (H - I)w$ implies an infinite number of solutions since $(H - I)w$ is non-null and w is arbitrary, including solutions that have negative elements which violate the non-negative constraints. Thus the generalized inverse method, while providing the form of all solutions to $A^*x^* = b$, offers no aid in obtaining an optimal solution to the L.P. problem.

the *vector of basis variables*, and (4) may be written in present notation as

$$x^* = \begin{bmatrix} x_B \\ x_2 \end{bmatrix} = \begin{bmatrix} B^{-1}b - B^{-1}A_2x_2 \\ x_2 \end{bmatrix}. \tag{5}$$

Now suppose we set $x_2 = 0$, i.e. set to zero the non-basic variables in x^*. Then (5) reduces to

$$x^* = \begin{bmatrix} x_B \\ 0 \end{bmatrix} = \begin{bmatrix} B^{-1}b \\ 0 \end{bmatrix}. \tag{6}$$

Thus the vector x_B used to obtain $x^* = [x_B' \ 0']'$ results in a solution x^* to the equations $A^*x^* = b$. This solution is associated with a unique corner-point: since k of the x_S's are set equal to zero ($m - k$ are associated with x_B), k of the inequalities $Ax \le b$ are forced to hold as equalities; furthermore, $n - k$ of the x's are set equal to zero, forcing the associated non-negativity constraints to hold as equalities. Thus the real variables in x^* represent the solution to n equations formed by forcing n of the constraints $Ax \le b$, $x \ge 0$ to hold as equalities, and the unique solution to this set of equations is by definition a corner-point. Any basis B has associated with it a unique corner-point, which may or may not be feasible depending on whether or not it satisfies the remaining constraints.

Example. In problem (1) there are six bases, and their relation to corner-points are as follows:

Basis	Basis Variables	$x_B = B^{-1}b$	Non-basic Variables	Inequalities Forced to Hold as Equalities	Corner-Point
$\begin{bmatrix} 1 & 0 \\ 0 & 1 \end{bmatrix}$	x_{S1}, x_{S2}	$\begin{bmatrix} x_{S1} \\ x_{S2} \end{bmatrix} = \begin{bmatrix} 8 \\ 7 \end{bmatrix}$	$x_1 = 0$ $x_2 = 0$	$x_1 = 0$ $x_2 = 0$	$\begin{bmatrix} x_1 \\ x_2 \end{bmatrix} = \begin{bmatrix} 0 \\ 0 \end{bmatrix}$
$\begin{bmatrix} 1 & 2 \\ 0 & 3 \end{bmatrix}$	x_{S1}, x_2	$\begin{bmatrix} x_{S1} \\ x_2 \end{bmatrix} = \begin{bmatrix} \frac{10}{3} \\ \frac{7}{3} \end{bmatrix}$	$x_1 = 0$ $x_{S2} = 0$	$x_1 = 0$ $2x_1 + 3x_2 = 7$	$\begin{bmatrix} x_1 \\ x_2 \end{bmatrix} = \begin{bmatrix} 0 \\ \frac{7}{3} \end{bmatrix}$
$\begin{bmatrix} 3 & 0 \\ 2 & 1 \end{bmatrix}$	x_1, x_{S2}	$\begin{bmatrix} x_1 \\ x_{S2} \end{bmatrix} = \begin{bmatrix} \frac{8}{3} \\ \frac{5}{3} \end{bmatrix}$	$x_{S1} = 0$ $x_2 = 0$	$3x_1 + 2x_2 = 8$ $x_2 = 0$	$\begin{bmatrix} x_1 \\ x_2 \end{bmatrix} = \begin{bmatrix} \frac{8}{3} \\ 0 \end{bmatrix}$
$\begin{bmatrix} 3 & 2 \\ 2 & 3 \end{bmatrix}$	x_1, x_2	$\begin{bmatrix} x_1 \\ x_2 \end{bmatrix} = \begin{bmatrix} 2 \\ 1 \end{bmatrix}$	$x_{S1} = 0$ $x_{S2} = 0$	$3x_1 + 2x_2 = 8$ $2x_1 + 3x_2 = 7$	$\begin{bmatrix} x_1 \\ x_2 \end{bmatrix} = \begin{bmatrix} 2 \\ 1 \end{bmatrix}$
$\begin{bmatrix} 3 & 1 \\ 2 & 0 \end{bmatrix}$	x_1, x_{S1}	$\begin{bmatrix} x_1 \\ x_{S1} \end{bmatrix} = \begin{bmatrix} \frac{7}{2} \\ -\frac{5}{2} \end{bmatrix}$	$x_2 = 0$ $x_{S2} = 0$	$x_2 = 0$ $2x_1 + 3x_2 = 7$	$\begin{bmatrix} x_1 \\ x_2 \end{bmatrix} = \begin{bmatrix} \frac{7}{2} \\ 0 \end{bmatrix}$
$\begin{bmatrix} 2 & 0 \\ 3 & 1 \end{bmatrix}$	x_2, x_{S2}	$\begin{bmatrix} x_2 \\ x_{S2} \end{bmatrix} = \begin{bmatrix} 4 \\ -5 \end{bmatrix}$	$x_1 = 0$ $x_{S1} = 0$	$x_1 = 0$ $3x_1 + 2x_2 = 8$	$\begin{bmatrix} x_1 \\ x_2 \end{bmatrix} = \begin{bmatrix} 0 \\ 4 \end{bmatrix}$

The last two bases are associated with infeasible corner-points and will not be examined by the revised simplex method.

Before detailing the steps of the revised simplex method in matrix notation we present the minimization problem and some related topics.

2. THE MINIMIZATION PROBLEM

a. Formulation

The matrix form of the minimization problem of L.P. is as follows:

$$\text{Minimize} \quad c'x$$
$$\text{subject to} \quad Ax \gtreqless b, \tag{7}$$
$$x \geq 0.$$

We first consider the case in which the inequalities are all "greater than"; the other possibilities are discussed in Section 3. Minimization problems typically occur in situations where the objective is to attain some specified level of performance at minimum cost. The following illustration examines such a situation.

Illustration 2. Stigler (1945) studied a diet problem where the objective was to minimize the cost of a diet, while the constraints involved the amounts of various nutrients required in the diet. In a simplified example, suppose that a certain diet needs protein and carbohydrates in amounts of 70 and 150 grams respectively; further, suppose only two foods are considered, one which costs 0.2¢ per gram and contains 0.5 gram of protein and 0.2 gram of carbohydrates per gram, and one which costs 0.03¢ per gram and contains 0.02 gram of protein and 0.6 gram of carbohydrates per gram. How many grams of each food should be purchased in order to meet the diet requirements at minimum cost?

As in the previous illustration, the objective function and the constraints can be written as linear functions of the decision variables, where the decision variables are how many grams of each good to purchase. The equations will be written in the standard form of L.P., but the objective now is to minimize cost:

$$\text{Minimize} \quad z = 0.2x_1 + 0.03x_2$$
$$\text{subject to} \quad 0.5x_1 + 0.02x_2 \geq 70,$$
$$0.2x_1 + 0.6\ x_2 \geq 150, \tag{8}$$
$$x_1 \geq 0,$$
$$x_2 \geq 0.$$

In addition to minimizing the objective function, (8) also differs from (1) in that both inequalities are "greater than or equal to." L.P. can handle this type of inequality, although a slight modification is necessary. Inequalities of this form imply a feasible region lying above the equations formed from the inequalities, as illustrated in Figure 2.

The feasible (shaded) region in Figure 2 continues out without limit, but since we are minimizing cost the optimal solution will be near the origin.

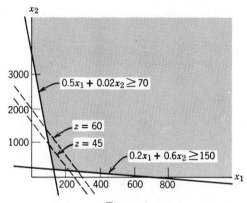

Figure 2

Using the "pencil technique" to obtain a graphical solution as before, the lines $z = 60$ and $z = 45$ are illustrated in Figure 2. Since we are minimizing, we want to move along lines of decreasing values of the objective function; i.e., toward the origin. Continuing to move the pencil parallel to the two lines, the line that touches the feasible region last occurs at $x_1 = 131.8$ and $x_2 = 206.1$. The minimum cost is 32.54¢ and it occurs at a feasible corner-point.

As in the maximization problem, an optimal solution of a minimization problem will always occur at a corner-point of the feasible region. In Figure 2 there are three feasible corner-points: $(750, 0)$, $(0, 3500)$ and $(131.8, 206.1)$. With slight modification, the revised simplex method of solution can be utilized to solve this type of minimization problem.

b.　The initial solution with "greater-than-or-equal-to" constraints

In the case of a "greater-than-or-equal-to" constraint, when we write the equality form using A^* as in (3), the "slack" variables must have a negative coefficient to take up the slack, and they are often called "surplus" variables. In Illustration 2 we have:

$$
\begin{aligned}
\text{Minimize} \quad & z = 0.2x_1 + 0.03x_2 \\
\text{subject to} \quad & 0.5x_1 + 0.02x_2 - x_{S1} \phantom{- x_{S2}} = 70, \\
& 0.2x_1 + 0.6\ x_2 - x_{S2} = 150, \\
& x_1,\ x_2,\ x_{S1},\ x_{S2} \phantom{- x_{S2}} \geq 0.
\end{aligned}
\tag{9}
$$

For example, if $0.5x_1 + 0.02x_2 = 90$, we need -20 to force the equality to hold; thus the negative coefficient of x_{S1} is appropriate.

Problem (9) is analogous to problem (3), but the A^* matrix in the minimization case does not contain an identity matrix, which is needed as a starting basis in the revised simplex procedure. In the maximization case, the identity matrix is present and corresponds to the feasible corner-point at the origin; in the minimization case, the origin is typically not a feasible solution (see the example above), but an identity matrix is nonetheless needed to start the procedure. We thus proceed to create it and subsequently insure that it is "disallowed."

To obtain an initial identity matrix in Illustration 2 (and in general with greater-than-or-equal-to constraints), we add *artificial variables* x_{d1} and x_{d2} to (9) as follows:

$$
\begin{aligned}
\text{Minimize} \quad & z = 0.2x_1 + 0.03x_2 \\
\text{subject to} \quad & 0.5x_1 + 0.02x_2 - x_{S1} \qquad + x_{d1} \qquad\quad = 70, \\
& 0.2x_1 + 0.6\ x_2 \qquad\quad - x_{S2} \qquad\quad + x_{d2} = 150, \\
& x_1,\ x_2,\ x_{S1},\ x_{S2},\ x_{d1},\ x_{d2} \qquad\qquad\ \geq 0.
\end{aligned}
\tag{10}
$$

There is now an identity matrix, and the revised simplex procedure can proceed; but by starting at the origin the procedure is starting at a corner-point which is not in the feasible region. The artificial variables are called that because they are not allowed in the final solution—they must be driven out. If they are zero, the remaining equations allow solutions only in the feasible region. Thus although we can start at the origin using the artificial variables we must be sure that they are eliminated from the final solution. This can be done in two ways: either by systematically driving them out or by pricing them in such a way that they are certain to be eliminated. The first procedure is called a Phase-I portion of a solution; the second is known as the M-pricing technique. Both procedures do exactly the same thing; they drive out the artificial variables before anything else is done. Phase-I is discussed by Hadley (1962); we will describe the M-pricing technique because of its relative simplicity.

If cost is being minimized, the M-pricing technique assigns a very large positive cost (M) to each artificial variable, while if profit is being maximized it assigns a very large negative profit to each artificial variable. In Illustration 2 a cost coefficient of 10,000¢ is sufficient to guarantee their removal. Thus (10) becomes:

$$
\begin{aligned}
\text{Minimize } & z = 0.2x_1 + 0.03x_2 \qquad\qquad\ + 10,000x_{d1} + 10,000x_{d2} \\
\text{subject to} \quad & 0.5x_1 + 0.02x_2 - x_{S1} \qquad + \qquad x_{d1} \qquad\qquad = 70, \\
& 0.2x_1 + 0.6\ x_2 \qquad\quad - x_{S2} \qquad\quad + \qquad x_{d2} = 150, \\
& x_1,\ x_2,\ x_{S1},\ x_{S2},\ x_{d1},\ x_{d2} \qquad\qquad\qquad\ \geq 0.
\end{aligned}
\tag{11}
$$

(11) can be written in the form of (3), where:

$$A* = \begin{bmatrix} 0.5 & 0.02 & -1 & 0 & +1 & 0 \\ 0.2 & 0.6 & 0 & -1 & 0 & +1 \end{bmatrix},$$

$$c*' = [0.2 \quad 0.03 \quad 0 \quad 0 \quad 10{,}000 \quad 10{,}000],$$

and
$$x*' = [x_1 \quad x_2 \quad x_{S1} \quad x_{S2} \quad x_{d1} \quad x_{d2}].$$

This form is ready for solution. It has an initial identity solution, and it guarantees that the procedure will terminate in the feasible region with an optimal solution.

3.   RELATED TOPICS

a.   Mixed constraints

In Sections 1 and 2, L.P. problems were examined where the constraints were either all "less than or equal to" or all "greater than or equal to." The procedure for handling mixed constraints is to combine the procedures used before. That is, when the constraints are written in inequality form, add slack variables with coefficients of $+1$ to the less-than constraints and surplus variables with coefficients of -1 to the greater-than constraints. Then add artificial variables with coefficients of $+1$ and a very high cost coefficient to the greater-than constraints. The initial solution consists of some slack variables at zero cost and some artificial variables at a very high cost, and the revised simplex proceeds to drive out the artificial variables.

b.   Equality constraints

A constraint that is initially in the form of an equality is handled by adding an artificial variable and using the M-pricing technique. While the initial solution will contain the artificial variable, the revised simplex procedure will eventually drive the artificial variable to zero, thus forcing the equality to hold.

c.   Changing the direction of an inequality

The elements of the b vector must be non-negative for the revised simplex method. If any b_i, $i = 1, \ldots, m$, is negative when the problem is initially formulated, the corresponding constraint can be multiplied through by -1, thus changing the signs of the elements of the constraint matrix A associated with that constraint and the direction of the inequality, as well as the sign of b, as desired. This does not affect the meaning of the constraint at all; the feasible region is unchanged. For example, a greater-than constraint multiplied by -1 becomes a less-than constraint, but the problem and the resulting solution are unchanged.

d. Changing a minimization (min) problem to a maximization (max) problem

Most computer programs are written to handle either minimization or maximization problems, but not both, so it is worthwhile to know how to change from a minimization problem to a maximization problem. This is done simply by multiplying the objective function by -1 and changing the word minimize to maximize; everything else remains unchanged. That is, minimizing $z = c'x$ subject to $Ax \gtreqless b$ and $x \geq 0$ is equivalent to maximizing $-c'x$ subject to $Ax \gtreqless b$, $x \geq 0$. Changing a maximization to a minimization is done in an opposite manner.

e. Unconstrained variables

L.P. procedures are designed under the assumption that all variables are non-negative. If the problem to be solved is such that one or more of the variables can be negative, i.e., these variables are unconstrained, L.P. procedures can still be utilized. For every unconstrained x_i, substitute two new variables $x_i^+ - x_i^- = x_i$ into the original formulation of the problem and proceed to solve the reformulated problem in terms of the difference between the non-negative variables $(x_i^+ - x_i^-)$ rather than the unrestricted x_i. When the solution has been obtained, replace $x_i^+ - x_i^-$ by x_i.

A Comprehensive Example. Consider the problem

$$\text{Minimize} \quad z = 2x_1 + 3x_2 + x_3$$

$$\text{subject to} \quad x_1 + x_2 - x_3 \leq 6,$$
$$2x_1 \quad\quad + x_3 \geq 1,$$
$$x_2 + x_3 = 4,$$
$$2x_1 - x_2 \quad\quad \leq -3,$$
$$x_1 \quad\quad\quad \geq 0,$$
$$x_2 \quad\quad \geq 0,$$
$$x_3 \text{ unrestricted.}$$

Let us change the form of the problem to that of a maximization and transform it so that it is amenable to the revised simplex procedure as follows:

$$\text{Maximize } z = -2x_1 - 3x_2 - x_3^+ + x_3^- + 0x_{S1} + 0x_{S2} + 0x_{S3} - Mx_{d1} - Mx_{d2} - Mx_{d3}$$

$$\text{subject to} \quad x_1 + x_2 - x_3^+ + x_3^- + x_{S1} \quad\quad\quad\quad\quad\quad = 6,$$
$$2x_1 \quad\quad + x_3^+ - x_3^- \quad\quad - x_{S2} \quad\quad + x_{d1} \quad\quad\quad = 1,$$
$$x_2 + x_3^+ - x_3^- \quad\quad\quad\quad\quad\quad + x_{d2} \quad\quad = 4,$$
$$-2x_1 + x_2 \quad\quad\quad\quad\quad\quad - x_{S3} \quad\quad\quad + x_{d3} = 3,$$
$$x_1, x_2, x_3^+, x_3^-, x_{S1}, x_{S2}, x_{S3}, x_{d1}, x_{d2}, x_{d3} \quad\quad\quad \geq 0.$$

The starting basis contains the variables $(x_{S1}, x_{d1}, x_{d2}, x_{d3})$ associated with an identity matrix of order 4. M is a large enough number so as to drive x_{d1}, x_{d2} and x_{d3} out of the basis eventually.

f.   The dual problem

For any L.P. problem written in the form

$$\text{Maximize}\quad z = c'x$$
$$\text{subject to}\quad Ax \leq b, \tag{12}$$
$$x \geq 0,$$

there exists a corresponding problem

$$\text{Minimize}\quad z = b'y$$
$$\text{subject to}\quad A'y \geq c, \tag{13}$$
$$y \geq 0.$$

The first of these problems is called the *primal problem* and the second is called the *dual problem*. If the original problem is (12), its dual is (13); if the original is (13), its dual is (12). The dual of the dual is the primal (see Exercise 6). The dual problem occupies an important position in L.P. theory and applications; it is discussed fully by Hadley (1962) and Dantzig (1963). The importance of the dual arises from the fact that the optimal solution to the dual problem, y_0, an $m \times 1$ column vector if A is $m \times n$, has as its elements the marginal value of each of the constraint limits, b_i. In most business and economic situations the b_i's represent scarce resources at the decision-maker's disposal; knowledge of how much profits would increase or costs would decrease with an additional unit of each of the various scarce resources is valuable information.

Example 1. The maximization problem (1) of Illustration 1 was:

$$\text{Maximize}\quad z = 20x_1 + 20x_2$$
$$\text{subject to}\qquad 3x_1 + 2x_2 \leq 8,$$
$$2x_1 + 3x_2 \leq 7,$$
$$x_1 \qquad\quad \geq 0,$$
$$x_2 \geq 0.$$

The dual of this problem is:[1]

$$\text{Minimize}\quad z = 8y_1 + 7y_2$$
$$\text{subject to}\qquad 3y_1 + 2y_2 \geq 20,$$
$$2y_1 + 3y_2 \geq 20, \tag{14}$$
$$y_1 \qquad\quad \geq 0,$$
$$y_2 \geq 0.$$

---

[1] In this particular example the technological coefficients matrix A equals A'. This is not generally the case.

Example 2. The minimization problem (8) of Illustration 2 was:

$$\text{Minimize} \quad z = 0.2x_1 + 0.03x_2$$

subject to
$$0.5x_1 + 0.02x_2 \geq 70,$$
$$0.2x_1 + 0.6\ x_2 \geq 150,$$
$$x_1 \qquad\qquad \geq 0,$$
$$x_2 \geq 0.$$

The dual of this problem is:

$$\text{Maximize} \quad z = 70y_1 + 150y_2$$

subject to
$$0.5\ y_1 + 0.2y_2 \leq 0.2,$$
$$0.02y_1 + 0.6y_2 \leq 0.03,$$
$$y_1 \qquad\qquad \geq 0,$$
$$y_2 \geq 0.$$

The dual variables are also known as *shadow prices* or *simplex multipliers*.

Illustration. The dual problem (14) associated with Illustration 1 is graphically represented and solved in Figure 3; the optimal solution is (4, 4).

As stated above, the elements of the optimal solution to the dual represent the marginal values or shadow prices of the constraint limits b_i. For example, since y_1 equals 4, an additional unit of machine I should be worth 4 units of profit (the first constraint in the primal problem represents available capacity on machine I). This is indeed the case; when machine I has 9 hours available instead of 8, the optimal solution to the primal is $(\frac{3}{5}, \frac{13}{5})$ for a total profit

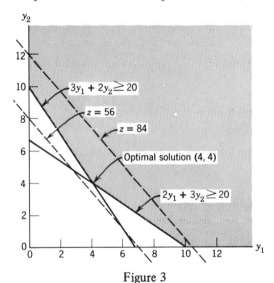

Figure 3

of 64, or 4 units larger than the initial profit of 60. (The reader may verify these results graphically by redrawing Figure 1 to represent 9 units of capacity for machine I.)

Note that y_2 also equals 4; this means that an additional unit of capacity of machine II will also increase profit by 4 units. Thus if the management of the machine shop had an opportunity to rent additional machine capacity at a cost of \$5/hour for machine I and \$3/hour for machine II, the first opportunity should be rejected and the second opportunity should be accepted, at least for some number of units.

How many units of capacity of machine II should be rented if the cost is \$3/hour? The shadow prices obtained from the dual are valid only within certain ranges of capacities. Thus as additional capacity of machine II is rented, at some point the basis will change, with a resulting change in the shadow price. The study of the range over which the shadow prices remain constant is called *sensitivity analysis*, while the behavior of a shadow price as capacity is continually varied is called *parametric programming*. For further discussion of these topics see Hadley (1962) or Bierman, Bonini and Hausman (1969).

The fundamental relation between the primal and the dual problems is that if x_0^* and y_0^* are optimal solutions to these problems respectively, then[1]

$$c'x_0 = c^{*'}x_0^* = z_0 = b^{*'}y_0^* \, b'y_0; \tag{15}$$

i.e., at their respective optima, the objective functions of the primal and the dual have identical values. Let us rearrange the elements of c^*, x^*, b^* and y^* such that $x_0^{*'} = [x_B \quad x]$, $c^{*'} = [c_B \quad 0]$ and $y_0^{*'} = [y_0 \quad y_S]$ where x_B are the basis variables, x are the non-basis variables (thus equal to zero), y_0 are the ordinary variables in the dual, y_S are the surplus variables associated with (13) in revised simplex form and c^* and b^* have been altered in a corresponding fashion. Then $c_B'x_B = b'y_0$ by (15); but $x_B = B^{-1}b$ from (6). Hence $c_B'B^{-1}b = b'y_0$, which is true for all non-negative b so that $y'_0 = c_B'B^{-1}$. Thus the value of the dual variables associated with (13) at its optimum can be obtained directly from the optimal basis B available from the revised simplex solution.

The ordinary dual variables $y'_0 = c_B'B^{-1}$, as mentioned above, can be interpreted as the marginal value (the shadow price) of an additional unit of each of the resources denoted by b_i. If the optimal solution to (12) contains some slack variables, the value of additional units of those resources should be zero. Suppose the ith constraint is non-binding; i.e., that slack variable is positive at the optimum. Suppose further that the column associated with that slack variable is the ith column[2] of B. That column will

---

[1] See Hadley (1962), pp. 228–230.
[2] The column in question could be in any position in B, with the same results holding; see Exercise 15.

have all zeros except for the ith element, which is equal to 1, and it can be shown (see Exercise 15) that the ith column of B^{-1} will be the same. Hence the ith element of $c'_B B^{-1}$ is c_{Bi}. But since the variable is a slack variable, its cost coefficient is zero, so that y_i in $y'_0 = c'_B B^{-1}$ is zero. Hence the dual variable y_i associated with a non-binding constraint (constraint i) is zero.

Much more could be said about the dual variables, in terms of both their relation to the primal problem and their economic interpretation and usage. For example, Baumol and Fabian (1964) discuss how the dual variables can be utilized in controlling the operations of a decentralized firm whose profit maximizing problem can be stated as an L.P. problem. Gale (1960) and Lancaster (1968) use a similar approach for pricing conditions for a competitive equilibrium. Extended discussions of duality theory and interpretation can be found in Hadley (1962), Dantzig (1963) and Dorfman, Samuelson and Solow (1958), while Bierman, Bonini and Hausman (1969) illustrate the use of dual variables in numerous decision-making situations.

4. APPLICATIONS OF LINEAR PROGRAMMING

L.P. provides a general solution method to an important class of problems in business and economics. In this section we illustrate two such applications and cite others.

Illustration 3. A Media Selection Problem. The problem of deciding where to advertise has been formulated as an L.P. problem by Bass and Lonsdale (1966). In a simplified version of their problem, suppose we can advertise only on the local radio station or in a monthly magazine. Further, suppose we are interested in two types of consumers; housewives in families with over \$10,000 income and housewives in families with under \$10,000 income. It is believed that a person in the first category will purchase twice as much as a person in the second category, and we want to maximize purchasing stimulus given that belief. One unit of advertising on the radio costs \$90, and reaches (on the average) 2 persons in the first category and 7 in the second. One unit of advertising in the magazine costs \$120; it reaches 6 persons in the first category and 3 in the second.[1] The manager requires at least 6 units on the radio for public relations purposes and no more than 12 units in the magazine (i.e. no more than 1 unit per month). Finally, the advertising budget is \$1,740 per year.

We formulate the problem as an L.P. problem. Let

x_1 = number of units of radio advertising, and

x_2 = number of units of magazine advertising.

---

[1] In this illustration it is assumed that the magazine's audience has no overlap with the radio's audience.

Then the problem is to maximize purchasing stimulus subject to management and budget constraints:

$$\text{Maximize} \quad z = [2(2) + 1(7)]x_1 + [2(6) + 1(3)]x_2$$

$$\text{subject to} \quad x_1 \qquad\qquad\qquad \geq 6,$$

$$x_2 \leq 12,$$

$$90x_1 + \qquad\qquad 120x_2 \leq 1{,}740,$$

$$x_1 \qquad\qquad\qquad \geq 0,$$

$$x_2 \geq 0.$$

Adding a surplus variable, two slack variables and an artificial variable, we obtain:

$$\text{Maximize} \quad z = 11x_1 + 15x_2 \qquad\qquad - 10{,}000x_d$$

$$\text{subject to} \quad x_1 \quad\; - x_{S1} \qquad\quad + \qquad x_d = 6,$$

$$x_2 \quad + x_{S2} \qquad\qquad\qquad = 12,$$

$$90x_1 + 120x_2 \qquad + x_{S3} \qquad = 1{,}740,$$

$$x_1 \qquad\qquad\qquad\qquad\qquad \geq 0,$$

$$x_2 \qquad\qquad\qquad\qquad\qquad \geq 0.$$

In this form, the revised simplex can be used to solve the problem.

Illustration 4. "Transportation" Problems. One special type of L.P. problem is the "transportation" problem, sometimes called the Hitchcock-Koopmans problem after the persons who first solved it. Dantzig's original simplex method was in fact formulated to solve this problem, even though it solves the more general L.P. problem as well. Any problem which has m sources, n destinations and a constant unit cost to produce at any source and to ship between any source and destination can be solved as a transportation L.P. There are special techniques for this problem's solution, but we will only give an example of a problem that falls into this class.

Suppose a manufacturer produces a product at 3 plants, with unit production costs of \$0.50, \$0.60 and \$0.45 respectively. He also has 4 market service warehouses with demands of 50, 55, 40 and 60 units respectively which must be satisfied. Transportation costs per unit from any plant to any warehouse are given in Table 1. Plant A can produce at most 40 units; plant B, 70 units; plant C, 140 units. This problem can be formulated as an L.P. transportation problem.

Let x_{A1} = the number of units produced in plant A and shipped to warehouse 1; the cost per unit of x_{A1} will be the production cost at A plus the shipping cost from A to 1, i.e. $0.50 + 0.05 = 0.55$. Let x_{B2} = the number,

TABLE 1. TRANSPORTATION COSTS PER UNIT

To:	Warehouse 1	Warehouse 2	Warehouse 3	Warehouse 4
From:				
Plant A	0.05	0.17	0.12	0.01
Plant B	0.17	0.06	0.13	0.04
Plant C	0.06	0.26	0.12	0.10

B to 2; the cost is $0.60 + 0.06 = 0.66$, etc; there are 12 such variables. The objective is to minimize cost. The L.P. is:

$$\text{Minimize } z = 0.55x_{A1} + 0.77x_{B1} + 0.51x_{C1} + 0.67x_{A2} + 0.66x_{B2}$$
$$+ 0.71x_{C2} + 0.62x_{A3} + 0.73x_{B3}$$
$$+ 0.57x_{C3} + 0.51x_{A4} + 0.64x_{B4} + 0.55x_{C4}$$

subject to $x_{A1} + x_{B1} + x_{C1} = 50$ (i.e. the total shipped to 1 is 50),

$$x_{A2} + x_{B2} + x_{C2} = 55,$$
$$x_{A3} + x_{B3} + x_{C3} = 40,$$
$$x_{A4} + x_{B4} + x_{C4} = 60,$$
$$x_{A1} + x_{A2} + x_{A3} + x_{A4} \leq 40 \quad \text{(i.e. the total from A is } \leq 40\text{),}$$
$$x_{B1} + x_{B2} + x_{B3} + x_{B4} \leq 70,$$
$$x_{C1} + x_{C2} + x_{C3} + x_{C4} \leq 140,$$
$$x_{ij} \geq 0, \ i = A, B, C \text{ and } j = 1, 2, 3, 4.$$

The constraints indicate that the demands must be satisfied and that plant capacities cannot be exceeded. Notice the form of the A matrix when this problem is written in matrix form:

$$A = \begin{bmatrix} 1 & 1 & 1 & 0 & 0 & 0 & 0 & 0 & 0 & 0 & 0 & 0 \\ 0 & 0 & 0 & 1 & 1 & 1 & 0 & 0 & 0 & 0 & 0 & 0 \\ 0 & 0 & 0 & 0 & 0 & 0 & 1 & 1 & 1 & 0 & 0 & 0 \\ 0 & 0 & 0 & 0 & 0 & 0 & 0 & 0 & 0 & 1 & 1 & 1 \\ 1 & 0 & 0 & 1 & 0 & 0 & 1 & 0 & 0 & 1 & 0 & 0 \\ 0 & 1 & 0 & 0 & 1 & 0 & 0 & 1 & 0 & 0 & 1 & 0 \\ 0 & 0 & 1 & 0 & 0 & 1 & 0 & 0 & 1 & 0 & 0 & 1 \end{bmatrix}.$$

This form is always present for transportation problems. Slack and surplus variables can be added to form the A^* matrix, and artificial variables can be used if necessary to obtain an initial solution.

An interesting property of A above is that every square sub-matrix has a determinant of $+1$, -1 or 0 (see Chapter 4, Exercise 10). As will be shown in Section 5, the revised simplex method continually takes the inverse of the current basis to obtain the solution. The only division involved is by the value of the determinant, but if that is always $+1$ or -1 (0 will not be used since the inverse does not exist for any such sub-matrix), then if we start with all integers we will always have all integers. This is an important aspect of transportation L.P.'s; they guarantee integer solutions, even though L.P.'s in general do not. There are algorithms to obtain the best integer solution to any L.P., but their computational performance is not yet proven for most problems [see Hadley (1964) for a discussion]. Thus this aspect of transportation L.P.'s is of interest.

Some production scheduling problems, personnel assignment problems and machine scheduling problems take the form of the transportation L.P. It is thus a useful general form, and usually few computations are required.

The above illustrations are only two of the applications of L.P. To mention some more, L.P. has been applied to capital budgeting problems by Weingartner (1963) and Charnes, Cooper and Miller (1959); a large class of production scheduling problems have been formulated as L.P.'s, and airline scheduling with random demand has been solved as an L.P. problem. The economic theory of equilibrium has been examined in this format, as have two-person constant sum games, mentioned earlier in the Exercises of Chapter 2. This list is by no means exhaustive, but it does give an indication of some of the applications and the types of problems amenable to L.P. The reader is referred to Dorfman, Samuelson and Solow (1958) for further discussion of economic theory and L.P. and to Bierman, Bonini and Hausman (1969), Hadley (1962) or Dantzig (1963) for a discussion of applications to business problems.

5.   THE REVISED SIMPLEX METHOD

Although L.P. computer programs are generally available to solve L.P. problems, it is instructive for the student to study the operation of the simplex method. The revised simplex method of solution differs in some computational details from the ordinary simplex, but the basic concepts are identical. We choose to present the revised simplex method since many computer programs used to solve L.P. problems utilize this method.

a. The revised simplex-maximization form

The revised simplex method begins with the basis $B = I$, the last m columns of A^*; this basis corresponds to the corner-point $x = 0$ (the null vector representing the origin), which is feasible since we assumed $b_i \geq 0$ for all i. Then a column (variable) is selected to enter the basis according to which of the non-basic variables (i.e., the variables currently zero) in x^* will increase the value of the objective function. Then a column (variable) is selected for removal from the existing basis in such a way that feasibility is maintained when the new $x_B = B^{-1}b$ is obtained; x_B represents the variables in the basis, and all other variables are zero. The substitution is then performed and a new basis, B_{new}, and a corresponding new basic solution, $x_{B,\text{new}}$, are obtained. Corresponding to the new basis is a new feasible corner-point, and, as will be seen subsequently, the value of z at the new corner-point is no lower than the value of z at the previous corner-point (and, in fact, is usually higher). The process is repeated (iterated) so long as a new substitution (feasible corner-point) can be found that will increase (or at least not decrease) the value of the objective function; when no such substitutions are possible the current corner-point is the optimal solution. Any feasible corner-point is attained at most once and since there are only a finite number of feasible corner-points, this process finds the optimal solution in a finite number of iterations.[1]

To introduce the revised simplex method, the objective function and the constraints $A^*x^* = b$ of (3) are written together in partitioned matrix form as:

$$\begin{bmatrix} 1 & -c^{*\prime} \\ 0 & A^* \end{bmatrix} \begin{bmatrix} z \\ x^* \end{bmatrix} = \begin{bmatrix} 0 \\ b \end{bmatrix}, \tag{16}$$

for the moment ignoring the $x^* \geq 0$ constraints. In (16) we have rewritten the objective function as $z - c^{*\prime}x^* = 0$ and the equations $A^*x^* = b$ have been augmented by this equation. The equations in (16) are the standard augmented form of the L.P. problem for the revised simplex method.

As stated above, the initial solution is obtained by setting the variables associated with the identity matrix equal to the corresponding b values. In Illustration 1 above,

$$x_{S1} = 8,$$

$$x_{S2} = 7.$$

The columns of A^* which form this solution are seen in the following

---

[1] While it is possible to construct examples which "cycle," i.e., examples in which a given set of corner-points are reattained periodically with no increase in the objective function, the authors are unaware of such a phenomenon ever occurring in a practical problem.

equations:

$$\begin{bmatrix} 1 & 0 \\ 0 & 1 \end{bmatrix} \begin{bmatrix} x_{S1} \\ x_{S2} \end{bmatrix} = \begin{bmatrix} 8 \\ 7 \end{bmatrix}. \tag{17}$$

The first step in any such problem uses the basis $B = I$, with the solution being $x_B = B^{-1}b = I^{-1}b = b$. Since b is non-negative and the remaining non-basic variables are zero, this solution satisfies the non-negativity constraints.

Before proceeding, some notation essential to the revised simplex method is given. Consider augmenting the matrix B: place the negative of the profit coefficients associated with the basis variables above the corresponding columns of B and add a first column with a one in the first element and the m remaining elements zero. This augmented matrix is denoted as B^*, using the star notation to indicate an augmented matrix. Thus

$$B^* = \begin{bmatrix} 1 & -c'_B \\ 0 & B \end{bmatrix}, \tag{18}$$

where c_B is the vector of profit coefficients for the variables in the current basis. For the illustration,

$$B^* = \begin{bmatrix} 1 & 0 & 0 \\ \hline 0 & 1 & 0 \\ 0 & 0 & 1 \end{bmatrix}, \quad \text{and for this first basis } B^{*-1} = \begin{bmatrix} 1 & 0 & 0 \\ 0 & 1 & 0 \\ 0 & 0 & 1 \end{bmatrix} \text{ also.}$$

This augmented basis now enables us to calculate the current value of z, the objective function: $z = c'_B x_B$. Thus we can obtain that value at each stage of the revised simplex. Still using star notation to indicate an augmented matrix (or vector) it can be shown that

$$x_B^* = \begin{bmatrix} z \\ x_B \end{bmatrix} = B^{*-1} \begin{bmatrix} 0 \\ b \end{bmatrix} = B^{*-1}b^*; \tag{19}$$

in our example,

$$x_B^* = \begin{bmatrix} 1 & 0 & 0 \\ 0 & 1 & 0 \\ 0 & 0 & 1 \end{bmatrix} \begin{bmatrix} 0 \\ 8 \\ 7 \end{bmatrix} = \begin{bmatrix} 0 \\ 8 \\ 7 \end{bmatrix}$$

for the first step. Derivation of (19) comes from the following result:[1]

$$\text{If} \qquad B^* = \begin{bmatrix} 1 & -c'_B \\ 0 & B \end{bmatrix} \qquad \text{then} \qquad B^{*-1} = \begin{bmatrix} 1 & c'_B B^{-1} \\ 0 & B^{-1} \end{bmatrix}. \tag{20}$$

b. Steps of the revised simplex method

The revised simplex method can be summarized in the following six steps.

1. Select the identity matrix to form the first basis. Then $B^{-1} = I$, and $x_B = b$.

2. Using, in turn, all columns a_j^* of A^* not in the basis, compute the scalar $c'_B B^{-1} a_j^* - c_j$. (Recall that $c'_B B^{-1}$ is available in the first row of B^{*-1}.) Choose the most negative value of $c'_B B^{-1} a_j^* - c_j$, and denote the corresponding column as column k; it is the next column (variable) that will enter the basis. If no value is negative then the current solution is optimal.

3. For column k chosen in step 2 compute the vector α_k^*:

$$\alpha_k^* = B^{*-1} \begin{bmatrix} -c_k \\ a_k^* \end{bmatrix} = \begin{bmatrix} -c_k + c'_B B^{-1} a_k^* \\ \alpha_{2,k} \\ . \\ . \\ . \\ \alpha_{m+1,k} \end{bmatrix}.$$

4. The column (variable) currently in the basis that is to be removed is obtained by finding the value of i which achieves the minimization

$$\min_{i=2,\ldots,m+1} \left\{ \frac{x_{B,i-1}}{\alpha_{ik}} , \text{ for } \alpha_{ik} > 0 \right\}$$

(where x_{Bj} is the jth variable in the current basis). Label the column found by this minimization as column l.

5. The new augmented basis inverse and solution vector (denoted B_{new}^{*-1} and $x_{B,\text{new}}^*$) are computed using the previous values as follows:

$$x_{B,\text{new}}^* = F x_B^* \tag{21}$$

$$B_{\text{new}}^{*-1} = F B^{*-1} \tag{22}$$

where the matrix F, shown below, is an identity matrix with the lth column changed to contain the ratios $-\alpha_{ik}/\alpha_{lk}$ except for the lth element, where

---

[1] See the partitioned form of the inverse, Section 5.7.

$1/\alpha_{lk}$ is used:

$$
F = \begin{bmatrix}
1 & 0 & 0 & 0 & \cdots & 0 & (c_k - c_B'B^{-1}a_k^*)/\alpha_{lk} & 0 & 0 & \cdots & 0 \\
0 & 1 & 0 & 0 & \cdots & 0 & -\alpha_{2k}/\alpha_{lk} & 0 & 0 & \cdots & 0 \\
0 & 0 & 1 & 0 & \cdots & 0 & -\alpha_{3k}/\alpha_{lk} & 0 & 0 & \cdots & 0 \\
\cdot & \cdot & \cdot & \cdot & & \cdot & \cdot & & & & \cdot \\
\cdot & \cdot & \cdot & \cdot & & \cdot & \cdot & & & & \cdot \\
\cdot & \cdot & \cdot & \cdot & & \cdot & \cdot & & & & \cdot \\
0 & 0 & 0 & 0 & \cdots & 1 & -\alpha_{l-1,k}/\alpha_{lk} & 0 & 0 & \cdots & 0 \\
0 & 0 & 0 & 0 & \cdots & 0 & (+1/\alpha_{lk}) & 0 & 0 & \cdots & 0 \\
0 & 0 & 0 & 0 & \cdots & 0 & -\alpha_{l+1,k}/\alpha_{lk} & 1 & 0 & \cdots & 0 \\
0 & 0 & 0 & 0 & \cdots & 0 & -\alpha_{l+2,k}/\alpha_{lk} & 0 & 1 & \cdots & 0 \\
0 & 0 & 0 & 0 & \cdots & 0 & -\alpha_{l+3,k}/\alpha_{lk} & 0 & 0 & \cdots & 0 \\
\cdot & \cdot & \cdot & \cdot & & \cdot & \cdot & & & & \cdot \\
\cdot & \cdot & \cdot & \cdot & & \cdot & \cdot & & & & \cdot \\
\cdot & \cdot & \cdot & \cdot & & \cdot & \cdot & & & & \cdot \\
0 & 0 & 0 & 0 & \cdots & 0 & -\alpha_{m+1,k}/\alpha_{lk} & 0 & 0 & \cdots & 1
\end{bmatrix}
\quad (23)
$$

6. Given $x_{B,\text{new}}^*$ and B_{new}^{*-1}, return to step 2 (dropping the "new" subscripts from x_B^* and B^{*-1}).

We now briefly interpret the steps outlined above. Step 1 provides a feasible solution ($x = 0$) from which to proceed. Step 2 checks the marginal contribution of one unit of each non-basic variable. If the cost of what it takes away from other variables ($c_B'B^{-1}a_j^*$) is smaller than the profit (c_j), the term $c_BB^{-1}a_j^* - c_j$ will be negative, and the variable should be brought into the basis. When there are no such variables remaining we cannot improve the value of the objective function and the procedure ends.

Given a new column (variable) to enter the basis from step 2, we need to know which column to delete from the basis. Since the marginal profit associated with each unit increment in the entering variable is positive, this marginal profit should be set at the highest possible level. This level is determined by the extent to which each element of the existing solution x_B^* is changed (driven toward negative values) as the new variable enters the basis and takes on positive values, remembering that $A^*x_{B,\text{new}}^* = b$. The kth column of A^* (the entering column) can be written as a linear combination of the columns in the existing basis; i.e., since

$$
\begin{bmatrix} \alpha_{2,k} \\ \cdot \\ \cdot \\ \cdot \\ \alpha_{m+1,k} \end{bmatrix} = B^{-1}a_k^*, \qquad a_k^* = B \begin{bmatrix} \alpha_{2,k} \\ \cdot \\ \cdot \\ \cdot \\ \alpha_{m+1,k} \end{bmatrix} = [\beta_1 \cdots \beta_m] \begin{bmatrix} \alpha_{2,k} \\ \cdot \\ \cdot \\ \cdot \\ \alpha_{m+1,k} \end{bmatrix} = \sum_{i=2}^{m+1} \alpha_{i,k}\beta_{i-1}
$$

where β_i is the ith column of B. Hence, from the last m elements of α_k in step 3 we obtain the substitution rate of the entering variable for each of the elements in the current solution x_B^*; i.e. the α_k vector indicates how much of each basic variable must be removed to allow one unit of the entering variable to enter. Step 4 keeps all the x_B's positive by choosing as the departing variable that x in the basis that is first driven to zero as the entering variable takes on positive values. It does this by checking the value of the basis variable divided by the corresponding element of α_k, thus obtaining a measure of how soon each basic variable is driven to zero as the new variable is introduced.

Step 5 computes what is called the product form of the augmented basis inverse B_{new}^{*-1}, and the new solution $x_{B,\text{new}}^*$. Step 6 returns the process to step 2 and eventually an optimal solution is reached.

Example. The actual computations and the table used in the revised simplex method are now demonstrated for Illustration 1:

$$\text{Maximize} \quad z = 20x_1 + 20x_2$$

$$\begin{aligned} \text{subject to} \quad 3x_1 + 2x_2 + x_{S1} \qquad\quad &= 8, \\ 2x_1 + 3x_2 \qquad\quad + x_{S2} &= 7, \\ x_1, x_2, x_{S1}, x_{S2} &\geq 0. \end{aligned}$$

Step 1. The selection of the identity portion of $A^* = \begin{bmatrix} 3 & 2 & 1 & 0 \\ 2 & 3 & 0 & 1 \end{bmatrix}$ is

accomplished with $x_{S1} = 8$, $x_{S2} = 7$.

Variables in Current Augmented Basis	B^{*-1}	x_B^*
z	$\begin{bmatrix} 1 & 0 & 0 \\ 0 & 1 & 0 \\ 0 & 0 & 1 \end{bmatrix}$	0
x_{S1}		8
x_{S2}		7

Step 2. Using columns 1 and 2 of A^*, we compute $c_B' B^{-1} a_j^* - c_j$, where $c_B' B^{-1} = [0 \quad 0]$ can be read from the above table. Thus $c_B' B^{-1} a_1^* - c_1 = -20$ and $c_B' B^{-1} a_2^* - c_2 = -20$. There is a tie, and we choose the variable x_1 to enter the basis. (Either one could be chosen.)

Step 3.

$$\alpha_1 = B^{*-1} \begin{bmatrix} -c_1 \\ a_1^* \end{bmatrix} = \begin{bmatrix} 1 & 0 & 0 \\ 0 & 1 & 0 \\ 0 & 0 & 1 \end{bmatrix} \begin{bmatrix} -20 \\ 3 \\ 2 \end{bmatrix} = \begin{bmatrix} -20 \\ 3 \\ 2 \end{bmatrix} = \begin{bmatrix} -c_1 + c_B' B^{-1} a_1^* \\ \alpha_{2,1} \\ \alpha_{3,1} \end{bmatrix}.$$

Step 4.

$$\min\left(\frac{x_{B1}}{\alpha_{2,1}}, \frac{x_{B2}}{\alpha_{3,1}}\right) = \min\{\tfrac{8}{3}, \tfrac{7}{2}\} = \tfrac{8}{3} \quad \text{for } x_{B1}, \quad \text{giving } l = 2,$$

where x_{B1}, x_{B2} are read from the table and $\alpha_{2,1}$, $\alpha_{3,1}$ were computed in step 3. Thus we should remove the first basic variable, which is x_{S1}, and the corresponding second column of B^*; $\alpha_{lk} = \alpha_{21} = 3$.

Step 5. The F matrix is as follows, placing the appropriate values in the second column of an identity matrix:

$$F = \begin{bmatrix} 1 & \frac{20}{3} & 0 \\ 0 & \frac{1}{3} & 0 \\ 0 & -\frac{2}{3} & 1 \end{bmatrix}.$$

Then

$$x^*_{B,\text{new}} = Fx^*_B = \begin{bmatrix} 1 & \frac{20}{3} & 0 \\ 0 & \frac{1}{3} & 0 \\ 0 & -\frac{2}{3} & 1 \end{bmatrix} \begin{bmatrix} 0 \\ 8 \\ 7 \end{bmatrix} = \begin{bmatrix} \frac{160}{3} \\ \frac{8}{3} \\ \frac{5}{3} \end{bmatrix},$$

and

$$B^{*-1}_{\text{new}} = FB^{*-1} = \begin{bmatrix} 1 & \frac{20}{3} & 0 \\ 0 & \frac{1}{3} & 0 \\ 0 & -\frac{2}{3} & 1 \end{bmatrix} \begin{bmatrix} 1 & 0 & 0 \\ 0 & 1 & 0 \\ 0 & 0 & 1 \end{bmatrix} = \begin{bmatrix} 1 & \frac{20}{3} & 0 \\ 0 & \frac{1}{3} & 0 \\ 0 & -\frac{2}{3} & 1 \end{bmatrix}.$$

We have now solved the problem using variables x_1 and x_{S2} in the basis; variable x_{S1} has been removed. The objective function has a value of $\frac{160}{3}$, representing $\frac{8}{3}$ units of x_1 at a profit of 20 per unit. There is no slack in the first equation, and the second equation has slack of $\frac{5}{3}$ units; furthermore, no variables are negative.

Step 6. Return to step 2 and check non-basic[1] variables.

Variables in Current Augmented Basis	B^{*-1}			x^*_B
z	1	$\frac{20}{3}$	0	$\frac{160}{3}$
x_1	0	$\frac{1}{3}$	0	$\frac{8}{3}$
x_{S2}	0	$-\frac{2}{3}$	1	$\frac{5}{3}$

Step 2. $c'_B B^{-1}$ is $[\frac{20}{3} \quad 0]$,

---

[1] Actually we need only examine column 2, since a vector that is removed on one iteration will never return on the next. (It may, however, enter again at some later stage.)

so that
$$c'_B B^{-1} a_2^* - c_2 = -\tfrac{20}{3} < 0,$$

and $c'_B B^{-1} a_{S1}^* - c_{S1} = +\tfrac{20}{3}$; column 2 (variable x_2) enters the basis.

Step 3.

$$\alpha_2 = B^{*-1} \begin{bmatrix} -c_2 \\ a_2^* \end{bmatrix} = \begin{bmatrix} 1 & \tfrac{20}{3} & 0 \\ 0 & \tfrac{1}{3} & 0 \\ 0 & -\tfrac{2}{3} & 1 \end{bmatrix} \begin{bmatrix} -20 \\ 2 \\ 3 \end{bmatrix} = \begin{bmatrix} -\tfrac{20}{3} \\ \tfrac{2}{3} \\ \tfrac{5}{3} \end{bmatrix}.$$

Step 4.

$$\min \left(\frac{x_{B1}}{\alpha_{2,2}}, \frac{x_{B2}}{\alpha_{3,2}} \right) = \min \left(\frac{\tfrac{8}{3}}{\tfrac{2}{3}}, \frac{\tfrac{5}{3}}{\tfrac{5}{3}} \right) = 1 \quad \text{for } x_{B2}, \quad \text{giving } l = 3;$$

thus x_{S2} and the corresponding 3rd column of B^* will be deleted from the basis. $\alpha_{lk} = \alpha_{3,2} = \tfrac{5}{3}$, and F will be formed as follows:

$$F = \begin{bmatrix} 1 & 0 & \dfrac{\tfrac{20}{3}}{\tfrac{5}{3}} \\ 0 & 1 & \dfrac{-\tfrac{2}{3}}{\tfrac{5}{3}} \\ 0 & 0 & \dfrac{1}{\tfrac{5}{3}} \end{bmatrix} = \begin{bmatrix} 1 & 0 & 4 \\ 0 & 1 & -0.4 \\ 0 & 0 & 0.6 \end{bmatrix}.$$

Step 5.
$$x_{B,\text{new}}^* = F x_B^* = \begin{bmatrix} 1 & 0 & 4 \\ 0 & 1 & -0.4 \\ 0 & 0 & 0.6 \end{bmatrix} \begin{bmatrix} \tfrac{160}{3} \\ \tfrac{8}{3} \\ \tfrac{5}{3} \end{bmatrix} = \begin{bmatrix} 60 \\ 2 \\ 1 \end{bmatrix},$$

$$B_{\text{new}}^{*-1} = F B^{*-1} = \begin{bmatrix} 1 & 0 & 4 \\ 0 & 1 & -0.4 \\ 0 & 0 & 0.6 \end{bmatrix} \begin{bmatrix} 1 & \tfrac{20}{3} & 0 \\ 0 & \tfrac{1}{3} & 0 \\ 0 & -\tfrac{2}{3} & 1 \end{bmatrix} = \begin{bmatrix} 1 & 4 & 4 \\ 0 & 0.6 & -0.4 \\ 0 & -0.4 & 0.6 \end{bmatrix}.$$

Step 6. The previous graphical solution indicates that the optimal solution $(x_1 = 2, x_2 = 1, z = 60)$ has been reached. The revised simplex method now terminates by giving all positive values when returning to step 2.

Step 2.
$$c'_B B^{-1} = [4 \quad 4],$$

so that
$$c'_B B^{-1} a_{S1}^* - c_{S1} = 4,$$

and
$$c'_B B^{-1} a_{S2}^* - c_{S2} = 4.$$

Since neither value is negative the objective function cannot be improved, and the optimal solution has been reached. The final table is shown below.

Variables in Current Augmented Basis	B^{*-1}			x_B^*
z	1	4	4	60
x_1	0	0.6	-0.4	2
x_2	0	-0.4	0.6	1

The final augmented basis is formed by the variables z, x_1 and x_2; z will always remain in the augmented basis and correspondingly the first column always contains a one and m zeros.

From (20) we know that the sub-matrix $\begin{bmatrix} 0.6 & -0.4 \\ -0.4 & 0.6 \end{bmatrix}$ of B^{*-1} equals B^{-1}, where B is the part of A^* formed by the columns associated with variables x_1 and x_2; $B = \begin{bmatrix} 3 & 2 \\ 2 & 3 \end{bmatrix}$. And clearly the sub-matrix $\begin{bmatrix} 0.6 & -0.4 \\ -0.4 & 0.6 \end{bmatrix}$ is the inverse of B.[1] Also, [4 4] should be $c_B B^{-1}$, and since $c_B' = [20\ \ 20]$ we see that this is true.

The revised simplex algorithm was started with an identity solution. Then we moved from that corner-point to an adjacent feasible corner-point by introducing one new column (variable) and removing one column (variable). We proceeded in this fashion, moving to an adjacent feasible corner-point, until no further moves were profitable. (Refer to Figure 1 to see the corner-points covered: $x_1 = 0$, $x_2 = 0$; $x_1 = \frac{8}{3}$, $x_2 = 0$; $x_1 = 2$, $x_2 = 1$.) The process was then terminated. It always terminates in a finite number of steps because there is only a finite number of feasible corner-points and no corner-point is ever reached twice.

c. **The revised simplex-minimization form**

Steps 1, 3, 4, 5 and 6 of the revised simplex algorithm as given in Section 5b are identical for a minimization problem. With the appending of the artificial variables, the identity solution is now available for step 1, and steps 3, 4, 5 and 6 are not concerned with the objective function. Step 2 is altered, however, as follows:

2′. Using all columns a_j^* of A^* not in the basis, compute the scalar $c_B' B^{-1} a_j^* - c_j$, where c_j is the cost coefficient associated with that variable

---

[1] In fact, L.P. provides an alternative way to get an inverse of a matrix. The method is discussed further in Exercises 8 and 9.

(column). Choose the most *positive*, i.e., $\max[c_B'B^{-1}a_j^* - c_j > 0]$ to enter the basis, and denote that column as column k; it is the variable (column) that will enter the basis. If no values are positive the current solution is then optimal.

Step $2'$ checks to see that the objective function decreases at each iteration. Since we are minimizing, that is the desired direction. When no values of $c_B'B^{-1}a_j^* - c_j$ are positive there is no way to lower the objective function, and the current solution is optimal.

The minimization form of the revised simplex method can now be demonstrated for Illustration 2. The initial solution is $x_{d1} = 70$, $x_{d2} = 150$, and the first table is given below. Note that $c_B'B^{-1} = c_B'I = [10{,}000 \quad 10{,}000]$ instead of $[0 \quad 0]$ as before.

Variables in Current Augmented Basis	B^{*-1}			x_B^*
z	1	10,000	10,000	2,200,000
x_{d1}	0	1	0	70
x_{d2}	0	0	1	150

The value of z is obtained as $10{,}000(70) + 10{,}000(150)$; i.e. $z = c_B'x_B$, the cost of the basis variables times the amounts used. Next, step $2'$ is performed.

$2'$.
$$c_B'B^{-1} = [10{,}000 \quad 10{,}000]$$
$$c_B'B^{-1}a_1^* - c_1 = 7{,}000 - 0.2 \ > 0$$
$$c_B'B^{-1}a_2^* - c_2 = 6{,}200 - 0.03 \ > 0$$
$$c_B'B^{-1}a_{S1}^* - c_{S1} = -10{,}000 - 0 < 0$$
$$c_B'B^{-1}a_{S2}^* - c_{S2} = -10{,}000 - 0 < 0.$$

Variable x_1 will enter the basis first. Note that the procedure immediately removes one of the artificials, but it does not introduce surplus variables until one or more real variables are positive.

3.
$$\alpha_k = \alpha_1 = B^{*-1}\begin{bmatrix} -c_1 \\ a_1^* \end{bmatrix} = \begin{bmatrix} 1 & 10{,}000 & 10{,}000 \\ 0 & 1 & 0 \\ 0 & 0 & 1 \end{bmatrix}\begin{bmatrix} -0.2 \\ 0.5 \\ 0.2 \end{bmatrix} = \begin{bmatrix} 6{,}999.8 \\ 0.5 \\ 0.2 \end{bmatrix}.$$

4. $\displaystyle\min_{i=2,3}\left(\frac{x_{B,i-1}}{\alpha_{i,1}}\right) = \min\left(\frac{70}{0.5}, \frac{150}{0.2}\right) = \frac{70}{0.5}$ for x_{B1}, giving $l = 2$;

$\alpha_{lk} = \alpha_{2,1} = 0.5$. Thus x_{d1} and the corresponding second column of B^* will be removed.

5. $\quad x^*_{B,\text{new}} = \begin{bmatrix} 1 & -13{,}999.6 & 0 \\ 0 & 1/0.5 & 0 \\ 0 & -0.2/0.5 & 1 \end{bmatrix} \begin{bmatrix} 2{,}200{,}000 \\ 70 \\ 150 \end{bmatrix} = \begin{bmatrix} 1{,}220{,}028 \\ 140 \\ 122 \end{bmatrix}$,

$\quad B^{*-1}_{\text{new}} = \begin{bmatrix} 1 & -13{,}999.6 & 0 \\ 0 & 1/0.5 & 0 \\ 0 & -0.2/0.5 & 1 \end{bmatrix} \begin{bmatrix} 1 & 10{,}000 & 10{,}000 \\ 0 & 1 & 0 \\ 0 & 0 & 1 \end{bmatrix}$

$\quad = \begin{bmatrix} 1 & -3{,}999.6 & 10{,}000 \\ 0 & 2 & 0 \\ 0 & -0.4 & 1 \end{bmatrix}$.

The new basis includes x_1 and x_{d2}. It is still not in the feasible region. The new table is:

Variables in Current Augmented Basis	B^{*-1}			x^*_B
z	1	−3,999.6	10,000	1,220,028
x_1	0	2	0	140
x_{d2}	0	−0.4	1	122

6. Return to 2′.

2′. $\qquad\qquad c'_B B^{-1} = [-3{,}999.6 \quad 10{,}000]$

$\qquad c'_B B^{-1} a^*_2 - c_2 = 5{,}920 - 0.03 > 0$

$\qquad c'_B B^{-1} a^*_{S1} - c_{S1} = 3{,}999.6 > 0$

$\qquad c'_B B^{-1} a^*_{S2} - c_{S2} = -10{,}000 < 0.$

Choose x_2 to enter the basis.

3. $\qquad\qquad \alpha_2 = B^{*-1} \begin{bmatrix} -0.03 \\ 0.02 \\ 0.6 \end{bmatrix} = \begin{bmatrix} 5{,}919.978 \\ 0.04 \\ 0.592 \end{bmatrix}$.

4. $\min\left\{\dfrac{140}{0.04}, \dfrac{122}{0.592}\right\} = \dfrac{122}{0.592}$, and x_{d2} is the variable to be removed, as was desired.

5. $\quad x^*_{B\text{new}} = \begin{bmatrix} 1 & 0 & -9,999.95 \\ 0 & 1 & -0.07 \\ 0 & 0 & 1.69 \end{bmatrix} \begin{bmatrix} 1,220,028 \\ 140 \\ 122 \end{bmatrix} = \begin{bmatrix} 32.54 \\ 131.8 \\ 206.2 \end{bmatrix}$

$B^{*-1}_{\text{new}} = \begin{bmatrix} 1 & 0 & -9,999.95 \\ 0 & 1 & -0.07 \\ 0 & 0 & 1.69 \end{bmatrix} \begin{bmatrix} 1 & -3,999.6 & 10,000 \\ 0 & 2 & 0 \\ 0 & -0.4 & 1 \end{bmatrix}$

$\qquad = \begin{bmatrix} 1 & 0.38 & 0.036 \\ 0 & 2 & -0.07 \\ 0 & -0.67 & 1.69 \end{bmatrix}.$

When B^{-1} and $c'_B B^{-1}$ are computed they check with the corresponding values in B^{*-1}_{new}. Accuracy is of prime importance for z and B^{*-1}_{new}, and six significant figures appear necessary in this illustration. Computers handle this easily, and even for desk calculation the solution x_B does not involve large numbers. Then z can be computed separately.

The optimal solution has now been reached; it may be verified by returning to step $2'$.

2'. $\qquad\qquad\qquad c_B B^{-1} = [0.38 \quad 0.036]$

$$c_B B^{-1} a^*_{S1} - c_{S1} = -0.38 < 0$$
$$c_B B^{-1} a^*_{S2} - c_{S2} = -0.036 < 0.$$

The artificial variables need not be checked because they will not re-enter the basis owing to their high cost. Since no value of $(c_B B^{-1} a^*_j - c_j) > 0$, the current solution is optimal.

We are now at a corner-point in the feasible region, as is always the case when the artificial variables are driven out. If it were not the optimal solution we would proceed as before to move to an adjacent corner-point, decreasing the value of the objective function until the optimal solution is reached. In this case the first feasible corner-point reached is optimal; this is not usually so. In general we are at a feasible corner-point when the artificial variables are gone and we proceed from there, using step 2 in a maximization problem or step $2'$ in a minimization problem.

6. CONCLUSIONS

Linear programming is an important tool for resource allocation under constraints. It is important for theoretical problems and for applications, because it is a general method of solution for a very large class of problems.

There are several solution techniques; we have examined one of the more efficient methods, the revised simplex method. Although graphical analysis can be used to describe linear programming, matrix algebra is of great help in presenting both the general problem and methods of solution. We have provided a brief introduction to the area without attempting to be complete. There are many side issues, other algorithms, possible complications and theoretical points that have been left unmentioned, for which the reader is referred to such texts as Hadley (1962) and Dantzig (1963) for a full discussion.

Several very efficient computer programs are available to solve very large problems. Hand solution of even small problems is tedious, but the existence of computer programs makes the application of L.P. economically feasible. In addition to regular computer programs, there are special techniques for handling large problems (e.g. "decomposition," which breaks a large problem into several small problems, and transportation L.P. codes which take advantage of the special form of A). Thus the computational problem is not an overriding consideration in L.P. problems, and large problems can be solved at a relatively small cost, enhancing the value of this widely applicable technique.

7. EXERCISES

1. Solve the following problems graphically:

   (a) Maximize $z = 2x_1 + 5x_2$ (b) Minimize $z = 4x_1 + 2x_2$

   subject to $x_1 + 3x_2 < 16$,   subject to $x_1 \qquad\ \geq\ 6$,

   $4x_1 + x_2 \leq 20$, $x_1 + x_2 \geq 10$,

   $x_1 \leq 4$, $x_1, x_2 \qquad \geq\ 0$.

   $x_1, x_2 \geq 0$.

2. Referring to Exercise 1, indicate on a graph the solution space, objective function line, corner-points and the best corner-point solution.

3. (a) Formulate the dual problem of either (a) or (b) in Exercise 1 and solve it graphically.

   (b) Solve the dual of the other problem from 1 by the revised simplex method.

4. For (b) of Exercise 1 verify that the solution to the primal problem gives the solution to the dual; i.e., show that the solution in 3 can be derived directly from the solution in 1.

5. Solve the following problems by the revised simplex method:

   (a) Maximize $z = x_1 + 3x_2 + 10x_3$

   subject to $x_1 + 4x_2 \qquad\qquad \leq\ 7$,

   $x_2 + 3x_3 \leq\ 8$,

   $3x_1 + x_2 + x_3 \leq 17$,

   $x_1, x_2, x_3 \geq\ 0$.

(b) Minimize $z = 7x_1 - 2x_2$

subject to
$$x_1 - x_2 \geq 0,$$
$$7x_1 + 4x_2 \geq 20,$$
$$3x_1 + 4x_2 \leq -6,$$
$$x_1, x_2 \geq 0,$$

6. Show that the dual of any dual is the associated primal.

7. Write the following L.P. problems in a form suitable for solution by the revised simplex method. (Do not solve them.)

(a) Maximize $z = x_1 + 3x_2 - x_3$

subject to
$$x_1 + x_2 + x_3 = 10,$$
$$3x_1 \qquad - \tfrac{1}{2}x_3 \geq 17,$$
$$x_2 + x_3 \leq 7,$$
$$x_1, x_2, x_3 \geq 0.$$

(b) Minimize $z = 2x_1 + x_2$

subject to
$$x_1 \qquad \leq 7,$$
$$x_2 \geq 6,$$
$$x_1 + x_2 = 17,$$
$$x_1, x_2 \geq 0.$$

8. Consider the following L.P. problem:

Maximize $z = 10{,}000x_1 + 100x_2 + x_3 - 10{,}000x_{S1} - 10{,}000x_{S2} - 10{,}000x_{S3}$

subject to
$$x_1 - 7x_2 + 8x_3 + x_{S1} \qquad\qquad\qquad = 2,$$
$$2x_1 + x_2 \qquad\qquad + x_{S2} \qquad = 3,$$
$$x_1 + x_2 - x_3 \qquad\qquad\qquad + x_{S3} = 1,$$
$$x_1, x_2, x_3, x_{S1}, x_{S2}, x_{S3} \geq 0.$$

Discuss why this problem can be used to obtain the inverse of

$$A = \begin{bmatrix} 1 & -7 & 8 \\ 2 & 1 & 0 \\ 1 & 1 & -1 \end{bmatrix}$$

and obtain A^{-1} by this means.

9. Using the method of Exercise 8, write an L.P. problem whose solution yields the inverse of the following matrix:

$$A = \begin{bmatrix} -1 & 0 & 17 & 1 \\ 6 & 5 & 4 & 3 \\ 2 & 1 & 0 & -1 \\ 12 & -11 & 0 & 0 \end{bmatrix}.$$

Do not solve the problem. Note that $x_1 = x_2 = x_3 = x_4 = 1$ gives a positive b vector that can be used.

10. In Exercises 5(a) and (b) and 7(a), write the vectors and matrices appropriate to putting the problem in matrix form. Do this for the $Ax \leq b$ (or $Ax \geq b$) and the $A^*x^* = b$ formats.

11. Solve Illustration 3 for the best media allocation.

12. Consider the following problem and its dual:

$$\text{Maximize} \quad z = x_1 + x_2 + 7x_3 + 4x_4 + 5x_5$$

$$\text{subject to} \quad 2x_1 + x_2 + 2x_3 + x_4 + x_5 \leq 18,$$

$$x_2 + 3x_4 - x_5 \leq 18,$$

$$x_i \geq 0, \quad i = 1, \ldots, 5.$$

Determine the value of the objective function at the optimum using graphical analysis only.

13. Write an L.P. problem that could represent a realistic situation (in the form of the illustrations) and solve the problem.

14. In Exercise 11 how many extra units of the objective function can be obtained if the manager relaxes (a) the first constraint, (b) the second constraint, (c) the budget constraint, each by one unit?

15. (a) Show that if a basis B has as its ith column the ith column of the associated identity matrix I_m in A^* [see equation (3)], then B^{-1} has its ith column identical to that of B and hence y_i in $y' = c'_B B^{-1}$ is zero. (Note that c_{Bi}, the profit coefficient of a slack variable, is zero.)

   (b) Show that if B has as its jth column the ith column of the associated identity matrix I_m in A^*, then B^{-1} has as its ith column the vector containing all zeros except for a 1 in the jth position, and, as a consequence, that y_i in $y' = c'_B B^{-1}$ is zero. (Note that c_{Bj}, the profit coefficient of the slack variable y_i whose column happens to be the jth column of B, is zero.)

REFERENCES

Bass, F. M., and R. T. Lonsdale (1966). An exploration of linear programming in media selection. *Journal of Marketing Research*, **3**, 179–188.

Baumol, W. J., and T. Fabian (1964). Decomposition, pricing for decentralization and external economies. *Management Science*, **11**, 1–32.

Bierman, H., Jr., C. P. Bonini, and W. H. Hausman (1969). *Quantitative Analysis for Business Decisions*. Third Edition; Irwin, Homewood, Ill.

Charnes, A., W. W. Cooper, and M. H. Miller (1959). Application of linear programming to financial budgeting and the cost of funds. *Journal of Business*, **22**, 20–46.

Dantzig, G. B. (1951). Maximization of a set of linear functions of variables subject to linear inequalities. In T. C. Koopmans (ed.), *Activity Analysis of Production and Allocation*. Wiley, New York.

Dantzig, G. B. (1963). *Linear Programming and Extensions.* Princeton University Press.

Dorfman, R., P. A. Samuelson, and R. M. Solow (1958). *Linear Programming and Economic Analysis.* McGraw-Hill, New York.

Gale, D. (1960). *The Theory of Linear Economic Models.* McGraw-Hill, New York.

Hadley, G. (1962). *Linear Programming.* Addison-Wesley, Reading, Mass.

Hadley, G. (1964). *Nonlinear and Dynamic Programming.* Addison-Wesley, Reading, Mass.

Lancaster, K. (1968). *Mathematical Economics.* Macmillan, New York.

Stigler, G. J. (1945). The cost of subsistence. *Journal of Farm Economics*, **27**, 303–314.

Weingartner, H. M. (1963). *Mathematical Programming and the Analysis of Capital Budgeting Problems.* Prentice-Hall, Englewood Cliffs, N.J.

CHAPTER 10

REGRESSION ANALYSIS

Regression analysis is a useful tool in the analysis of business and economic problems, and its techniques are well suited to being described in matrix terminology. The matrix descriptions of regression clarify both the underlying theoretical framework of the analysis and the necessary computing procedures.

The outline of regression analysis presented here is self-contained, but a previous acquaintance with statistics, particularly hypothesis testing, is assumed. Such references as Kendall and Stuart (1958, 1961) and Mood and Graybill (1963) will provide the reader with greater statistical detail than space here permits, while Graybill (1961), Johnston (1963), Goldberger (1964) and Searle (1966) will give him more extensive matrix descriptions than presented here.

1. INTRODUCTION

In predicting how an economy's rate of inflation varies according to its rate of unemployment, one possible model would be that when the unemployment rate is zero the rate of inflation is b_0 per cent; and when the unemployment rate is x per cent the rate of inflation is changed from b_0 per cent to $(b_0 + b_1 x)$ per cent. This macroeconomic model, similar to a model suggested by Phillips (1958), implies that the expected value of y_i, the rate of inflation in time period i, is of the form

$$b_0 + b_1 x_i$$

where x_i is the unemployment rate in period i. Although this is the anticipated form of the rate of inflation, it is unlikely in practice to be its exact form. On assuming linearity, discrepancies between $b_0 + b_1 x_i$ and the actual value of y_i might arise from a number of causes:

(i) Factors other than the rate of unemployment affect the rate of inflation; e.g. mobility of the labor force.

(ii) The economic system, over and above the total effect of all relevant factors, has a basic and unpredictable element of randomness.

(iii) There may be measurement errors in y_i.

To recognize these discrepancies the model is written as

$$y_i = b_0 + b_1 x_i + e_i \tag{1}$$

where e_i is a random error (or disturbance) term arising from one or more of the causes stated above.

Having introduced the random error term e_i into the model, we must now specify certain characteristics of its probability distribution in order to be able to use the model. The simplest and usual assumptions are that the mean value of e_i is zero, that its variance is constant (and therefore independent of i and of x_i) and that the values of e_i over the different time periods are independent of one another. Using E to denote expectation [$E(e_i)$ is the "expected value" of e_i], these assumptions are

$$E(e_i) = 0, \qquad \text{for all } i$$

and
$$E(e_i e_j) = \begin{cases} 0, & \text{for } i \neq j \\ \sigma^2, & \text{for } i = j. \end{cases} \tag{2}$$

Using (2) and taking the expected value of (1) gives

$$E(y_i) = E(b_0 + b_1 x_i + e_i) = b_0 + b_1 x_i + E(e_i)$$

so that
$$E(y_i) = b_0 + b_1 x_i. \tag{3}$$

This is the expected value of y_i as introduced at the outset.

Equations (1) through (3) specify the model to be examined. The variable denoted by y is usually called the *dependent variable* (because it depends on x) and the x-variable is correspondingly referred to as the *independent variable*.[1] Suppose that we have a sample of n observations on y_i and x_i. For each pair of observations (y_i, x_i) equation (1) is assumed to exist. The n equations are

$$y_1 = b_0 + b_1 x_1 + e_1$$
$$y_2 = b_0 + b_1 x_2 + e_2$$
$$\cdot$$
$$\cdot$$
$$\cdot$$
$$y_n = b_0 + b_1 x_n + e_n.$$

---

[1] Note that in (3), $E(x_i)$ was taken to be x_i; it is assumed that the x-values are specified in advance and are not random variables.

By writing the vectors

$$y' = [y_1 \quad y_2 \quad \cdots \quad y_n],$$
$$e' = [e_1 \quad e_2 \quad \cdots \quad e_n],$$
$$b' = [b_0 \quad b_1]$$

and the matrix

$$X = \begin{bmatrix} 1 & x_1 \\ 1 & x_2 \\ \cdot & \cdot \\ \cdot & \cdot \\ \cdot & \cdot \\ 1 & x_n \end{bmatrix}$$

these equations can be rewritten in vector form as

$$y = Xb + e.$$

Note, however, that these equations are unlike any previous set discussed in this book, owing to the nature of the error terms represented by e; by definition these error terms make the equations consistent for any values of b_0 and b_1. But the error terms are unknown, so that a solution for b cannot be derived as $b = X^{-1}(y - e)$, nor as $b = G(y - e)$ where G is a generalized inverse of X. It is therefore impossible to determine b_0 and b_1 exactly. They can, however, be *estimated* from the data, and although the estimates are not the exact values of b_0 and b_1 they provide useful information about them.

Several procedures are available for making estimates of b_0 and b_1. The most frequently used is that known as the least squares procedure, and it is the one described here. Regardless of how we derive them or of what their values are, let us denote the estimates of b_0 and b_1 by $\hat{b}_0$ and $\hat{b}_1$ respectively. Then, for a given value of x_i the corresponding estimated value of y_i, denoted by $\hat{y}_i$, would be

$$\hat{y}_i = \hat{b}_0 + \hat{b}_1 x_i.$$

The difference between this estimated value and the observed value of y_i corresponding to x_i is $y_i - \hat{y}_i$; this difference is called the *estimated error term* or residual, $\hat{e}_i$. Thus

$$\hat{e}_i = y_i - \hat{y}_i.$$

With this formulation the least squares procedure for deriving the estimates $\hat{b}_0$ and $\hat{b}_1$ is to choose them in such a way that the sum of squares of the estimated error terms (SSE) is minimized.[1] This means $\hat{b}_0$ and $\hat{b}_1$ are chosen

---

[1] We do not justify this criterion here but simply note that (*a*) it is well established and widely used in statistical methodology and (*b*) if the cost or loss function associated with estimated errors is quadratic, this criterion minimizes the expected cost of these errors.

so as to minimize

$$\text{SSE} = \sum_{i=1}^{n} \hat{e}_i^2 = \sum_{i=1}^{n} (y_i - \hat{y}_i)^2 = \sum_{i=1}^{n} (y_i - \hat{b}_0 - \hat{b}_1 x_i)^2. \tag{4}$$

The minimization is achieved, and $\hat{b}_0$ and $\hat{b}_1$ are derived, by minimizing the last expression in (4) with respect to $\hat{b}_0$ and $\hat{b}_1$, using differential calculus to do so. Before performing the minimization we make a change in notation, rewriting (1) as

$$y_i = b_0 x_{i0} + b_1 x_{i1} + e_i$$

where $x_{i0} = 1$ and $x_{i1} = x_i$ for all i, and y_i and e_i are unchanged. This minor change in the notation for the x's simplifies the subsequent discussion of situations involving several x-variables.

Incorporating the above change of notation into (4) gives

$$\text{SSE} = \sum_{i=1}^{n} \hat{e}_i^2 = \sum_{i=1}^{n} (y_i - \hat{b}_0 x_{i0} - \hat{b}_1 x_{i1})^2,$$

minimization of which involves equating to zero the first partial derivatives with respect to $\hat{b}_0$ and $\hat{b}_1$. Thus we set

$$\frac{\partial}{\partial \hat{b}_0} \sum_i \hat{e}_i^2 = -2 \sum_{i=1}^{n} x_{i0}(y_i - \hat{b}_0 x_{i0} - \hat{b}_1 x_{i1}) = 0$$

and $$\frac{\partial}{\partial \hat{b}_1} \sum_i \hat{e}_i^2 = -2 \sum_{i=1}^{n} x_{i1}(y_i - \hat{b}_0 x_{i0} - \hat{b}_1 x_{i1}) = 0,$$

which simplify to

$$\Sigma\, y_i x_{i0} = \hat{b}_0 \, \Sigma\, x_{i0}^2 + \hat{b}_1 \, \Sigma\, x_{i0} x_{i1}$$

and $$\Sigma\, y_i x_{i1} = \hat{b}_0 \, \Sigma\, x_{i0} x_{i1} + \hat{b}_1 \, \Sigma\, x_{i1}^2,$$

where all summations are over the range $i = 1, 2, \ldots, n$. These equations are called the *normal equations*. The solutions to these two simultaneous linear equations are the values of $\hat{b}_0$ and $\hat{b}_1$ which minimize[1] $\text{SSE} = \sum_{i=1}^{n} \hat{e}_i^2$. Since $x_{i0} = 1$ for $i = 1, \ldots, n$, we have $\Sigma\, x_{i0}^2 = n$, $\Sigma\, y_i x_{i0} = \Sigma\, y_i$ and $\Sigma\, x_{i0} x_{i1} = \Sigma\, x_{i1}$. Hence the normal equations reduce to

$$\hat{b}_0 n + \hat{b}_1 \, \Sigma\, x_{i1} = \Sigma\, y_i$$

and $$\hat{b}_0 \, \Sigma\, x_{i1} + \hat{b}_1 \, \Sigma\, x_{i1}^2 = \Sigma\, y_i x_{i1}, \tag{5}$$

---

[1] The two partial derivatives are the first-order (necessary) conditions for a minimum. It can be shown that the solution also satisfies the second-order conditions for a minimum.

with the solutions

$$\hat{b}_1 = \frac{\Sigma\, y_i x_{i1} - (\Sigma\, y_i)(\Sigma\, x_{i1})/n}{\Sigma\, x_{i1}^2 - (\Sigma\, x_{i1})^2/n}$$

and

$$\hat{b}_0 = (\Sigma\, y_i - \hat{b}_1 \Sigma\, x_{i1})/n.$$

Example. Suppose observations on the variables of our illustration are as shown in Table 1.

TABLE 1. DATA OF AN EXAMPLE

i	y_i	x_{i0}	x_{i1}
1	2.0	1	5.0
2	3.0	1	4.2
3	3.2	1	4.3
4	4.0	1	3.8
5	3.6	1	4.4
6	2.1	1	5.2
7	1.7	1	6.0
8	1.2	1	6.5
9	1.8	1	5.9
10	2.4	1	4.0
Sums	$\sum_{i=1}^{10} y_i = 25.0$	$\sum_{i=1}^{10} x_{i0} = 10 = n$	$\sum_{i=1}^{10} x_{i1} = 49.3$
Sums of Squares	$\sum_{i=1}^{10} y_i^2 = 69.94$		$\sum_{i=1}^{10} x_{i1}^2 = 251.03$
Sum of Products		$\sum_{i=1}^{10} y_i x_{i1} = 116.54$	

The system of normal equations is

$$10.00\hat{b}_0 + \ \ 49.30\hat{b}_1 = \ \ 25.00,$$
$$49.30\hat{b}_0 + 251.03\hat{b}_1 = 116.54,$$

and the solution is $\hat{b}_0 = 6.65$ and $\hat{b}_1 = -0.841$. Thus the estimate of our model, obtained by least squares, is

$$\hat{y}_i = 6.65 - 0.841x_{i1}.$$

If this were the true model we would conclude that no inflation is expected when the unemployment rate is $6.65/0.841 = 7.91\%$, and that with each 1% drop in the unemployment rate 0.841% inflation is expected.

Note that when $x_i = 0$ in equation (3), $E(y_i) = b_0$. This value is known as the *intercept*; it is the expected value of y_i in the model when x_i is zero. The model is correspondingly called an *intercept model*. Conversely, a *no-intercept model* is one in which there is no intercept; i.e. in which $E(y_i) = 0$ when $x_i = 0$. Such a model, analogous to (3), would be $E(y_i) = b_1 x_i$.

In the intercept model, as has already been indicated, the intercept b_0 is estimated by using $x_{i0} = 1$ for $i = 1, 2, \ldots, n$, i.e. by having an artificial first variable that is unity for each observation. In the no-intercept model this variable is not needed and does not enter into the estimation procedure; the only variables used are those actually observed.

The relationship envisaged in the previous equations is linear in the x's, and the analysis we are to describe, referred to so far as regression analysis, is more precisely called *linear regression*, or (for more than one x-variable) *multiple linear regression* analysis. Sometimes the underlying relationship envisaged between variables may not be a linear one, although the assumption of linearity is common and is the only relationship considered here. However, non-linear functions that are linear in the parameters to be estimated (the b's) can be studied by linear regression analysis. Thus $E(y) = b_1 x_1^p + b_2 x_2^q$ cannot be analyzed by linear regression if p and q are unknown parameters to be estimated; but if they are known then $E(y)$ is a linear function of the variables x_1^p and x_2^q and hence b_1 and b_2 can be estimated. Likewise the equation $E(y) = b_1 x + b_2 x^2 + b_3 x^3$ is not linear in x, but it is linear in the three variables x, x^2 and x^3. In other cases non-linear functions can sometimes be transformed into linear functions of transformed variables. For example, Meltzer (1963) specifies an aggregate demand for money equation as $M = ar^\alpha W^\beta$, where M is money demand, r is an interest rate, W is a measure of wealth of the nation and a, α and β are parameters to be estimated. By taking logarithms he obtains the equation

$$\log M = \log a + \alpha \log r + \beta \log W$$

which is linear in the logarithms of the original variables, and hence is in a form suited to linear regression analysis.

2. MULTIPLE LINEAR REGRESSION: THE CASE OF k x-VARIABLES

The procedure for estimating the coefficients of more than two x-variables, say k of them, is a direct extension of the preceding section.

a. No-intercept model

The equations for this model are

$$y_i = b_1 x_{i1} + b_2 x_{i2} + \cdots + b_k x_{ik} + e_i$$

for $i = 1, 2, \ldots, n$ observations. As a vector equation they can be written as before,

$$y = Xb + e, \tag{6}$$

where $\qquad y' = [y_1 \quad y_2 \quad \cdots \quad y_n], \qquad b' = [b_1 \quad b_2 \quad \cdots \quad b_k],$

$$e' = [e_1 \quad e_2 \quad \cdots \quad e_n]$$

and
$$X = \begin{bmatrix} x_{11} & x_{12} & \cdots & x_{1k} \\ x_{21} & x_{22} & \cdots & x_{2k} \\ \cdot & & & \\ \cdot & & & \\ \cdot & & & \\ x_{n1} & x_{n2} & \cdots & x_{nk} \end{bmatrix}.$$

We wish, as before, to find least squares estimators of the elements of b. If they are represented by $\hat{b}' = [\hat{b}_1 \quad \hat{b}_2 \quad \cdots \quad \hat{b}_k]$, the corresponding estimate of the vector y is

$$\hat{y} = X\hat{b}$$

and the vector of estimated errors is

$$\hat{e} = y - \hat{y} = y - X\hat{b}.$$

The sum of squares of these estimated errors is

$$\text{SSE} = \sum_{i=1}^{n} \hat{e}^2 = \hat{e}'\hat{e} = (y - X\hat{b})'(y - X\hat{b}) = (y' - \hat{b}'X')(y - X\hat{b})$$

$$= y'y - 2\hat{b}'X'y + \hat{b}'X'X\hat{b}$$

after utilizing $y'X\hat{b} = \hat{b}'X'y$ (a scalar).

The vector $\hat{b}$ is derived by minimizing $\hat{e}'\hat{e}$; i.e., by minimizing $y'y - 2\hat{b}'X'y + \hat{b}'X'X\hat{b}$ through equating to zero the partial derivatives with respect to the elements of $\hat{b}$. Using a vector of differential operators,[1] this gives

$$\frac{\partial(\hat{e}'\hat{e})}{\partial \hat{b}} = -2X'y + 2X'X\hat{b} = 0.$$

These equations, equivalent to

$$X'X\hat{b} = X'y,$$

are the normal equations for multivariate regression. Provided $(X'X)^{-1}$ exists they have the solution

$$\hat{b} = (X'X)^{-1}X'y, \tag{7}$$

in which $\hat{b}$ is the least squares estimator of b.

---

[1] See Section 13.5.

Now $X'X$ is square, and from the nature of X it is the (symmetric) matrix of sums of squares and cross-products of the n observations on the x-variables:

$$X'X = \begin{bmatrix} \Sigma\, x_{i1}^2 & \Sigma\, x_{i1}x_{i2} & \cdots & \Sigma\, x_{i1}x_{ik} \\ \Sigma\, x_{i1}x_{i2} & \Sigma\, x_{i2}^2 & \cdots & \Sigma\, x_{i2}x_{ik} \\ \cdot & \cdot & & \cdot \\ \cdot & \cdot & & \cdot \\ \cdot & \cdot & & \cdot \\ \Sigma\, x_{i1}x_{ik} & \Sigma\, x_{i2}x_{ik} & \cdots & \Sigma\, x_{ik}^2 \end{bmatrix}.$$

$X'y$ is likewise the vector of sums of products of the observations on the x-variables and the y-variables:

$$X'y = \begin{bmatrix} \Sigma\, x_{i1}y_i \\ \Sigma\, x_{i2}y_i \\ \cdot \\ \cdot \\ \cdot \\ \Sigma\, x_{ik}y_i \end{bmatrix}.$$

Thus the vector of estimators $\hat{b} = (X'X)^{-1}X'y$ is the vector of sums of products of the x's and y's pre-multiplied by the inverse of the matrix of sums of squares and products of the x's.

Example. Suppose we have the following five sets of observations from the no-intercept model $y_i = b_1x_{i1} + b_2x_{i2} + b_3x_{i3} + e_i$:

i	y_i	x_{i1}	x_{i2}	x_{i3}
1	24	1	1	3
2	26	1	0	1
3	20	2	1	1
4	20	3	2	1
5	23	1	2	0

The b's are estimated as

$$\hat{b} = \begin{bmatrix} \hat{b}_1 \\ \hat{b}_2 \\ \hat{b}_3 \end{bmatrix} = (X'X)^{-1}X'y$$

where

$$X = \begin{bmatrix} 1 & 1 & 3 \\ 1 & 0 & 1 \\ 2 & 1 & 1 \\ 3 & 2 & 1 \\ 1 & 2 & 0 \end{bmatrix} \quad \text{and} \quad y = \begin{bmatrix} 24 \\ 26 \\ 20 \\ 20 \\ 23 \end{bmatrix}.$$

Hence $(X'X)^{-1} = \begin{bmatrix} 16 & 11 & 9 \\ 11 & 10 & 6 \\ 9 & 6 & 12 \end{bmatrix}^{-1} = \frac{1}{90} \begin{bmatrix} 28 & -26 & -8 \\ -26 & 37 & 1 \\ -8 & 1 & 13 \end{bmatrix}$

and $X'y = \begin{bmatrix} 173 \\ 130 \\ 138 \end{bmatrix},$

so $\hat{b} = \frac{1}{90} \begin{bmatrix} 28 & -26 & -8 \\ -26 & 37 & 1 \\ -8 & 1 & 13 \end{bmatrix} \begin{bmatrix} 173 \\ 130 \\ 138 \end{bmatrix} = \begin{bmatrix} 4 \\ 5 \\ 6 \end{bmatrix}.$

Note that $(X'X)^{-1}$ must exist for $\hat{b} = (X'X)^{-1}X'y$ to exist; and $(X'X)^{-1}$ *will* exist provided X has full column rank (i.e. provided the k vectors of x-variables are linearly independent; see Section 6.12b).

b. Intercept model

In (1) we wrote the model as $y_i = b_0 + b_1 x_i + e_i$, where b_0 is the intercept. We continue to use the symbol b_0 for the intercept to distinguish it from $b_1, b_2, \ldots, b_k$, the coefficients of the x's. Then the intercept model for multiple regression is

$$y_i = b_0 + b_1 x_{i1} + \cdots + b_k x_{ik} + e_i.$$

If, as before, we introduce an artificial variable $x_{i0} = 1$ for all $i = 1, 2, \ldots, n$, the model can be written as

$$y_i = b_0 x_{i0} + b_1 x_{i1} + \cdots + b_k x_{ik} + e_i. \tag{8}$$

In vector notation this is

$$y = X^* b^* + e \tag{9}$$

where y and e are as before, where X^* is X augmented by a first column of 1's and where b^* is b with b_0 added as a first element. We now define the symbol **1** to be a vector (in this case of order n) of elements that are all unity; i.e.

$$\mathbf{1}' = [1 \quad 1 \quad 1 \quad \cdots \quad 1].$$

(This is the only symbol for which bold-face type will be used; it distinguishes **1** from 1.) Then X^* and b^* may be written

$$X^* = [\mathbf{1} \quad X] \quad \text{and} \quad b^{*'} = [b_0 \quad b']. \tag{10}$$

Now, apart from these changes in formulation the model in (9) is exactly the same form as that of (6) and so, as in (7), the vector of least squares estimates is obtained as before:[1]

$$\hat{b}^* = (X^{*\prime}X^*)^{-1}X^{*\prime}y.$$

Example. Suppose the parameters of the model

$$y_i = b_0 + b_1 x_{i1} + b_2 x_{i2} + e_i$$

are to be estimated from the following six sets of observations:

i	y_i	x_{i1}	x_{i2}
1	10	1	0
2	17	4	6
3	13	2	4
4	14	2	3
5	12	1	1
6	15	3	5

Then
$$X^* = \begin{bmatrix} 1 & 1 & 0 \\ 1 & 4 & 6 \\ 1 & 2 & 4 \\ 1 & 2 & 3 \\ 1 & 1 & 1 \\ 1 & 3 & 5 \end{bmatrix} \quad \text{and} \quad \hat{b}^* = \begin{bmatrix} \hat{b}_0 \\ \hat{b}_1 \\ \hat{b}_2 \end{bmatrix}$$

and it will be found that

$$\hat{b}^* = (X^{*\prime}X^*)^{-1}X^{*\prime}y$$

$$= \frac{1}{112}\begin{bmatrix} 129 & -105 & 37 \\ -105 & 161 & -77 \\ 37 & -77 & 41 \end{bmatrix}\begin{bmatrix} 81 \\ 189 \\ 283 \end{bmatrix} \qquad (11)$$

$$= \begin{bmatrix} 1075/112 \\ 133/112 \\ 47/112 \end{bmatrix}. \qquad (12)$$

---

[1] In virtually all regression situations, the augmenting of X by a first column of 1's augments the column rank of X by 1, thus not affecting the existence or lack thereof of $(X'X)^{-1}$ and $\hat{b}$.

3.   PROPERTIES OF LEAST SQUARES ESTIMATORS

a.   Assumptions

We have seen that when the model is $y = Xb + e$, whether it represents the no-intercept model or the intercept model, the least squares estimator of b is $\hat{b} = (X'X)^{-1}X'y$. Properties of this estimator arise from assumptions made about the elements of e.

As already mentioned, it is frequently assumed that the elements of e are a random sample from a distribution having zero mean and variance σ^2, with the covariance between any two elements being zero. Using E to denote expectation we have $E(e_i) = 0$ and $E(e_i^2) = \sigma^2$ for all i and $E(e_i e_j) = 0$ for $i \neq j$. In terms of the vector e this means

$$E(e) = 0 \tag{13}$$

and using this, the variance-covariance matrix of the e's (see Section 3.6) is

$$\text{var}(e) = E[e - E(e)][e - E(e)]' = E(ee') = \sigma^2 I. \tag{14}$$

These assumptions about the elements of e enable us to find the expected value and variance of the estimator $\hat{b}$; an estimator of σ^2 can also be found. As we shall see in Section 4, in order to consider significance tests relating to $\hat{b}$, we need additional assumptions concerning the exact form of the probability distribution of the elements of e.

b.   Expected values and variances

The expected value of y is Xb. For, with $y = Xb + e$, the expected value of y is $E(y) = E(Xb + e) = Xb + E(e)$ and because $E(e) = 0$ from (13),

$$E(y) = Xb. \tag{15}$$

An estimator is *unbiased* if its expected value is equal to the parameter being estimated. The estimator $\hat{b}$ is unbiased; i.e. $E(\hat{b}) = b$. Since $\hat{b} = (X'X)^{-1}X'y$,

$$E(\hat{b}) = E[(X'X)^{-1}X'y] = (X'X)^{-1}X'E(y) = (X'X)^{-1}X'Xb = b, \tag{16}$$

utilizing (15) for $E(y)$.

We will show that the variance-covariance matrix of $\hat{b}$ is $(X'X)^{-1}\sigma^2$. By definition, the variance-covariance matrix of $\hat{b}$ is

$$\text{var}(\hat{b}) = E[\hat{b} - E(\hat{b})][\hat{b} - E(\hat{b})]' = E(\hat{b} - b)(\hat{b} - b)',$$

from (16). Now because $y = Xb + e$,

$$\hat{b} - b = (X'X)^{-1}X'(Xb + e) - b = (X'X)^{-1}X'e.$$

Therefore
$$\begin{aligned}
\operatorname{var}(\hat{b}) &= E[(X'X)^{-1}X'ee'X(X'X)^{-1}] \\
&= (X'X)^{-1}X'E(ee')X(X'X)^{-1} \\
&= (X'X)^{-1}X'\sigma^2 IX(X'X)^{-1} \qquad \text{from (14)} \\
&= (X'X)^{-1}\sigma^2. \qquad\qquad\qquad\qquad\qquad (17)
\end{aligned}$$

Thus the inverse matrix $(X'X)^{-1}$ used for obtaining $\hat{b}$ also determines the variances and covariances of the elements[1] of $\hat{b}$; the variances and covariances are required for tests of significance presented in Section 4.

c. Estimating the error variance

The sum of squared deviations of the observed y's from their estimated values has been referred to as the error sum of squares:

$$\operatorname{SSE} = \sum_{i=1}^{n}(y_i - \hat{y}_i)^2 = (y - \hat{y})'(y - \hat{y}) = \hat{e}'\hat{e}.$$

A convenient computing form for this is to replace $\hat{y}$ by $X\hat{b}$ and get

$$\begin{aligned}
\operatorname{SSE} &= (y - X\hat{b})'(y - X\hat{b}) \\
&= y'y - 2\hat{b}'X'y + \hat{b}'X'X\hat{b} \\
&= y'y - 2\hat{b}'X'y + \hat{b}'(X'y)
\end{aligned}$$

because $X'X\hat{b} = X'y$ (the normal equations). Hence

$$\operatorname{SSE} = y'y - \hat{b}'X'y. \qquad\qquad (18)$$

This result is useful because $y'y$ is the total sum of squares of the y-observations and $\hat{b}'X'y$ is the sum of products of elements of $\hat{b}$ with the right-hand sides of the normal equations $X'X\hat{b} = X'y$ from which $\hat{b}$ is derived.

Another useful form for SSE is obtained by replacing $\hat{b}$ in (18) by $(X'X)^{-1}X'y$; then

$$\operatorname{SSE} = y'[I - X(X'X)^{-1}X']y \qquad\qquad (19)$$

in which the matrix $[I - X(X'X)^{-1}X']$ is idempotent (see Section 7.3b), i.e.,

$$[I - X(X'X)^{-1}X']^2 = I - X(X'X)^{-1}X'. \qquad\qquad (20)$$

The value of the expression for SSE in (19) is that it leads to an unbiased estimator of σ^2, the variance associated with the random error terms of the model. Using $y = Xb + e$ in (19) gives

$$\operatorname{SSE} = (b'X' + e')[I - X(X'X)^{-1}X'](Xb + e),$$

---

[1] Least squares estimators are often called best linear unbiased estimators. They are linear in the y_i's from (7), unbiased from (16), and are best in the sense that they can be shown to have the smallest variance of any such estimator.

and because $X'[I - X(X'X)^{-1}X']$ is null (as is its transpose) this gives

$$\text{SSE} = e'[I - X(X'X)^{-1}X']e.$$

With $\text{var}(e) = \sigma^2 I$ it can be shown[1] from this that

$$E(\text{SSE}) = (n - k)\sigma^2.$$

Hence $\qquad\qquad\qquad \hat{\sigma}^2 = \dfrac{\text{SSE}}{n - k} \qquad\qquad\qquad (21)$

is an unbiased estimator of σ^2, where n is the number of observations and k is the number of x-variables.

d. The intercept model

With the single exception of (21), the above results apply to the model $y = Xb + e$ whether it represents a no-intercept model or an intercept model. However, in the case of the latter, certain results pertaining to the intercept can be stated explicitly rather than leaving them as an intrinsic part of $\hat{b}$. Thus, in writing the intercept model as $y = X^*b^* + e$ as in (9), X^* and b^* are defined in (10) in terms of X and b. Manipulation of the estimators $\hat{b}^* = (X^{*'}X^*)^{-1}X^{*'}y$ in terms of X and $\hat{b} = (X'X)^{-1}X'y$ leads to explicit results for $\hat{b}_0$, the intercept, and more importantly, to results for $\hat{b}$ that are independent of the intercept. These results are important because they are the means by which most computer programs handle regression analysis.

The results are simply stated here; for derivations see Exercise 11. First we define $\bar{y}$ as the mean of the y-observations and

$$w' = [\bar{x}_1 \quad \bar{x}_2 \quad \cdots \quad \bar{x}_k], \qquad\qquad (22)$$

the vector of means of the x-observations. Then it can be shown that

$$\hat{b} = (X'X - nww')^{-1}(X'y - n\bar{y}w), \qquad\qquad (23)$$

$$\text{var}(\hat{b}) = (X'X - nww')^{-1}\sigma^2, \qquad\qquad (24)$$

$$\text{cov}(\hat{b}, \bar{y}) = 0, \qquad\qquad (25)$$

$$\hat{b}_0 = \bar{y} - w'\hat{b}, \qquad\qquad (26)$$

$$\text{var}(\hat{b}_0) = [1/n + w'(X'X - nww')^{-1}w]\sigma^2, \qquad\qquad (27)$$

$$\text{cov}(\hat{b}_0, \hat{b}) = -(X'X - nww')^{-1}w\sigma^2, \qquad\qquad (28)$$

and $\qquad \text{SSE} = y'y - \hat{b}^{*'}X^{*'}y = y'y - n\bar{y}^2 - \hat{b}'(X'y - n\bar{y}w). \qquad (29)$

---

[1] Since e is a vector of random variables with $E(e) = 0$ and $E(ee') = \sigma^2 I$, Theorem 4.19 of Graybill gives $E(e'Ae) = r\sigma^2$ where $A^2 = A$ and the rank of A is r. (See Section 13.3.)

Expressions (23), (24) and (29) are particularly important. By the nature of the vector w defined in (22), nww' is the matrix whose elements are correction factors (for the means) for the sums of squares and products of the x's that are the elements of $X'X$. Hence $X'X - nww'$ is the matrix of corrected sums of squares and products of the x's.[1] Similarly, $X'y - n\bar{y}w$ is the vector of corrected sums of products of the x's with the y's. Thus (23) and (24) are exactly the same form as $\hat{b} = (X'X)^{-1}X'y$ and $\mathrm{var}(\hat{b}) = (X'X)^{-1}\sigma^2$ obtained in (7) and (17), only using corrected sums of squares and products instead of uncorrected sums. Similarly, (29) is the same form as SSE $= y'y - b'X'y$ in (18), only using corrected values. Thus it is possible to use the symbolic forms $\hat{b} = (X'X)^{-1}X'y$, $\mathrm{var}(\hat{b}) = (X'X)^{-1}\sigma^2$ and SSE $= y'y - \hat{b}'X'y$ in both the no-intercept model and the intercept model. In both cases they give estimators (and their covariance matrix) of the regression coefficients; and also the error sum of squares, SSE. In the no-intercept case $X'X$ and $X'y$ represent uncorrected sums of squares and products, whereas in the intercept case they represent corrected sums.

In using (18) or (29) to estimate σ^2 for the intercept model we have to amend equation (21), which applies to a no-intercept model. In terms of such a model the intercept model is, by the addition of the artificial variable x_{i0} in (8), a no-intercept model with $k + 1$ variables. Hence for the intercept model, using (18) or (29) for SSE,

$$\hat{\sigma}^2 = \frac{\text{SSE}}{n - k - 1}. \tag{30}$$

Example (*continued*). In (12) we found $\hat{b}^*$ directly as $(X^{*'}X^*)^{-1}X^{*'}y$. We now find its elements $\hat{b}_0$ and $\hat{b}$ by the formulae just discussed. From the data we have

$$\begin{aligned}
\Sigma\, y_i &= 81, & \Sigma\, x_{i1} &= 13, & \Sigma\, x_{i2} &= 19, \\
\Sigma\, y_i^2 &= 1123, & \Sigma\, x_{i1}^2 &= 35, & \Sigma\, x_{i2}^2 &= 87, \\
\Sigma\, y_i x_{i1} &= 189, & \Sigma\, x_{i1}x_{i2} &= 54, & \text{and} \quad \Sigma\, y_i x_{i2} &= 283.
\end{aligned}$$

Hence
$$w' = [\bar{x}_1 \quad \bar{x}_2] = [13/6 \quad 19/6]$$

$$(X'X - nww')^{-1} = \begin{bmatrix} 35 - 13^2/6 & 54 - 13(19)/6 \\ 54 - 13(19)/6 & 87 - 19^2/6 \end{bmatrix}^{-1}$$

$$= \begin{bmatrix} 41/6 & 77/6 \\ 77/6 & 161/6 \end{bmatrix}^{-1} = \begin{bmatrix} 161/112 & -77/112 \\ -77/112 & 41/112 \end{bmatrix} \tag{31}$$

and
$$X'y - n\bar{y}w = \begin{bmatrix} 189 - 13(81)/6 \\ 283 - 19(81)/6 \end{bmatrix} = \begin{bmatrix} 81/6 \\ 159/6 \end{bmatrix}. \tag{32}$$

---

[1] I.e., the matrix $X'X - nww'$ would be the same as $X'X$ if the observations contained in X were computed as deviations from their means.

Therefore, from (23), (31) and (32)

$$\hat{b} = \begin{bmatrix} 161/112 & -77/112 \\ -77/112 & 41/112 \end{bmatrix} \begin{bmatrix} 81/6 \\ 159/6 \end{bmatrix} = \begin{bmatrix} 133/112 \\ 47/112 \end{bmatrix}; \tag{33}$$

and from (26) and (33),

$$\hat{b}_0 = 81/6 - [13/6 \quad 19/6] \begin{bmatrix} 133/112 \\ 47/112 \end{bmatrix} = 1{,}075/112.$$

These values correspond to those of $\hat{b}^*$ derived in (12).

Similarly, from (27) and (31)

$$\text{var}(\hat{b}_0) = \left\{ \tfrac{1}{6} + [13/6 \quad 19/6]\tfrac{1}{112} \begin{bmatrix} 161 & -77 \\ -77 & 41 \end{bmatrix} \begin{bmatrix} 13/6 \\ 19/6 \end{bmatrix} \right\} \sigma^2 = \tfrac{129}{112}\sigma^2$$

and from (28) and (31)

$$\text{cov}(\hat{b}_0, \hat{b}) = \frac{-1}{112} \begin{bmatrix} 161 & -77 \\ -77 & 41 \end{bmatrix} \begin{bmatrix} 13/6 \\ 19/6 \end{bmatrix} \sigma^2 = \begin{bmatrix} -105/112 \\ 37/112 \end{bmatrix} \sigma^2$$

with (24) and (31) giving

$$\text{var}(\hat{b}) = \tfrac{1}{112} \begin{bmatrix} 161 & -77 \\ -77 & 41 \end{bmatrix} \sigma^2.$$

These results correspond identically to $\text{var}(\hat{b}^*) = (X^{*\prime}X^*)^{-1}\sigma^2$, similar to (17), using $(X^{*\prime}X^*)^{-1}$ shown in (11).

Finally, from (29), using (33) and (32)

$$\text{SSE} = 1{,}123 - (81)^2/6 - [133/112 \quad 47/112] \begin{bmatrix} 81/6 \\ 159/6 \end{bmatrix}$$

$$= 177/6 - 3{,}041/112$$

$$= 263/112 \tag{34}$$

and hence from (30)

$$\hat{\sigma}^2 = (263/112)/3 = 263/336. \tag{35}$$

4.   TESTS OF SIGNIFICANCE

Up to now the only assumptions made about the error terms of the model (the elements of e) have been that they are random variables having some distribution in which $E(e) = 0$ and $E(ee') = \sigma^2 I$; no assumption has been made about the exact form of the distribution. If it is plausible to assume that the distribution is normal, i.e. that the error terms are normally distributed with zero means and variance-covariance matrix $\sigma^2 I$, then tests of significance familiarly associated with regression analysis can be performed.

a. Adequacy of the model

Equations (6), (13) and (14) constitute the model as so far described. Having obtained estimators $\hat{b}$ from (7) or (23) and (26) the question might be asked: "How adequate is the model for the data?" Since the estimators $\hat{b}$ lead to estimated (or predicted) values of the dependent variable, namely $\hat{y} = X\hat{b}$, this question can be answered in terms of a statistic that measures the correlation between the observed y-values and the predicted $\hat{y}$-values. This correlation is sometimes called the multiple correlation, denoted by R. In practice, more frequent use is made of the square of this correlation, R^2, referred to as the coefficient of determination; it measures the fraction of the total variability in the observed y-values that has been accounted for by fitting the model.

Since R is a correlation, the coefficient of determination, R^2, always has a value between 0 and 1. The closer it is to 1 the better the model explains the data; and a test based on the value of R^2 and the properties of the F-distribution[1] can be used to test the significance of the model. Expressions for R^2 and the F-statistic are shown in Table 2.

TABLE 2. R^2 AND F FOR REGRESSION MODELS WITH n SETS OF OBSERVATIONS AND k INDEPENDENT x-VARIABLES

Statistic	No-intercept Model	Intercept Model
Multiple correlation (R)	$R = \dfrac{\Sigma\, y_i \hat{y}_i}{\sqrt{\Sigma\, y_i^2 (\Sigma\, \hat{y}_i^2)}}$	$R = \dfrac{\Sigma\, (y_i - \bar{y})(\hat{y}_i - \bar{\hat{y}})}{\sqrt{\Sigma\, (y_i - \bar{y})^2 [\Sigma\, (\hat{y}_i - \bar{\hat{y}})^2]}}$
Error sum of squares, SSE	$\text{SSE} = y'y - \hat{b}'X'y$ [Eq. (18)]	$\text{SSE} = y'y - n\bar{y}^2 - \hat{b}'(X'y - n\bar{y}w)$ [Eq. (29)]
Coefficient of determination	$R^2 = 1 - \dfrac{\text{SSE}}{y'y}$	$R^2 = 1 - \dfrac{\text{SSE}}{y'y - n\bar{y}^2}$
F-statistic	$F = \dfrac{(n-k)R^2}{k(1-R^2)}$	$F = \dfrac{(n-k-1)R^2}{k(1-R^2)}$
Degrees of freedom for F	$(k, n-k)$	$(k, n-k-1)$

The expression for R^2 in the no-intercept model is derived by writing R^2 as $(y'\hat{y})^2/(y'y)(\hat{y}'\hat{y})$, replacing $\hat{y}$ by $X\hat{b} = X(X'X)^{-1}X'y$, and simplifying (see Exercise 8). Derivation of R^2 in the intercept model is contained in Section 9.10 of Searle (1966).

---

[1] See, for example, Goldberger (1964, p. 109).

The F-values shown in Table 2 have the F-distribution with degrees of freedom (D.F.) as shown and provide tests of the adequacy of the models concerned.

b. Analysis of variance

The calculation of the F-tests discussed above is usually summarized in an analysis of variance table.

The sum of squares of the observed y's is $y'y$, and the corrected sum of squares is $y'y - n\bar{y}^2$. Using these as the total sums of squares (SST) in the two models respectively we write

$$\text{SST} = y'y \quad \text{in the no-intercept model}$$
$$\text{and} \qquad \text{SST} = y'y - n\bar{y}^2 \quad \text{in the intercept model.} \tag{36}$$

Now $\text{SSE} = \Sigma\,(y_i - \hat{y}_i)^2$ is the sum of squares of the deviations of the observed y's from their predicted values. Therefore the difference between SST and SSE, namely

$$\text{SSR} = \text{SST} - \text{SSE},$$

represents that portion of SST which is attributable to having fitted the regression. Using (36) for SST and the values of SSE shown in Table 2 gives

$$\text{SSR} = \hat{b}'X'y \quad \text{in the no-intercept model}$$
$$\text{and} \qquad \text{SSR} = \hat{b}'(X'y - n\bar{y}w) \quad \text{in the intercept model.} \tag{37}$$

SSR is usually called the sum of squares due to regression. In the no-intercept model it refers to the whole model; in the intercept model it refers to regression on the x-variables, over and above fitting the mean.

Splitting SST into two portions in this fashion, SSR and SSE, is the underlying process of the analysis of variance technique; it is usually summarized in an analysis of variance table. Two tables, that for the no-intercept model and that for the intercept model, are shown in Table 3.

Although development of these tables started from the definition of SSE, from which SSR was defined as SST − SSE, computing procedures are

TABLE 3. ANALYSES OF VARIANCE FOR FITTING k x-VARIABLES

Source of Variation	No-intercept Model		Intercept Model[1]	
	D.F.	Sum of Squares	D.F.	Sum of Squares
Regression on k x's	k	$\text{SSR} = \hat{b}'X'y$	k	$\text{SSR} = \hat{b}'(X'y - n\bar{y}w)$
Error	$n - k$	$\text{SSE} = \text{SST} - \text{SSR}$	$n - k - 1$	$\text{SSE} = \text{SST} - \text{SSR}$
Total	n	$\text{SST} = y'y$	$n - 1$	$\text{SST} = y'y - n\bar{y}^2$

[1] See Section 4e for a detailed development.

more easily based on SSR, deriving SSE as SST − SSR. This is because SSR, as shown in (37), is easy to calculate. Recalling the normal equations on which the vector of estimators $\hat{b}$ from (7) and (23) is based, we have

$$X'X\hat{b}' = X'y \qquad \text{in the no-intercept model}$$

and $\qquad (X'X - nww')\hat{b}' = X'y - nyw \qquad \text{in the intercept model.}$

(38)

Thus in (37) we see that SSR is the sum of products of the elements of $\hat{b}$ with the right-hand sides of the equations from which $\hat{b}$ is derived. Hence either expression in (37) is easily calculated. For this reason Table 3 shows SSR as given in (37) and SSE as SST − SSR.

From Tables 2 and 3 one can derive R^2 as

$$R^2 = \frac{\text{SSR}}{\text{SST}} \tag{39}$$

in both models. Thus it is that R^2, the coefficient of determination, represents the fraction of the total sum of squares that has been accounted for by fitting the regression (in the intercept case, above and beyond fitting the mean). Also, since from (21) and (30),

$$\hat{\sigma}^2 = \frac{\text{SSE}}{n - k} \qquad \text{in the no-intercept model}$$

and $\qquad \hat{\sigma}^2 = \frac{\text{SSE}}{n - k - 1} \qquad \text{in the intercept model,}$

(40)

it can also be seen from Tables 2 and 3 that

$$F = \frac{\text{SSR}}{k\hat{\sigma}^2} \tag{41}$$

in both models.

The section of Table 3 pertaining to the no-intercept model is basic and is the point from which development of the section pertaining to the intercept model originates. This development is outlined in Section 4e. Since in many instances the intercept model is the one most frequently used, further discussion is confined to this model.

Example (*continued*). We continue the example started earlier, treating it as an intercept model with two ($k = 2$) x-variables. To calculate Table 3 we need

$$SSR = \hat{b}'(X'y - n\bar{y}w)$$

$$= [133/112 \quad 47/112] \begin{bmatrix} 81/6 \\ 159/6 \end{bmatrix}, \qquad \text{from (32) and (33)}$$

$$= 3{,}041/112$$

and
$$SST = y'y - n\bar{y}^2 = 1,123 - 81^2/6$$
$$= 177/6.$$

Hence
$$SSE = SST - SSR = 177/6 - 3,041/112$$
$$= 263/112, \quad \text{as obtained earlier in (34).}$$

Thus Table 3 for this example is as shown in Table 4. From this, (39),

TABLE 4. ANALYSIS OF VARIANCE OF EXAMPLE
(USING INTERCEPT MODEL)

Source of Variation	D.F.	Sum of Squares
Regression	2	SSR = 3,041/112
Error	3	SSE = 263/112
Total	5	SST = 177/6

(40) and (41) give

$$R^2 = \frac{3,041}{112} \bigg/ \frac{177}{6} = 0.92$$

$$\hat{\sigma}^2 = \frac{263}{112} \bigg/ (6 - 2 - 1) = \frac{263}{336} = 0.78$$

and
$$F = \frac{3,041}{112} \bigg/ \frac{2(263)}{336} = \frac{3(3,041)}{2(263)} = 17.3.$$

The F-statistic has the F-distribution with 2 and 3 D.F., the tabulated value of which at the 5% level is 9.55. Hence, since $17.3 > 9.55$, we conclude that at the 5% level of significance the hypothesis that $b_1 = b_2 = 0$ must be rejected.

c. Sub-sets of x-variables

Suppose that a sub-set of p of our x-variables is of secondary interest compared to the other $k - p$. We might then wish to test if those p variables are contributing significantly to the regression, over and above the fitting of the other $k - p$ variables. To do this we fit two models: one, for all k x's, calculating the regression sum of squares which we shall call SSR_k; and another, for just the $k - p$ x-variables, for which we calculate SSR_{k-p}. Then to test whether the p variables from among the k variables are contributing significantly to the regression, we calculate

$$F = \frac{(n - k - 1)(SSR_k - SSR_{k-p})}{p(SST - SSR_k)}. \tag{42}$$

This has the F-distribution with p and $n - k - 1$ degrees of freedom. The calculation of F can be summarized in an analysis of variance table (Table 5).

Calculation of the three terms required for this analysis is as follows. (i) SST $= y'y - n\bar{y}^2$, and (ii) $\text{SSR}_k = \hat{b}'X'y$ in the usual way. The third term, SSR_{k-p}, is derived implicitly, through a single expression for the difference (iii) $\text{SSR}_k - \text{SSR}_{k-p}$ used in the second line of Table 5. Consider the

TABLE 5. ANALYSIS OF VARIANCE FOR INTERCEPT MODEL:
TESTING p x-VARIABLES FROM AMONG k x-VARIABLES

Source of Variation	D.F.	Sum of Squares
Regression on $k - p$ variables	$k - p$	SSR_{k-p}
Regression on p variables over and above that on the $k - p$ variables	p	$\text{SSR}_k - \text{SSR}_{k-p}$
Error	$n - k - 1$	$\text{SST} - \text{SSR}_k$
Total	$n - 1$	SST

p x-variables whose significance is being tested. Suppose the b's corresponding to these are the last p b's in b. Then partition $\hat{b}$ as

$$\hat{b}' = [\hat{b}'_{k-p} \quad \hat{b}'_p],$$

where $\hat{b}_p$ represents the estimators under consideration. Also, partition the inverse $(X'X)^{-1}$ conformably with this partitioning of $\hat{b}$. To do this we will—for notational simplicity—write T for $(X'X)^{-1}$ and so, in partitioned form, we have

$$(X'X)^{-1} = T = \begin{bmatrix} T_{rr} & T_{rp} \\ T_{pr} & T_{pp} \end{bmatrix}$$

where $r = k - p$ and the subscripts on the sub-matrices of T denote their order; also, $T_{pr} = (T_{rp})'$. Then, for Table 5 we have the result[1]

$$\text{SSR}_k - \text{SSR}_{k-p} = \hat{b}'_p T_{pp}^{-1} \hat{b}_p. \tag{43}$$

This expression is readily computed: we already have $\hat{b}_p$, as part of $\hat{b}$; and from $(X'X)^{-1}$ we get T_{pp} and invert it.[2] This is considerably less effort than obtaining SSR_{k-p} directly, for that requires inverting a matrix of order $k - p$ whereas (43) demands inverting only a $p \times p$ matrix. When k is much larger than p, as is usually the case, the advantage is clear, and especially

---

[1] Proven in Searle (1966, p. 246), who uses $k + m$ and k respectively where we have used k and $k - p$ here.

[2] It is this use of part of the inverse of $X'X$ that gives the calculation its name of "invert part of the inverse" rule.

so when one wishes to test several different sub-sets of p x-variables. Furthermore, although described here in terms of the p x-variables being the last p in the sequence $x_1, x_2, \ldots, x_k$, it does, in fact, apply to any p x-variables: take the $\hat{b}$'s corresponding to the p x-variables and call them the vector $\hat{b}_p$. In $(X'X)^{-1}$ take the $p \times p$ matrix corresponding to the same p x-variables; call it T_{pp} and invert it, T_{pp}^{-1}. Then for this $\hat{b}_p$ and this T_{pp}, (43) applies.

Example (*continued*).

Suppose in our example, which has two x-variables, we wish to test the significance of fitting the first x-variable over and above the fitting of the second. Then $k = 2$, and $p = 1$ refers to the first variable. SSR_k is the regression sum of squares for fitting both variables and its value is the SSR shown in Table 4; i.e. $SSR_k = 3{,}041/112$.

To obtain $SSR_k - SSR_{k-p}$ by (43) we need $\hat{b}_p$, the estimated regression coefficient for the first variable; from (33) this is $\hat{b}_p = 133/112$. And T_{pp} is the leading element in $(X'X)^{-1}$ where $X'X$ is here a matrix of corrected sums of squares and products. Hence from (31) $T_{pp} = 161/112$ and so in (43)

$$SSR_k - SSR_{k-p} = \frac{133}{112}\left(\frac{161}{112}\right)^{-1}\left(\frac{133}{112}\right) = \frac{133^2}{112(161)}$$

Therefore with $SST - SSR_k = 263/112$ as in Table 4, and $n = 6$, substitution in (42) gives

$$F_{1,3} = \frac{(6 - 2 - 1)133^2}{(1)112(161)}\frac{112}{263}$$

which simplifies to

$$F_{1,3} = \frac{3(133^2)}{161(263)} = 1.25. \tag{44}$$

Since the tabulated value of F with 1 and 3 degrees of freedom at the 5% level of significance is 10.13, the hypothesis that the first variable is not contributing significantly to the regression over and above the second cannot be rejected.

From the above, SSR_{k-p} can be calculated as

$$SSR_{k-p} = SSR_k - \frac{133^2}{112(161)} = \frac{3041}{112} - \frac{133^2}{112(161)}$$

which simplifies to

$$SSR_{k-p} = \frac{159^2}{966}$$

This, it can be shown, is identical to the regression sum of squares for fitting just the $k - p$ variables (namely the second variable). The (intercept) model for this is $y_i = b_0 + b_2 x_{i2} + e_i$, and the equation for fitting it comes from the second part of (43). It reduces to

$$(\Sigma\, x_{i2}^2 - n\bar{x}_2^2)\tilde{b}_2 = \Sigma\, x_{i2}y_i - n\bar{x}_2\bar{y}$$

where $\tilde{b}_2$ is the estimate of b_2 in this case. This equation, from the basic data given earlier, is

$$(87 - 19^2/6)\tilde{b}_2 = 283 - 19(81)/6$$

i.e.
$$(161/6)\tilde{b}_2 = 159/6$$

and so
$$\tilde{b}_2 = 159/161.$$

Hence from (37), which in this case reduces to

$$\text{SSR} = \tilde{b}_2(\Sigma\, x_{i2}y_i - n\bar{x}_2\bar{y}),$$

we have the regression sum of squares which shall be identified as SSR_{k-p}:

$$\text{SSR}_{k-p} = (159/161)159/6 = 159^2/966$$

as obtained earlier.

d. Tests of individual variables

The significance tests based on the F-statistic from the analysis of variance table (see Section 4a above) test the null hypothesis that all the co-efficients $b_1, b_2, \ldots, b_k$ are zero; in this sense the F-test is a test of the entire model's ability to explain the variation in the dependent variable. In contrast to this, attention can also be focused on just one of the k independent x-variables to test whether, *after* fitting the other $k - 1$ variables, the coefficient of a particular x-variable is zero.

This test is a special case of the previous section, namely when $p = 1$. The F-statistic of (42) then becomes

$$F = \frac{(n - k - 1)(\text{SSR}_k - \text{SSR}_{k-1})}{\text{SST} - \text{SSR}_k} \tag{45}$$

with 1 and $n - k - 1$ degrees of freedom.

Suppose that it is the ith x-variable whose contribution to the regression is being considered. Denote the corresponding value of (45) by F_i. Then it can be shown that

$$F_i = t_i^2 \tag{46}$$

where
$$t_i = \frac{\hat{b}_i}{\sqrt{\widehat{\text{var}}(\hat{b}_i)}}. \tag{47}$$

In this expression $\hat{b}_i$ is the estimated coefficient corresponding to the ith x-variable and obtained from $\hat{b}$, and $\widehat{\text{var}}(\hat{b}_i)$ is the estimated variance of $\hat{b}_i$.

Thus, if d_i is the ith diagonal element of $(X'X - nww')^{-1}$, then $\mathrm{var}(\hat{b}_i) = d_i\sigma^2$ from (24), and with $\hat{\sigma}^2$ as given in (30), $\widehat{\mathrm{var}}(\hat{b}_i) = d_i\hat{\sigma}^2$.

The t-value shown in (47) has the t-distribution with $n - k - 1$ degrees of freedom[1] and provides a test of significance of fitting the ith variable after fitting the other $k - 1$ variables. Because of (46) it is equivalent to the F-test based on (45).

Example (*continued*). As in (32), (33) and (34),

$$\hat{b} = \begin{bmatrix} 133/112 \\ 47/112 \end{bmatrix} \quad \text{and} \quad (X'X - nww')^{-1} = \tfrac{1}{112}\begin{bmatrix} 161 & -77 \\ -77 & 41 \end{bmatrix};$$

and SSE $= 263/112$ with $n - k - 1 = 3$. Hence

$$t_1 = \frac{\hat{b}_1}{\sqrt{\widehat{\mathrm{var}}(\hat{b}_1)}} = \frac{\hat{b}_1}{\sqrt{d_1\hat{\sigma}^2}} = \frac{\hat{b}_1}{\sqrt{d_1(\mathrm{SSE})/(n - k - 1)}}$$

$$= \frac{133/112}{\sqrt{(161/112)(263/336)}} = \frac{133\sqrt{3}}{\sqrt{161(263)}} = 1.12.$$

Since the tabulated value of t at the 5% level for $n - k - 1 = 3$ degrees of freedom is 3.182, we conclude that at the 5% level of significance the null hypothesis that $b_1 = 0$ cannot be rejected. Note that $t^2 = (1.12)^2 = 1.25 = F$ of equation (44), thus demonstrating (46); and the conclusion based on t is identical to that based on F.

e. Analysis of variance for the intercept model

The intercept model with k x-variables is equivalent, for analysis procedures, to a no-intercept model with $k + 1$ variables. Therefore, as in (18) the appropriate error sum of squares is SSE $= y'y - \hat{b}*'X*'y$, as shown in (29). The corresponding analysis of variance, similar to the first section of Table 3, is therefore as shown in Table 6.

TABLE 6.   ANALYSIS OF VARIANCE

Source of Variation	D.F.	Sums of Squares
Regression on k x-variables and an intercept	$k + 1$	$\hat{b}*'X*'y$
Error	$n - k - 1$	$y'y - \hat{b}*'X*'y$
Total	n	$y'y$

---

[1] See, for example, Goldberger (1964, p. 108).

The first line of Table 6 is the regression sum of squares attributable to fitting both the k x-variables and an intercept. Consider, however, fitting a model that has just an intercept: $y_i = b_0 + e_i$. It is equivalent to fitting just the mean, and the sum of squares due thereto is $n\bar{y}^2$. Subtracting this from the first line of Table 6, that line can be written as two lines:

$$\begin{array}{lcl} \text{Fitting the mean (intercept)} & 1 & n\bar{y}^2 \\ \text{Regression due to } k \text{ } x\text{-variables} & k & \hat{b}^{*\prime}X^{*\prime}y - n\bar{y}^2. \end{array} \qquad (48)$$

They total $\hat{b}^{*\prime}X^{*\prime}y$ with $k + 1$ degrees of freedom. Now (29) shows that

$$\hat{b}^{*\prime}X^{*\prime}y - n\bar{y}^2 = \hat{b}'(X'y - n\bar{y}w). \qquad (49)$$

Hence, using (48) and (49), and (29) again for SSE, Table 6 becomes Table 7. Removing the first line of this table by subtracting it from the last line gives the analysis of variance for the intercept model contained in Table 3.

TABLE 7. ANALYSIS OF VARIANCE

Source of Variation	D.F.	Sums of Squares
Mean	1	$n\bar{y}^2$
Regression on k x-variables	k	$\hat{b}'(X'y - n\bar{y}w)$
Error	$n - k - 1$	$\text{SSE} = y'y - n\bar{y}^2 - \hat{b}'(X'y - n\bar{y}w)$
Total	n	$y'y$

5. SUMMARY OF REGRESSION CALCULATIONS

The general expressions involved in estimating the linear regression on k x-variables are summarized and listed below in terms of suitable computing formulas.

Description of data:

     n = number of sets of observations.

     k = number of x-variables.

Estimation (Sections 2 and 3):

     $X'X = k \times k$ matrix of sums of squares and products of observed x's.

     $X'y = k \times 1$ vector of sums of products of observed x's and y's.

         Note: In the no-intercept model these sums are uncorrected; in the intercept model these sums are corrected for means.

$w = k \times 1$ vector of means of the x's.

$\bar{y} =$ mean of the y's.

$\hat{b} = (X'X)^{-1}X'y$: estimated regression coefficients.

$\text{SSE} = y'y - \hat{b}'X'y$: error sum of squares.

$$k^* = \begin{cases} k & \text{in the no-intercept model;} \\ k + 1 & \text{in the intercept model.} \end{cases}$$

$\hat{\sigma}^2 = \text{SSE}/(n - k^*)$: estimated error variance.

$\widehat{\text{var}}(\hat{b}) = (X'X)^{-1}\hat{\sigma}^2$: estimated variance-covariance matrix of $\hat{b}$.

Intercept model:

$\hat{b}_0 = \bar{y} - w'\hat{b}$: estimated intercept.

$\widehat{\text{var}}(\hat{b}_0) = [1/n + w'(X'X)^{-1}w]\hat{\sigma}^2$: estimated variance of $\hat{b}_0$.

$\widehat{\text{cov}}(\hat{b}_0, \hat{b}) = -(X'X)^{-1}w\hat{\sigma}^2$: estimated covariance of $\hat{b}_0$ with $\hat{b}$.

Tests of significance (Section 4):

$\text{SSR} = \hat{b}'X'y$: regression sum of squares.

$$\text{SST} = \begin{cases} y'y & \text{in the no-intercept model} \\ y'y - n\bar{y}^2 & \text{in the intercept model} \end{cases}: \text{total sum of squares.}$$

$R^2 = \text{SSR}/\text{SST}$: coefficient of determination.

$F_{k,n-k*} = \text{SSR}/k\hat{\sigma}^2$: F-value, for testing the model.

$d_i = i$th diagonal element of $(X'X)^{-1}$.

$t_i = \hat{b}_i/\hat{\sigma}\sqrt{d_i}$: t-value, on $n - k^*$ D.F., for testing significance of ith x-variable, after fitting all others.

$F_{p,n-k*} = (\text{SSR}_k - \text{SSR}_{k-p})/p\hat{\sigma}^2$: F-value, on p and $n - k^*$ D.F., for testing significance of p variables after fitting the other $k - p$.

We have now shown how the technique of regression analysis can be described as a series of matrix expressions. The calculations involved are easily programmed for a computer, and consequently programs for fitting regressions and deriving the associated statistics are available in most computing facilities. While the calculations become quite tedious when n and k are large, especially k, since the order of $X'X$ whose inverse is required, today's computers can handle values of k even as large as 100 in less than two minutes.

6. EXERCISES

1. Insofar as dividend payments are a component of personal income, an econometric model of the United States might well include an equation relating dividends to the level of economic activity. The most logical relationship is that dividends are a function of corporate profits; the most general linear relationship would be $D_t = b_0 + b_1 P_t$, where D_t and P_t are dividends and profits in year t. A study of dividend policies by Lintner (1956) indicates that there is a widespread acceptance of some specific ratio of dividends to earnings as reasonable. If this is true for individual companies, it should also be true in the aggregate.

   (a) What would this policy imply about the coefficient b_0 in the above equation?

   (b) The following data are taken from the 1968 *Economic Report of the President*:

Year	Corporate Profits after Taxes	Corporate Dividend Payments
1960	26.7	13.4
1961	27.2	13.8
1962	31.2	15.2
1963	33.1	16.5
1964	38.4	17.8
1965	45.2	19.8
1966	49.3	21.5
1967	47.2	22.8

   Estimate b_0 and b_1 from these data.

   (c) Test the implication stated in part (a). From your test does it appear that Lintner's observation is correct?

2. Solve equations (5) of the chapter.

3. Show that equations (5) are a special case of the normal equations $X'X\hat{b} = X'y$.

4. Show that the solutions to equations (5) are of the form $\hat{b} = (X'X)^{-1}X'y$.

5. In the example of Section 1 (involving 10 observations) multiply every y_i and every x_{i1} by 1.1 and recalculate the regression. Why is your value of $\hat{b}_1$ the same as in the example? And why does your value of $\hat{b}_0$ equal 7.315?

6. In the example of Section 1 add 5 to every y_i and 10 to every X_{i1} and recalculate the regression. Explain the relationship of your answers to those given in Section 1.

7. For the case of a single x-variable, explain the relationship between the estimated coefficients of the regression of $c_1y + k_1$ on $c_2x + k_2$ with those of the regression of y on x. Use Exercises 5 and 6 to illustrate your results.

8. Derive the expression for R^2 for the no-intercept model as given in Table 2.
9. Chow (1967) proposes the following model for computer rental fees:

$$r = b_0 t^{b_1} m^{b_2} a^{b_3} \quad \text{or} \quad \log_{10} r = \log_{10} b_0 + b_1 \log_{10} t + b_2 \log_{10} m + b_3 \log_{10} a,$$

where r = monthly rental fee for a computer,
       t = time for an addition cycle in the computer,
      m = memory size of computer,
      a = access time of computer.

(a) What signs would you expect b_1, b_2 and b_3 to carry?
(b) Suppose we have the following data[1] on five computers:

Computer	r (thousand $/mo.)	t (microseconds)	m [thousands of words (6 bit characters)]	a (microseconds)
1	6.5	690	4.8	10
2	8.3	3.5	16	1.75
3	0.875	675	4	8
4	22.5	12	131	0.75
5	47	0.8	262	0.75

Find $\hat{b}_0$, $\hat{b}_1$, $\hat{b}_2$ and $\hat{b}_3$.
(c) What is the coefficient of determination?
(d) Does the log model fit at the 5% level? Would you get a different R^2 if you calculated it in terms of the original model rather than the log form?
(e) Is b_2 significantly different from zero at the 5% level?
(f) What parameters would be affected if base e were used rather than base 10?
10. For X^* and w' of equations (10) and (22) respectively, show that $1'X^* = [n \quad w']$ if $1' = [1 \quad 1 \quad \cdots \quad 1]$.
11. Prove equations (23) through (29) by substituting from (10) in the equation below (10). *Hint:* You may want to use

$$\begin{bmatrix} n & p' \\ p & Q \end{bmatrix}^{-1} = \begin{bmatrix} 1/n & 0 \\ 0 & 0 \end{bmatrix} + \begin{bmatrix} n^{-2}p'Vp & -n^{-1}p'V \\ -n^{-1}Vp & V \end{bmatrix}$$

where $V = [Q - n^{-1}pp']^{-1}$ and n is a scalar.

REFERENCES

Chow, G. C. (1967). Technological change and the demand for computers. *The American Economic Review*, **57**, 1115–1130.
Goldberger, A. S. (1964). *Econometric Theory*. Wiley, New York.

---

[1] Data adapted from *Computer Characteristics Quarterly*, **VIII**, No. 2, Keydata and Adams Associates, Watertown, Mass.

Graybill, F. A. (1961). *An Introduction to Linear Statistical Models.* Vol. I, McGraw-Hill, New York.

Johnston, J. (1963). *Econometric Methods.* McGraw-Hill, New York.

Kendall, M. G., and A. Stuart (1958, 1961). *The Advanced Theory of Statistics.* Vols. I and II, Charles Griffin, London.

Lintner, J. (1956). Distribution of incomes of corporations among dividends, retained earnings and taxes. *The American Economic Review, Papers and Proceedings,* May, 97–113.

Meltzer, A. H. (1963). The demand for money: the evidence from the time series. *The Journal of Political Economy,* **71,** 219–246.

Mood, A. M., and F. A. Graybill (1963). *Introduction to the Theory of Statistics.* McGraw-Hill, New York.

Phillips, A. W. (1958). The relation between unemployment and the rate of change in money wage rates in the United Kingdom, 1862–1957. *Economica,* **25,** 283–299.

Searle, S. R. (1966). *Matrix Algebra for the Biological Sciences.* Wiley, New York.

CHAPTER 11

LINEAR MODELS

---

1. INTRODUCTION

Special kinds of x-variables known as dummy variables are of particular interest in regression analysis. Dummy variables are used in a regression model to indicate the presence or absence of various attributes; their values are always zero or unity. The technique of using them is known as regression on dummy variables or as the analysis of linear models. As we shall see, use of these variables causes the matrix $X'X$ to be singular and so the equation $X'X\hat{b} = X'y$ of Chapter 10 cannot be solved using $(X'X)^{-1}$. A convenient method of obtaining a solution is to use the ge..eralized inverse procedures discussed in Chapter 7. A simple illustration serves to introduce the general ideas.

Illustration. We use an abbreviated form of a model discussed by Henderson, Hind and Brown (1961) in reporting the effect on apple sales of different advertising strategies used in a point-of-purchase advertising campaign. In one city where six stores were involved in the campaign, three of them emphasized "uses of apples" in their advertising, two stores utilized "healthful qualities" and one store used no particular strategy. Sales data at the six stores are shown in Table 1.

For the entries in this table let y_{ij} denote the sales in the jth store using the ith advertising strategy, i taking on the values 1, 2 and 3 for uses, health and none respectively; and $j = 1, 2, \ldots, n_i$, where n_i is the number of observations in the ith strategy of advertising. The problem is to estimate the effect of advertising strategy on sales. To do this we assume that the observation y_{ij} is the sum of three parts:

$$y_{ij} = \mu + \alpha_i + e_{ij}, \tag{1}$$

TABLE 1.   APPLE SALES IN SIX STORES
ACCORDING TO ADVERTISING STRATEGY

	Uses	Health	None
	101	84	32
	105	88	
	94		
Totals	300	172	32

where μ represents the population mean of sales, α_i is the effect of the ith strategy on sales and e_{ij} is a random error term peculiar to the observation y_{ij}. As in regression it is assumed that the e_{ij}'s are independently distributed with zero mean, i.e., $E(e_{ij}) = 0$. Then $E(y_{ij}) = \mu + \alpha_i$. It is also assumed that each e_{ij} has the same variance, σ^2, so that the variance-covariance matrix of the vector of e-terms is $\sigma^2 I$. This model is clearly linear since it is based on the assumption that y_{ij} consists of the simple sum of its three component parts: μ, α_i and e_{ij}. Models of this type are often called 1-*way classification models* since there is one classification scale involved, namely, the type of advertising strategy used. We hereafter refer to the three strategies as three classes.

The problem is to estimate the μ and α_i terms and also σ^2, the variance of the error terms. Not all of the terms μ and α_i can be estimated satisfactorily; only certain linear functions of them can. This is not a matter for concern, though, because the number of linear functions that can be estimated is large and usually includes those of greatest interest; e.g. differences between the classes, such as $\alpha_1 - \alpha_2$. Sometimes, however, functions of interest cannot be estimated satisfactorily because of a paucity of data. On all occasions, though, a method is needed for ascertaining which functions can be estimated satisfactorily and which cannot. This is provided in what follows.

To develop the method of estimation we write down the six observations in terms of equation (1) of the model:

$$101 = y_{11} = \mu + \alpha_1 \qquad\qquad + e_{11}$$
$$105 = y_{12} = \mu + \alpha_1 \qquad\qquad + e_{12}$$
$$94 = y_{13} = \mu + \alpha_1 \qquad\qquad + e_{13}$$
$$84 = y_{21} = \mu \qquad + \alpha_2 \qquad + e_{21}$$
$$88 = y_{22} = \mu \qquad + \alpha_2 \qquad + e_{22}$$
$$32 = y_{31} = \mu \qquad\qquad + \alpha_3 + e_{31}$$

These equations are easily written in vector form as

$$
\begin{bmatrix} 101 \\ 105 \\ 94 \\ 84 \\ 88 \\ 32 \end{bmatrix} = \begin{bmatrix} y_{11} \\ y_{12} \\ y_{13} \\ y_{21} \\ y_{22} \\ y_{31} \end{bmatrix} = \begin{bmatrix} 1 & 1 & 0 & 0 \\ 1 & 1 & 0 & 0 \\ 1 & 1 & 0 & 0 \\ 1 & 0 & 1 & 0 \\ 1 & 0 & 1 & 0 \\ 1 & 0 & 0 & 1 \end{bmatrix} \begin{bmatrix} \mu \\ \alpha_1 \\ \alpha_2 \\ \alpha_3 \end{bmatrix} + \begin{bmatrix} e_{11} \\ e_{12} \\ e_{13} \\ e_{21} \\ e_{22} \\ e_{31} \end{bmatrix} \tag{2}
$$

or as
$$ y = Xb + e, \tag{3} $$

where y is the vector of observations, X is the matrix of 0's and 1's, $b' = [\mu \quad \alpha_1 \quad \alpha_2 \quad \alpha_3]$ is the vector of parameters to be estimated and e is the vector of error terms.

Note that equation (3) is exactly the same form as (6) in Section 10.2. In both cases the vectors y and e are of observations and errors, respectively, and in both cases b is a vector of parameters whose estimators are desired. Also, the properties of e are the same: $E(e) = 0$, so that $E(y) = Xb$ and $E(ee') = \sigma^2 I$. The only difference is in the form of X: in regression it is a matrix of observations on x-variables, whereas in (3) its elements are either 0 or 1, depending on the absence or presence of particular terms of the model in each y_{ij} observation. But, with respect to applying the principle of least squares for estimating the elements of b, there is no difference between equations (3) here and equation (6) of Section 10.2. Consequently the least squares procedure described in Chapter 10 gives $\hat{b}$ satisfying the normal equations

$$ X'X\hat{b} = X'y. \tag{4} $$

In ordinary regression analysis the solution $\hat{b}$ was obtained from this as

$$ \hat{b} = (X'X)^{-1}X'y, $$

$X'X$ being nonsingular and therefore having an inverse. Now, however, we shall see that $X'X$ has no inverse; nevertheless, the methods of generalized inverses can be used to advantage to obtain a solution.

2. THE NORMAL EQUATIONS AND THEIR SOLUTIONS

We seek a solution $\hat{b}$ to the normal equations in (4). But notice that for X, the matrix of 0's and 1's in (2), the matrix

$$X'X = \begin{bmatrix} 6 & 3 & 2 & 1 \\ 3 & 3 & 0 & 0 \\ 2 & 0 & 2 & 0 \\ 1 & 0 & 0 & 1 \end{bmatrix}$$

is singular (its first row equals the sum of its last three rows). This is generally the case; in analyzing a linear model of the general nature being discussed here, $X'X$ is almost invariably singular. Accordingly, its rank, r say, is less than its order, and $\hat{b}$ cannot be obtained as $(X'X)^{-1}X'y$. But using a generalized inverse of $X'X$ leads to a solution.

Let G be any generalized inverse of $X'X$, such that $X'XGX'X = X'X$. We assume here only that G satisfies this equation, whether it is calculated by the method of Section 7.3 or by some other method. In addition, we define $H = GX'X$. With these definitions the normal equations of (4) can be solved immediately from Theorem 2 of Section 7.4a as

$$\hat{b} = GX'y + (H - I)z, \tag{5}$$

where z is arbitrary. This means there is no unique solution $\hat{b}$; and by the nature of (4) with $X'X$ being singular, this is to be expected.

This lack of uniqueness in the solutions of the normal equations is discussed in many places, usually in terms of making use of the fact that for equations like (4) the addition of what is often called a "convenient restraint" or "obvious restriction" (such as $\alpha_1 + \alpha_2 + \alpha_3 = 0$) does lead to a single solution. But indeed this is convenient only for very particular cases of (4), and for the general case there *are* no "convenient" or "obvious" restrictions leading to unique solutions. Although a great deal has been written about obtaining solutions through adding extra equations ("conditions," "restraints" or "contraints") of this nature—and almost any text on experimental design refers to them—they are avoided here by considering the solution, given in (5), obtained from using the generalized inverse. We now examine the properties of $\hat{b}$ derived in this fashion.

3. PROPERTIES OF THE SOLUTIONS

a. Assumptions

As in equations (13) and (14) in Chapter 10, we assume $E(e) = 0$ and $\text{var}(e) = \sigma^2 I$ where $\text{var}(e)$ is the variance-covariance matrix of the unobservable random error terms.

b. Expected values and variances

Taking the expected value of equation (5) gives

$$E(\hat{b}) = GX'E(y) + (H - I)E(z)$$
$$= GX'Xb + (H - I)z$$
$$= Hb + (H - I)z, \quad \text{since } H = GX'X, \tag{6}$$

which is different from b, even if z is taken as a null vector. However, as is shown later, this does not create difficulties.

To find the variance-covariance matrix of $\hat{b}$ we note from (5) and (6) that

$$\hat{b} - E(\hat{b}) = GX'y - Hb = GX'(y - Xb) = GX'e.$$

Hence $$\text{var}(\hat{b}) = E[\hat{b} - E(\hat{b})][\hat{b} - E(\hat{b})]'$$
$$= E(GX'ee'XG') = GX'E(ee')XG' = GX'\sigma^2 IXG'$$
$$= GX'XG'\sigma^2. \tag{7}$$

In the special case that G is calculated by the method suggested in Chapter 7, G is symmetric and $GX'XG = G$; thus (7) reduces to $\text{var}(\hat{b}) = G\sigma^2$. However, these conditions on G are not true of all generalized inverses of $X'X$, so that (7) must remain as the general expression for $\text{var}(\hat{b})$. But this turns out to be no restriction at all on the usefulness of this approach.

c. Some matrix results

We subsequently make considerable use of matrix results that stem from the definition of G, namely

$$X'XGX'X = X'X.$$

Transposing this equation gives

$$X'XG'X'X = X'X,$$

which shows that G' is also a generalized inverse of $X'X$. Furthermore (by Exercise 13 of Chapter 7), GX' is a generalized inverse of X and therefore

$$XGX'X = X, \quad \text{i.e.,} \quad XH = X, \tag{8}$$

and $$XGX' \text{ is unique for all } G. \tag{9}$$

Also, transposing (8) gives

$$X'XG'X' = X'. \tag{10}$$

These results also mean that

$$(XGX')^2 = XGX' \tag{11}$$

and $$(I - XGX')^2 = (I - XGX'); \tag{12}$$

i.e. that XGX' and $(I - XGX')$ are idempotent (see Section 7.3b). Note that these results, being true for any generalized inverse G of $X'X$, are also true for G' because it, too, is a generalized inverse of $X'X$.

d. Estimating the error variance

In regression analysis, $\hat{y} = X\hat{b}$ is a vector of predicted values and from this can be computed the error sum of squares SSE, previously expressed in equation (18) of Chapter 10 as $y'y - \hat{b}'X'y$. We now investigate the situation in the case of the linear model. As before,

$$\text{SSE} = \sum_i \sum_j (y_{ij} - \hat{y}_{ij})^2$$
$$= (y' - \hat{y}')(y - \hat{y}).$$

But
$$\hat{y} = X\hat{b} = XGX'y + X(H - I)z,$$

and because of (8) this gives $\hat{y} = XGX'y$. Therefore

$$\text{SSE} = y'(I - XG'X')(I - XGX')y = y'(I - XGX')^2 y$$

by (9), and from (12) this becomes

$$\text{SSE} = y'(I - XGX')y. \tag{13}$$

Because of (9) SSE has the same, unique value no matter what generalized inverse of $X'X$ is used. Furthermore, from (8) $X(H - I)$ is null; thus SSE can be written from (13) as

$$\text{SSE} = y'y - y'X[GX'y + (H - I)z] = y'y - y'X\hat{b}$$
$$= y'y - \hat{b}'X'y. \tag{14}$$

Hence the error sum of squares is the total uncorrected sum of squares $y'y$ after subtracting from it the sum of products of the elements in $\hat{b}$ each multiplied by the corresponding right-hand side of the normal equations $X'X\hat{b} = X'y$. This is exactly the same result as in regression analysis; however, in regression $\hat{b}$ was unique, whereas here $\hat{b}$ represents just one of many solutions to the normal equations $X'X\hat{b} = X'y$. Here, for *any* solution $\hat{b}$ from (5), the error sum of squares is $y'y - \hat{b}'X'y$; this value is the same regardless of the solution used.

To derive the expected value of SSE we substitute $y = Xb + e$ into (13) to get

$$\text{SSE} = (b'X' + e')(I - XGX')(Xb + e)$$

which, after simplifying, becomes

$$\text{SSE} = e'(I - XGX')e \tag{15}$$

because of (8). Furthermore, because of (12) and the footnote at the end of Section 10.3c, it can be shown that the expected value of (15) is

$$E(\text{SSE}) = (n - r)\sigma^2,$$

$n - r$ being the rank of $I - XGX'$ where r is the rank of X and $X'X$. Thus an unbiased estimator of σ^2 is

$$\hat{\sigma}^2 = \frac{\text{SSE}}{n - r}. \tag{16}$$

4. ESTIMABLE FUNCTIONS

Equation (6) showed that $\hat{b}$ is not an unbiased estimator of b; however, we can derive unique, unbiased estimators of certain linear combinations of the elements of b from $\hat{b}$. This is achieved by using Theorem 5 of Section 7.6c which states that certain linear combinations of the elements of the solution $\hat{b}$ have a unique value no matter what solution for $\hat{b}$ is obtained from (5). These combinations are $q'\hat{b}$ where q' is such that

$$q'H = q';$$

i.e. for any q' satisfying this equation, $q'\hat{b}$ is invariant to whatever solution $\hat{b}$ is used. Furthermore, the expected value of $q'\hat{b}$ is, using (6),

$$E(q'\hat{b}) = q'Hb + q'(H - I)z = q'Hb = q'b$$

because $q'(H - I)$ is null.

In other words, provided $q'H = q'$, the linear function $q'\hat{b}$ is invariant to whatever solution for $\hat{b}$ is obtained from (5), and it is an unbiased estimator of $q'b$. This is precisely the property of estimability: linear functions of the elements of $\hat{b}$ that are invariant to $\hat{b}$ are the invariant, unbiased estimators of the same linear functions of the elements of b—and these are the *only* linear functions of the parameters that can be so estimated. Such functions are called *estimable functions*. Hence if q' is such that $q'H = q'$, then $q'\hat{b}$ is the invariant unbiased estimator of $q'b$, and $q'b$ is an estimable function.

We now note that *any* q' of the form $q' = w'H$, no matter what the vector w' is, satisfies $q'H = q'$ because $q'H$ then equals $w'H^2 = w'H = q'$, since $H^2 = H$. Thus for any arbitrary vector w',

$$q'b = w'Hb$$

is an estimable function; and its estimator, which is invariant to the choice of $\hat{b}$, is

$$\begin{aligned}
q'\hat{b} &= w'H\hat{b} \\
&= w'H[GX'y + (H - I)z] \\
&= w'HGX'y \\
&= w'GX'y \tag{17}
\end{aligned}$$

because of (10).

Expression (17) also equals $w'\hat{b}_\dagger$ where $\hat{b}_\dagger$ is taken as $GX'y$, the value of $\hat{b}$ obtained by putting $z = 0$ in (5).

The variance of the estimator $q'\hat{b}$ is

$$\begin{aligned}
\operatorname{var}(q'\hat{b}) &= q'\operatorname{var}(\hat{b})q, && \text{from (18) in Section 3.6}\\
&= q'GX'XG'q\sigma^2, && \text{from (7)}\\
&= q'GX'XGX'XG'q\sigma^2, && \text{from the definition of } G\\
&= q'HGH'q\sigma^2, && \text{from the definition of } H\\
&= q'Gq\sigma^2, && (18)
\end{aligned}$$

because $q'b$ is estimable and so $q'H = q'$. Note that (18) is unique for all G; see Exercise 3. This result is analogous to that pertaining when $(X'X)$ is nonsingular [see equation (17) in Chapter 10].

The covariance between the estimators $q_1\hat{b}$ and $q_2\hat{b}$ of two estimable functions can be derived in a similar fashion. It reduces to

$$\operatorname{cov}(q_1'\hat{b}, q_2'\hat{b}) = q_1'Gq_2\sigma^2.$$

Example (*continued*). The general results developed thus far are demonstrated by continuing with the example of the illustration in Section 1. The normal equations are

$$X'Xb = X'y,$$

where
$$X'X = \begin{bmatrix} 6 & 3 & 2 & 1 \\ 3 & 3 & 0 & 0 \\ 2 & 0 & 2 & 0 \\ 1 & 0 & 0 & 1 \end{bmatrix} \quad \text{and} \quad X'y = \begin{bmatrix} 504 \\ 300 \\ 172 \\ 32 \end{bmatrix}. \qquad (19)$$

A generalized inverse of $X'X$ is

$$G = \begin{bmatrix} 0 & 0 & 0 & 0 \\ 0 & \frac{1}{3} & 0 & 0 \\ 0 & 0 & \frac{1}{2} & 0 \\ 0 & 0 & 0 & 1 \end{bmatrix} \quad \text{for which } H = GX'X = \begin{bmatrix} 0 & 0 & 0 & 0 \\ 1 & 1 & 0 & 0 \\ 1 & 0 & 1 & 0 \\ 1 & 0 & 0 & 1 \end{bmatrix}.$$

Hence from (5), the solution to the normal equations is

$$\begin{aligned}
\hat{b} &= GX'y + (H - I)z \\
&= \begin{bmatrix} 0 & 0 & 0 & 0 \\ 0 & \frac{1}{3} & 0 & 0 \\ 0 & 0 & \frac{1}{2} & 0 \\ 0 & 0 & 0 & 1 \end{bmatrix}\begin{bmatrix} 504 \\ 300 \\ 172 \\ 32 \end{bmatrix} + \begin{bmatrix} -1 & 0 & 0 & 0 \\ 1 & 0 & 0 & 0 \\ 1 & 0 & 0 & 0 \\ 1 & 0 & 0 & 0 \end{bmatrix}\begin{bmatrix} z_1 \\ z_2 \\ z_3 \\ z_4 \end{bmatrix}
\end{aligned}$$

or
$$\hat{b} = \begin{bmatrix} -z_1 \\ 100 + z_1 \\ 86 + z_1 \\ 32 + z_1 \end{bmatrix}, \tag{20}$$

where z_1 is arbitrary, the first element of the arbitrary vector z.

From the basic data

$$y'y = \sum_i \sum_j y_{ij}^2 = 45{,}886 \tag{21}$$

and with this, SSE is given by (14) as $y'y - \hat{b}'X'y$. Since XGX' is unique we know that $\hat{b}'X'y = y'XGX'y$ has the same value no matter what solution we use for $\hat{b}$; we now demonstrate this uniqueness for our example. From (20),

$$\hat{b}'X'y = \begin{bmatrix} -z_1 & 100 + z_1 & 86 + z_1 & 32 + z_1 \end{bmatrix} \begin{bmatrix} 504 \\ 300 \\ 172 \\ 32 \end{bmatrix}$$

$$= z_1(-504 + 300 + 172 + 32) + 100(300) + 86(172) + 32(32)$$

$$= 45{,}816$$

independent of z_1. This will be the value of $\hat{b}'X'y$ for any solution $\hat{b}$, and so

$$\text{SSE} = 45{,}866 - 45{,}816 = 70. \tag{22}$$

Hence, using (22) in (16) gives

$$\hat{\sigma}^2 = 70/(6 - 3) = 70/3, \tag{23}$$

the rank of X being $r = 3$. The values given in (22) and (23) will be unaltered no matter what value is obtained for $\hat{b}$ in (20).

Estimable functions are $q'b$ where $q' = w'H$ for arbitrary w' and $b' = [\mu \quad \alpha_1 \quad \alpha_2 \quad \alpha_3]$, the vector of parameters of the model. Thus for $w' = [w_0 \quad w_1 \quad w_2 \quad w_3]$, estimable functions are

$$q'b = w'Hb = (w_1 + w_2 + w_3)\mu + w_1\alpha_1 + w_2\alpha_2 + w_3\alpha_3 \tag{24}$$

for which estimators are, from (17), $q'\hat{b} = w'GX'y$; from the nature of one-way classification models this can be expressed as

$$q'\hat{b} = w'GX'y = w_1\bar{y}_1. + w_2\bar{y}_2. + w_3\bar{y}_3. \tag{25}$$

with $\bar{y}_i.$ defined as $\bar{y}_i. = \left(\sum_{j=1}^{n_i} y_{ij} \right) \Big/ n_i.$

These equations illustrate an important aspect of the general results, namely that they apply for *any* vector w'; i.e. (24) and (25) hold true for any values that we care to give to w_1, w_2 and w_3. There are two consequences of this. First, by giving specific values to the w's we can obtain specific estimable functions from (24); for example, for $w_1 = w_2 = 1$ and $w_3 = 0$ we see that the function $2\mu + \alpha_1 + \alpha_2$ is estimable. Second, we can find out whether particular functions of the parameters that interest us are estimable—by seeing if w's can be found such that (24) reduces to the function of interest; e.g. is $\alpha_1 - \alpha_2$ estimable? Yes, because with $w_1 = 1$, $w_2 = -1$ and $w_3 = 0$, (24) reduces to $\alpha_1 - \alpha_2$. This, of course, is equivalent to seeing whether, for a particular value of $q'b$ that interests us, q' satisfies $q'H = q'$. Thus, for $q'b = \alpha_1 - \alpha_2$ in the example, $q' = [0 \quad 1 \quad -1 \quad 0]$ and we see that

$$q'H = [0 \quad 1 \quad -1 \quad 0] \begin{bmatrix} 0 & 0 & 0 & 0 \\ 1 & 1 & 0 & 0 \\ 1 & 0 & 1 & 0 \\ 1 & 0 & 0 & 1 \end{bmatrix} = [0 \quad 1 \quad -1 \quad 0] = q';$$

$$(26)$$

thus $\alpha_1 - \alpha_2$ is estimable. Similarly we find that μ is not estimable, for the q' appropriate to $q'b = \mu$ is $q' = [1 \quad 0 \quad 0 \quad 0]$, for which $q'H \neq q'$. Thus it is impossible to choose values for the w's that reduce (24) to be just μ. On all occasions the estimator of (24) is given by the corresponding value of (25). There are, of course, an unlimited number of sets of values that can be given to the w's and hence an unlimited number of estimable functions—but only $n - r = 3$ of them can be linearly independent. Examples are shown in Table 2. It can be shown that the sets of values given to the w's in Table 2 are linearly independent and therefore so are the corresponding estimable functions.

TABLE 2.　EXAMPLES OF ESTIMABLE FUNCTIONS AND THEIR ESTIMATORS

Example	Values of w's			Estimable Function	Estimator
	w_1	w_2	w_3	$(w_1 + w_2 + w_3)\mu + w_1\alpha_1$ $+ w_2\alpha_2 + w_3\alpha_3$	$w_1\bar{y}_1. + w_2\bar{y}_2. + w_3\bar{y}_3.$ $= 100w_1 + 86w_2 + 32w_3$
1	1	-1	0	$\alpha_1 - \alpha_2$	$\bar{y}_1. - \bar{y}_2. = 14$
2	0	1	-1	$\alpha_2 - \alpha_3$	$\bar{y}_2. - \bar{y}_3. = 54$
3	$\frac{1}{3}$	$\frac{1}{3}$	$\frac{1}{3}$	$\mu + \frac{1}{3}(\alpha_1 + \alpha_2 + \alpha_3)$	$(\bar{y}_1. + \bar{y}_2. + \bar{y}_3.)/3 = 72\frac{2}{3}$

The last line of the table shows why it is a "convenient restraint" to have $\alpha_1 + \alpha_2 + \alpha_3 = 0$; but it is certainly not necessary, neither for ascertaining what functions of the parameters are estimable, nor for obtaining the estimators of them. In this instance it provides an estimator of μ, but only in one particular restricted situation, namely that in which it is meaningful to define the α's such that they sum to zero. This may or may not be an appropriate form of definition, depending on the particular situation. However, even if it is appropriate, it is useful in the estimation process only if it enters into an estimable function as just exemplified; otherwise it is of little help. (In our example there seems to be no justification for assuming that $\alpha_1 + \alpha_2 + \alpha_3 = 0$ or any other constant.)

Finally, the variance of the estimator of any estimable function is, as given in (18)

$$\text{var}(q'\hat{b}) = q'Gq\sigma^2.$$

For example, with the particular q' given in (26) this is

$$\text{var}(\widehat{\alpha_1 - \alpha_2}) = (\tfrac{1}{3} + \tfrac{1}{2})\sigma^2 = (\tfrac{5}{6})\sigma^2.$$

5.   TESTS OF SIGNIFICANCE

a.   Analysis of variance

Use has already been made of the error sum of squares

$$\text{SSE} = y'y - \hat{b}'X'y$$

as shown in (14). Suppose this is subtracted from the total corrected[1] sum of squares

$$\text{SST} = y'y - n\bar{y}^2$$

just as was done in the case of regression. Then we have

$$\text{SSM} = \text{SST} - \text{SSE}$$
$$= \hat{b}'X'y - n\bar{y}^2$$

which can be described as the sum of squares due to fitting the model, analogous to SSR of regression analysis. This division of SST into two parts can be summarized in an analysis of variance table, Table 3, similar to the analyses[2] in Section 10.4.

---

[1] Corrected for the mean; i.e., measured as deviations from the mean. See Section 10.4b.

[2] There is one apparent inconsistency between this table and the corresponding second part of Table 3 in Section 10.4. There, the D.F. associated with SSE is $n - k - 1$ and here, in Table 3, it is $n - r$ where r is the rank of X. Actually, this is also the case in the earlier Table 3 (of Section 10.4), because there the rank of the X-matrix of the intercept model, namely X^*, is $k + 1$—see also Table 5 below. There is therefore no inconsistency; $r - 1$ corresponds to k in the same way.

We have, from Table 3,

$$R^2 = \frac{SSM}{SST} \tag{27}$$

as the coefficient of determination for fitting the model; this corresponds to equation (39) of Chapter 10. Similarly, in line with equation (41) of Chapter 10,

$$F_{r-1, n-r} = \frac{SSM}{(r-1)\hat{\sigma}^2}. \tag{28}$$

TABLE 3.   ANALYSIS OF VARIANCE FOR FITTING
A LINEAR MODEL

Source of Variation	D.F.	Sum of Squares
Fitting the model	$r - 1$	$SSM = \hat{b}' X' y - n\bar{y}^2$
Error	$n - r$	$SSE = SST - SSM$
Total	$n - 1$	$SST = y'y - n\bar{y}^2$

After assuming that the error terms in the model are normally distributed the value of F in (28) can be compared with tabulated values of the F-distribution, at $r - 1$ and $n - r$ degrees of freedom, to test the significance of the complete model.

b. Tests of general linear hypotheses

We now consider testing linear hypotheses, a linear hypothesis being a hypothesis that some linear function of the parameters equals some assigned constant. The discussion must be confined to linear hypotheses involving estimable functions, since these are the only ones that can be tested. Thus, in the example above, the function $\alpha_1 - \alpha_2$ is estimable and the hypothesis that $\alpha_1 - \alpha_2$ equals some constant, m_0 say, can be tested; but the hypothesis that $\mu = 0$ cannot be tested.

In general, a linear hypothesis that $q'b = m_0$, where m_0 is assigned and $q'b$ is estimable, is tested by amending the model $y = Xb + e$ to take account of the hypothesis. In this connection the model $y = Xb + e$ is usually referred to as the *full model*, and amendment in terms of the hypothesis yields what is called the *reduced model*, since it represents the full model reduced by the conditions of the hypothesis. Thus if $q'b = m_0$ led to b being changed into b^* the reduced model would be $y = X^*b^* + e$ where X^* is the form of X corresponding to b^*. In the example of this chapter $b' = [\mu \quad \alpha_1 \quad \alpha_2 \quad \alpha_3]$, and to test the hypothesis that $\alpha_1 = \alpha_2$, i.e. that $\alpha_1 - \alpha_2 = 0$ (which is testable because $\alpha_1 - \alpha_2$ is estimable), $b^{*\prime}$ would be $b^{*\prime} = [\mu \quad \alpha_1 \quad \alpha_3]$. The matrix X is

$$
X = \begin{bmatrix} 1 & 1 & 0 & 0 \\ 1 & 1 & 0 & 0 \\ 1 & 1 & 0 & 0 \\ 1 & 0 & 1 & 0 \\ 1 & 0 & 1 & 0 \\ 1 & 0 & 0 & 1 \end{bmatrix}, \quad \text{and} \quad X^* \text{ would be } X^* = \begin{bmatrix} 1 & 1 & 0 \\ 1 & 1 & 0 \\ 1 & 1 & 0 \\ 1 & 1 & 0 \\ 1 & 1 & 0 \\ 1 & 0 & 1 \end{bmatrix}.
$$

Many hypotheses cannot be represented by a single equation $q'b = m_0$ but require several such equations. These are usually thought of in vector form $Q'b = m$, where the rows of $Q'b$ are linearly independent estimable functions. In this way $Q'H = Q'$, and if Q' has s rows the rank of Q' is s. As an example, the hypothesis that $\alpha_1 = \alpha_2 = \alpha_3$ can be represented in the form $Q'b = m$ as

$$
\begin{bmatrix} 0 & 1 & -1 & 0 \\ 0 & 1 & 0 & -1 \end{bmatrix} \begin{bmatrix} \mu \\ \alpha_1 \\ \alpha_2 \\ \alpha_3 \end{bmatrix} = \begin{bmatrix} 0 \\ 0 \end{bmatrix}. \tag{29}
$$

The hypothesis $Q'b = m$ is known as the *general linear hypothesis*. The test of this hypothesis requires obtaining the sum of squares due to fitting the full model and that due to fitting the reduced model. The first of these is $\text{SSM} = \hat{b}'X'y - n\bar{y}^2 = y'XGX'y - n\bar{y}^2$ from (14) where G is a generalized inverse of $X'X$, and since this refers to the full model it shall be denoted by

$$
\text{SSM (full model)} = y'XGX'y - n\bar{y}^2.
$$

By analogy, the sum of squares for fitting the reduced model $y = X^*b^* + e$ (derived from the full model by applying to it the hypothesis $Q'b = m$) is

$$
\text{SSM (reduced model)} = y'X^*G^*X^{*\prime}y - n\bar{y}^2,
$$

where G^* is a generalized inverse of $X^{*\prime}X^*$. The difference between these two expressions is used to obtain

$$
F_{s,\,n-r} = [\text{SSM (full model)} - \text{SSM (reduced model)}]/s\hat{\sigma}^2,
$$

where s is the rank of Q' and $\hat{\sigma}^2 = \text{SSE}/(n - r)$ as in (16). Comparing F with tabulated values of the F-distribution for s and $n - r$ D.F. provides the test of the hypothesis.

While the numerator of F could be calculated as $y'XGX'y - y'X^*G^*X^{*\prime}y$, to do so would require $X^*G^*X^{*\prime}$ for each and every hypothesis that is to be tested. However, because in both models SSM + SSE = SST, the expression for F can also be written as

$$F_{s,n-r} = [\text{SSE (reduced model)} - \text{SSE (full model)}]/s\hat{\sigma}^2$$

and from this a computing formula can be developed which avoids altogether the necessity of obtaining X^*. The formula utilizes $\hat{b}$ as well as Q' and m, the specifications of the hypothesis, and it is relatively easy to compute. It is given in the following theorem, which enables us to test hypotheses in a straightforward manner once $\hat{b}$ is obtained.

Theorem.[1] When fitting the linear model $y = Xb + e$, the numerator sum of squares of the F-value used for testing the (testable) general linear hypothesis $Q'b = m$, for Q' consisting of s linearly independent rows, is $(Q'\hat{b} - m)'(Q'GQ)^{-1}(Q'\hat{b} - m)$ where $\hat{b} = GX'y$ is a solution to the normal equations $X'Xb = X'y$ and G is a generalized inverse of $X'X$.

This theorem means that the F-value for testing the hypothesis $Q'b = m$ can be calculated as

$$F_{s,n-r} = [(Q'\hat{b} - m)'(Q'GQ)^{-1}(Q'\hat{b} - m)]/s\hat{\sigma}^2. \tag{30}$$

With $Q'\hat{b}$ being the estimator of the estimable functions $Q'b$ in the full model it is apparent that once $\hat{b} = GX'y$ has been calculated F is readily obtainable.

Example (*continued*). From the basic data $n\bar{y}^2 = (504)^2/6 = 42,336$; and with $y'y = 45,866$ in (21),

$$\text{SST} = 45,886 - 42,336 = 3,550.$$

Hence, with SSE = 70 in (22) the analysis of variance of Table 3 is given in

TABLE 4.   ANALYSIS OF VARIANCE OF EXAMPLE

Source of Variation	D.F.	Sums of Squares
Fitting the model	2	3,480
Error	3	70
Total	5	3,550

Table 4. From (16), (27) and (28)

$$\hat{\sigma}^2 = 70/3,$$
$$R^2 = 3,480/3,550 = 0.98,$$

and

$$F_{2,3} = 3,480/2(70/3) = 522/7 = 74.57. \tag{31}$$

[1] Proof is given in Searle (1966, p. 277).

At the 5% level of significance the critical value of $F_{2,3}$ is 9.55, so the null hypothesis that the model has no explanatory power must be rejected at this level of significance. As shown in Table 2, the function $\alpha_1 - \alpha_2$ is estimable, and therefore a testable hypothesis is $\alpha_1 - \alpha_2 = 0$. Expressing this as $Q'b = 0$, Q' is a vector having rank $s = 1$: $Q' = [0 \quad 1 \quad -1 \quad 0]$. It will be found that $Q'GQ = 5/6$, $Q'\hat{b} = 14$, and hence from (30) the statistic for testing the hypothesis $\alpha_1 = \alpha_2$ is

$$F = \frac{14(5/6)^{-1}14}{1(70/3)} = \frac{1,176/5}{70/3} = 10.08$$

which is F-distributed with $(1, 3)$ degrees of freedom. At a 5% level of significance the critical value of $F_{1,3} = 10.13$, so the hypothesis $\alpha_1 - \alpha_2 = 0$ cannot be rejected at that level of significance.

Another hypothesis of interest would be $\alpha_1 = \alpha_2 = \alpha_3$; this would be written as $Q'b = 0$ as in equation (29) from which the F-statistic for testing the hypothesis is calculated from (30) as follows. From (20) and (29)

$$Q'\hat{b} = \begin{bmatrix} 0 & 1 & -1 & 0 \\ 0 & 1 & 0 & -1 \end{bmatrix} \begin{bmatrix} 0 \\ 100 \\ 86 \\ 32 \end{bmatrix} = \begin{bmatrix} 14 \\ 68 \end{bmatrix}$$

and

$$Q'GQ = \begin{bmatrix} 0 & 1 & -1 & 0 \\ 0 & 1 & 0 & -1 \end{bmatrix} \begin{bmatrix} 0 & 0 & 0 & 0 \\ 0 & \frac{1}{3} & 0 & 0 \\ 0 & 0 & \frac{1}{2} & 0 \\ 0 & 0 & 0 & 1 \end{bmatrix} \begin{bmatrix} 0 & 0 \\ 1 & 1 \\ -1 & 0 \\ 0 & -1 \end{bmatrix} = \begin{bmatrix} \frac{5}{6} & \frac{1}{3} \\ \frac{1}{3} & \frac{4}{3} \end{bmatrix}.$$

Hence with

$$(Q'GQ)^{-1} = \begin{bmatrix} \frac{4}{3} & -\frac{1}{3} \\ -\frac{1}{3} & \frac{5}{6} \end{bmatrix} \quad \text{and} \quad m = \begin{bmatrix} 0 \\ 0 \end{bmatrix}$$

equation (30) gives

$$F_{2,3} = [14 \quad 68] \begin{bmatrix} \frac{4}{3} & -\frac{1}{3} \\ -\frac{1}{3} & \frac{5}{6} \end{bmatrix} \begin{bmatrix} 14 \\ 68 \end{bmatrix} \Big/ 2(70/3) = 522/7 = 74.57.$$

This is the same value as in (31), because the hypothesis that $\alpha_1 = \alpha_2 = \alpha_3$ is identical to the hypothesis of no model (apart from a mean).

6.   SUMMARY OF LINEAR MODEL CALCULATIONS

We have shown that the algebra of linear models can, like that of linear regression, be described as a series of matrix expressions readily available for computer programming. A summary is therefore listed below of the formulae for deriving estimable functions, their estimators, variances of the estimators and the associated analysis of variance table.

Our summary is based on using b to denote the vector of parameters of the model.

Data:

n = number of observations.

y = vector of observations.

$\bar{y}$ = mean of observations.

X = matrix of coefficients of elements of b in the model.

r = rank of X.

Estimation:

$X'X$ = matrix of coefficients in the normal equations $X'X\hat{b} = X'y$.

$X'y$ = vector of totals in the normal equations.

G = any generalized inverse of $X'X$, satisfying $X'XGX'X = X'X$.

$H = GX'X$.

w' = vector of arbitrary elements.

$q'b = w'Hb$: estimable function.

$q'\hat{b} = w'GX'y$: estimator of estimable function.

$\text{var}(q'\hat{b}) = q'Gq\sigma^2$: variance of estimator.

$\text{cov}(\hat{q}'_1\hat{b}, \hat{q}'_2\hat{b}) = q'_1Gq_2\sigma^2$: covariance between estimators of two estimable functions.

$\text{SSE} = y'y - \hat{b}'X'y$: error sum of squares.

$\hat{\sigma}^2 = \text{SSE}/(n-r)$: estimated error variance.

Significance tests:

$\text{SST} = y'y - n\bar{y}^2$: total (corrected) sum of squares.

$\begin{aligned}\text{SSM} &= \hat{b}'X'y - n\bar{y}^2 \\ &= y'XGX'y - n\bar{y}^2\end{aligned}\Bigg\}$: sum of squares due to fitting the linear model.

$\text{SSE} = \text{SST} - \text{SSM}$: error sum of squares.

$R^2 = \text{SSM}/\text{SST}$: coefficient of determination.

$F_{r-1, n-r} = \text{SSM}/(r-1)\hat{\sigma}^2$: F-statistic of the full model.

$Q'b = m$: testable hypothesis of s linearly independent estimable functions.

$F_{s, n-r} = [(Q'\hat{b} - m)'(Q'GQ)^{-1}(Q'\hat{b} - m)]/s\hat{\sigma}^2$
     : F-statistic for hypothesis $Q'b = m$ with Q' having rank s.

7. CONCLUSIONS

In this chapter we have presented an example of the simplest type of linear model, but the general results developed apply *in toto* to all linear models. The type most frequently encountered is that involved in survey data, where there are likely to be many classifications of the data—rather than just one as we have illustrated—and where interactions between different groups of classes may also be considered. For example, a market research survey where customers are classified by age group, sex, education level and income group could be analyzed by the linear model techniques outlined here. Searle (1966) discusses additional cases and describes an example of a model that includes both an ordinary regression variable and dummy variables. In these and all other situations involving data that can be classified into groups in various ways, the concepts and procedures described here are directly applicable—namely those of estimable functions and their estimators and tests of hypotheses. Numerous texts deal with these topics in greater detail, such as Kempthorne (1952), Anderson (1958), Scheffé (1959) and Graybill (1961), although none of them make explicit use of the generalized inverse. Further consideration of this approach is to be found in Rao (1965) and Searle (1966), and a discussion of how these techniques relate to analyses of covariance in business and economic problems is given in Lin and White (1968).

8. EXERCISES

1. For the example in Section 4, test the hypothesis that $\alpha_1 = \alpha_3$ if it is a testable hypothesis.
2. Show that XGX' is always symmetric, where G is a generalized inverse of $X'X$, even when G is not symmetric.
3. Show that if $q'b$ is an estimable function then equation (18), $\text{var}(q'\hat{b}) = q'Gq\sigma^2$, is unique for all G (whether symmetric or not).
4. Consider the following table whose entries contain sales of three products under two advertising strategies:

Unit Sales

	Strategy 1	Strategy 2	Total
Product 1	y_{11}	y_{12}	$y_{1.}$
Product 2	y_{21}	y_{22}	$y_{2.}$
Product 3	y_{31}	y_{32}	$y_{3.}$
Total	$y_{.1}$	$y_{.2}$	$y_{..}$

(a) Write down the equation of the linear model for this situation.

(b) With observations in matrix form $y = Xb + e$, write X and b and obtain $X'y$ and $X'X$.

(c) Obtain a symmetric generalized inverse G of $X'X$ and calculate $H = GX'X$.

(d) Derive a general set of estimable functions.

5. For Exercise 4 above:

(a) Derive in general form the unique, unbiased estimators of the estimable functions you obtained in (d).

(b) Suppose you wished to test the hypothesis that there is no significant difference between the sales of Product 1 and the sales of Product 2. Is this a testable hypothesis (i.e., can you find an estimable function representing this hypothesis)?

(c) How many LIN estimable functions are there? Why?

6. It is possible to combine the concepts of regression analysis and regression on dummy variables. Consider the following sales data:

Five Sales Volumes

Advertising Strategy i	Product j	Sales y_{ij}	Previous Year Sales x_{ij}
1	1	8	1
1	2	3	2
1	3	7	1
2	1	5	3
2	2	4	2

(a) Using the model $y_{ij} = \mu + \alpha_i + \beta x_{ij} + e_{ij}$, write X and b of $y = Xb + e$.

(b) Obtain a symmetric generalized inverse G of $X'X$ and calculate $H = GX'X$.

(c) Derive a general set of estimable functions.

7. For Exercise 6 above:

(a) Derive the set of unique, unbiased estimators of the functions obtained in 6(c).

(b) What is the estimator of $(\alpha_1 - \alpha_2)$?

(c) What is the variance of that estimator?

(d) Calculate the customary estimator of σ^2.

8. If G is a symmetric generalized inverse of $X'X$ satisfying the first two of Penrose's four conditions (See Section 7.3a), and if

$$(i)\ a = GX'y, \qquad (ii)\ s = (y - Xa)'(y - Xa),$$

and $$(iii)\ b = a - GQ(Q'GQ)^{-1}(Q'a - m),$$

show that

(a) $s = y'y - a'X'y$;

(b) $Q'b = m$.

REFERENCES

Anderson, T. W. (1958). *Introduction to Multivariate Statistical Analysis*. Wiley, New York.

Graybill, F. A. (1961). *An Introduction to Linear Statistical Models*. Vol. I, McGraw-Hill, New York.

Henderson, P. L., J. F. Hind, and S. E. Brown (1961). Sales effects of two campaign themes. *Journal of Advertising Research*, **1**, 2–11.

Kempthorne, O. (1952). *Design and Analysis of Experiments*. Wiley, New York.

Lin, Chi-Yuan, and W. L. White (1968). Four procedures for testing linear hypotheses. *Industrial Management Review* (M.I.T.), **10**, 13–30.

Rao, C. R. (1965). *Linear Statistical Inference and its Applications*. Wiley, New York.

Scheffe, H. (1959). *The Analysis of Variance*. Wiley, New York.

Searle, S. R. (1966). *Matrix Algebra for the Biological Sciences*. Wiley, New York.

CHAPTER 12

CHARACTERISTIC ROOTS AND VECTORS

Characteristic roots and vectors have applications in many different situations. In this chapter, after describing the basic development of characteristic roots and vectors, we illustrate their application to systems of difference equations of a nature that arise in describing dynamic (changing) systems in business and economics, such as inventory models, macroeconomic models, models of stock price movements and market equilibrium models.

The equations that we shall be concerned with are those known as first-order difference equations. They usually relate to measurements of a variable (or a set of variables) made at different points in time. A simple example is $u_t = 3u_{t-1}$ where u_t and u_{t-1} may be scalars or vectors measured at times t and $t - 1$ respectively. If the u's are scalars we have a single difference equation; if they are vectors we have a set or *system* of difference equations. In the latter case the equation does not need to be as simple as suggested but might be of the form $u_t = Au_{t-1}$ where A is a square matrix and u_0 is the vector of initial values at time zero.

In Chapter 8 (Markov Chains) a particular type of difference equation was discussed: $u'_t = u'_{t-1}P$ where P is a transition probability matrix. The very special nature of transition probability matrices and their use and frequent occurrence in many practical problems have led to the equations $u'_t = u'_{t-1}P$ receiving special and widespread attention in the literature. Now, however, we consider the more general equations $u_t = Au_{t-1}$ for any square matrix A. To do so, suppose for a moment that for some value of t the vector u_t takes a particularly simple form: $u_t = \lambda u_{t-1}$ where λ is a scalar. We then have

$$u_t = Au_{t-1} = \lambda u_{t-1},$$

or, more particularly $\qquad Au_{t-1} = \lambda u_{t-1}.$

This is our starting point for considering characteristic roots and vectors. We shall see how it leads to a solution of the equations $u_t = Au_{t-1}$ that readily enables the behavior of u_t over time to be explored.

1. CHARACTERISTIC ROOTS

In starting from the equation $Au_{t-1} = \lambda u_{t-1}$ we drop the subscript $t-1$ and consider the equivalent equation

$$Au = \lambda u. \tag{1}$$

That is, consider a vector u which when pre-multiplied by a square matrix A is equal to itself multiplied by a scalar λ. If a scalar and a vector exist that satisfy this equation it can be rewritten as $Au - \lambda u = 0$, equivalent to

$$(A - \lambda I)u = 0. \tag{2}$$

(The symbol 0 here represents a null vector, and henceforth is used interchangeably as a null vector or null matrix and also as the scalar zero.) As seen in Chapter 7, an equation of this form has a non-null solution for the vector u only if the rank of $(A - \lambda I)$ is less than its order, in which case its determinant is zero; i.e.,

$$|A - \lambda I| = 0. \tag{3}$$

This equation establishes conditions under which equation (1) is true for a non-null vector u; namely, values of λ which satisfy (3) are such that (1) is also satisfied. Equation (3) is known as the *characteristic equation*. When A is of order n the characteristic equation is a polynomial in λ of degree n, and hence has n solutions $\lambda_1, \lambda_2, \ldots, \lambda_n$. For each of them (1) holds true, and so in general we would expect to find n vectors $u_1, u_2, \ldots, u_n$ corresponding to the n λ's. It is shown later that this is not always the case, but let us assume for the moment that it is. Then for each λ_i that is a solution to (3), (1) is true, i.e.,

$$Au_i = \lambda_i u_i, \quad \text{for} \quad i = 1, 2, \ldots, n. \tag{4}$$

The λ_i's are known as the *characteristic roots* of the matrix A, and the corresponding vectors u_i are the *characteristic vectors* of A. The word "characteristic" is not the only term that can be used in this context; the roots are referred to as latent roots, λ-roots, characteristic values or eigen-values and the corresponding vectors are known as latent vectors, λ-vectors or eigen-vectors. The terms "characteristic roots" and "characteristic vectors" are used throughout this book.

Example. The matrix

$$A = \begin{bmatrix} 1 & 4 \\ 9 & 1 \end{bmatrix}$$

has the characteristic equation, $|A - \lambda I| = 0$, as follows:

$$\left\{ \begin{bmatrix} 1 & 4 \\ 9 & 1 \end{bmatrix} - \begin{bmatrix} \lambda & 0 \\ 0 & \lambda \end{bmatrix} \right\} = 0;$$

i.e., $\det \begin{vmatrix} 1 - \lambda & 4 \\ 9 & 1 - \lambda \end{vmatrix} = 0.$ (5)

A characteristic equation is always of this form: λ is subtracted from every diagonal element of the determinant of A and the resulting determinant equated to zero. Expanding (5) gives

$$(1 - \lambda)^2 - 36 = 0,$$

and hence $\lambda = -5$ or 7. Methods for obtaining the characteristic vector u corresponding to each solution of the characteristic equation are discussed later, but if we let $u = \begin{bmatrix} 2 \\ -3 \end{bmatrix}$ and $\lambda = -5$ then we find that equation (1), $Au = \lambda u$, is satisfied; i.e.

$$\begin{bmatrix} 1 & 4 \\ 9 & 1 \end{bmatrix} \begin{bmatrix} 2 \\ -3 \end{bmatrix} = -5 \begin{bmatrix} 2 \\ -3 \end{bmatrix}.$$

Thus $\begin{bmatrix} 2 \\ -3 \end{bmatrix}$ is the characteristic vector corresponding to the characteristic root -5; similarly $\begin{bmatrix} 2 \\ 3 \end{bmatrix}$ is the characteristic vector corresponding to the characteristic root 7, because

$$\begin{bmatrix} 1 & 4 \\ 9 & 1 \end{bmatrix} \begin{bmatrix} 2 \\ 3 \end{bmatrix} = 7 \begin{bmatrix} 2 \\ 3 \end{bmatrix}.$$

Illustration. Chow (1957) raises the importance of the age distribution of automobiles in designing a measure of the automobile stock for aggregate demand analysis. We illustrate two models of automobile demand, both based on the existing stock and its age distribution.

Suppose automobiles are either 1, 2 or 3 years old (the ages can be extended without altering the technique). Vehicles described as 1 year old consist of replacements of 20% of last year's 2-year-old cars and 100% of last year's 3-year-olds; those that are 2 years old consist of all of last year's 1-year-olds; and the current 3-year-old cars consist of 80% of last year's 2-year-olds. With this description of the aging process of automobiles, let us write

$u'_t = [u_{1t} \quad u_{2t} \quad u_{3t}]$ as the age distribution vector in year t, where u_{it} is the number of vehicles of age i in year t. Then the relationship between the age distribution vector in one year and the next is described by the equation $u_t = Au_{t-1}$ where

$$A = \begin{bmatrix} 0 & 0.2 & 1 \\ 1 & 0 & 0 \\ 0 & 0.8 & 0 \end{bmatrix}.$$

Since all worn-out vehicles are replaced with new ones the total stock of automobiles (the sum of the components of u_t) does not change from year to year. However, we can investigate whether certain age distribution vectors perpetuate themselves under this model. If $u_{t+1} = Au_t = u_t$ then the vector u_t perpetuates itself; thus we can ask if a vector u (dropping the time subscript) exists which satisfies equation (1), $Au = \lambda u$, with $\lambda = 1$.

From (3) the characteristic equation is $|A - \lambda I| = -\lambda^3 + 0.2\lambda + 0.8 = 0$. The characteristic root $\lambda = 1$ satisfies this equation and the corresponding characteristic vector[1] $u' = [a \quad a \quad 0.8a]$, with a being a scalar, satisfies $Au = 1u$; i.e.,

$$Au = \begin{bmatrix} 0 & 0.2 & 1 \\ 1 & 0 & 0 \\ 0 & 0.8 & 0 \end{bmatrix} \begin{bmatrix} a \\ a \\ 0.8a \end{bmatrix} = 1 \begin{bmatrix} a \\ a \\ 0.8a \end{bmatrix} = u. \tag{6}$$

Hence if u_t is of the form $u'_t = [a \quad a \quad 0.8a]$ for arbitrary a, the age distribution vector will remain unchanged over time, i.e., $u_{t+1} = u_t$ and $u_{t+n} = u_t$ for any positive n. This means that a mix of cars involving an equal number of 1- and 2-year-old cars and 0.8 times that number of 3-year-olds will be a stable mix.

We noted above that this first model of automobile demand does not allow any growth in the total stock of automobiles. Let us now introduce growth through a second model. Suppose population and per capita income grow at a relatively constant rate over time; then we will assume that the number of new automobiles produced over and above replacement production is proportional to the existing stock. In the first model the total replacement production needed in year t is the first row of A multiplied by u_{t-1}, namely $0u_{1,t-1} + 0.2u_{2,t-1} + 1u_{3,t-1}$. Based on our assumption about the growth rate being proportional to total existing stock, the production needed for growth will be $\delta(u_{1,t-1} + u_{2,t-1} + u_{3,t-1})$ where δ (a scalar) is the growth proportionality factor. Then the total production of new cars in

---

[1] Derivation of characteristic vectors is discussed in Section 2.

year t will be $\delta u_{1,t-1} + (0.2 + \delta)u_{2,t-1} + (1 + \delta)u_{3,t-1}$, so that $u_t = Au_{t-1}$ now has

$$A = \begin{bmatrix} \delta & 0.2 + \delta & 1 + \delta \\ 1 & 0 & 0 \\ 0 & 0.8 & 0 \end{bmatrix}.$$

Now, because annual growth has been built into the model we can no longer expect u_t to ever equal u_{t-1}. But we might ask: Can u_t ever equal λu_{t-1}; i.e. can the age distribution vector in one year be, apart from a proportionality factor λ, the same as the age distribution vector the previous year? This would mean $u_t = Au_{t-1} = \lambda u_{t-1}$. Thus the question becomes: What vector u satisfies $Au = \lambda u$ and what is λ? As before, the characteristic equation is $|A - \lambda I| = 0$ which in this case reduces to

$$\lambda^3 - \lambda^2 \delta - \lambda(0.2 + \delta) - 0.8(1 + \delta) = 0.$$

This has a root $\lambda = 1 + \delta$, and the corresponding characteristic vector is $[(1 + \delta)a \quad a \quad 0.8a/(1 + \delta)]$, for arbitrary a. For example, suppose $\delta = 0.2$. Then

$$A = \begin{bmatrix} 0.2 & 0.4 & 1.2 \\ 1 & 0 & 0 \\ 0 & 0.8 & 0 \end{bmatrix},$$

and the characteristic equation is

$$\lambda^3 - 0.2\lambda^2 - 0.4\lambda - 0.96 = 0$$

with a root $\lambda = 1.2$; the corresponding characteristic vector is $[6a/5 \quad a \quad 2a/3]$. This means that if the initial age distribution vector is of the form $[6a/5 \quad a \quad 2a/3]$ for any value of a, the age distribution will remain in the ratio $(6/5):1:(2/3)$ for all future years, and the total stock will grow at an annual rate of 20%.

2. CHARACTERISTIC VECTORS

In the illustration concerning the age distribution of automobiles, $u_t = Au_{t-1} = A^2 u_{t-2} = \cdots = A^t u_0$ where u_0 is the initial age distribution vector. Hence calculations to obtain u_t given u_0 involve taking the tth power of the matrix A which, by repeated multiplication, can be tedious for large t.

Suppose, however, there is a matrix U and a diagonal matrix D such that $A = UDU^{-1}$. Then

$$A^2 = AA = UDU^{-1}UDU^{-1} = UD^2U^{-1}$$

and in general, for t being a positive integer,

$$A^t = (UDU^{-1})^t = UD^tU^{-1}.$$

Then $\qquad\qquad u_t = A^t u_0 = UD^t U^{-1} u_0$

can be computed, by using the powers of diagonal matrix D—a simple procedure once U and U^{-1} have been obtained from the characteristic vectors of A. Development of U is now discussed.

The first step involved in calculating the characteristic vectors of a matrix is to obtain its characteristic roots by solving the characteristic equation. Methods for determining the corresponding characteristic vectors then depend on certain conditions relative to the characteristic roots. Two situations are initially distinguishable: when the characteristic roots are all different, and when they are not all different.

a.   Characteristic roots all different

When the characteristic roots of a matrix are all different the corresponding characteristic vectors form a linearly independent set of vectors.[1] If the characteristic vectors are then used as columns of a matrix

$$U = [u_1 \quad u_2 \quad \cdots \quad u_n],$$

U is square of order n and, because its columns are independent, $r(U) = n$, U is nonsingular and U^{-1} exists. Therefore, by writing

$$Au_i = \lambda_i u_i, \qquad \text{for} \quad i = 1, 2, \ldots, n,$$

as

$$A[u_1 \quad u_2 \quad \cdots \quad u_n] = [u_1 \quad u_2 \quad \cdots \quad u_n] \begin{bmatrix} \lambda_1 & 0 & \cdots & & & 0 \\ 0 & \lambda_2 & & & & \cdot \\ & & & & & \cdot \\ \cdot & & & & & \cdot \\ \cdot & & & & \cdot & 0 \\ \cdot & & & & & \cdot \\ 0 & \cdots & & & 0 & \lambda_n \end{bmatrix}$$

because all u_i are column vectors, we obtain

$$AU = UD$$

where D is the n-order diagonal matrix of the characteristic roots $\lambda_1, \lambda_2, \ldots, \lambda_n$. Hence

$$A = UDU^{-1} \quad \text{and} \quad D = U^{-1}AU. \tag{7}$$

D is often referred to as the *canonical form of A under similarity*.

----

[1] See Searle (1966, p. 167) for proof.

Example (*continued*). It has already been shown that the matrix $A = \begin{bmatrix} 1 & 4 \\ 9 & 1 \end{bmatrix}$ has characteristic roots -5 and 7, and that the corresponding characteristic vectors are $\begin{bmatrix} 2 \\ -3 \end{bmatrix}$ and $\begin{bmatrix} 2 \\ 3 \end{bmatrix}$. The matrix U which has these vectors as columns is

$$U = \begin{bmatrix} 2 & 2 \\ -3 & 3 \end{bmatrix} \quad \text{with} \quad U^{-1} = \tfrac{1}{12} \begin{bmatrix} 3 & -2 \\ 3 & 2 \end{bmatrix},$$

and

$$U^{-1}AU = \tfrac{1}{12} \begin{bmatrix} 3 & -2 \\ 3 & 2 \end{bmatrix} \begin{bmatrix} 1 & 4 \\ 9 & 1 \end{bmatrix} \begin{bmatrix} 2 & 2 \\ -3 & 3 \end{bmatrix}$$

$$= \begin{bmatrix} -5 & 0 \\ 0 & 7 \end{bmatrix}$$

$$= D,$$

the diagonal matrix of the characteristic roots.[1]

As indicated earlier, equations (7) provide a useful method of deriving the powers of A once U and U^{-1} have been obtained. For then only the powers of D are required, and they are easy to calculate because D is diagonal. Thus from (7), for t being an integer,

$$A^t = UD^tU^{-1} \tag{8}$$

including, if it exists, $(A^{-1})^t \equiv A^{-t} = U(D^{-1})^tU^{-1} = UD^{-t}U^{-1}$.

Illustration. Suppose the state of a machine at period 0 is described by the vector $x_0' = [x_{01} \quad x_{02}]$ where x_{01} is the probability that the machine is working properly at period 0 and x_{02} is the probability that it needs adjustment. It has been shown (Sections 2.5d and 8.1) that the state of the machine n periods later equals the product of the initial state vector multiplied by the nth power of the transition probability matrix P:

$$x_n' = x_0'P^n.$$

If the matrix P is reduced to its canonical form under similarity, $P = UDU^{-1}$, then its nth power is

$$P^n = UD^nU^{-1}$$

where

$$D^n = \begin{bmatrix} \lambda_1^n & 0 \\ 0 & \lambda_2^n \end{bmatrix}.$$

---

[1] Characteristic roots can be complex or real numbers; see Section 7c.

This method of calculation may often be simpler than calculating P^n by successive matrix multiplications.

b. Characteristic vectors when roots are all different

For each λ_i that is a solution of the characteristic equation we have, by (4), $Au_i = \lambda_i u_i$, and hence

$$(A - \lambda_i I)u_i = 0. \qquad (9)$$

Now λ_i has been derived as a solution to the characteristic equation $|A - \lambda I| = 0$. Therefore $|A - \lambda_i I| = 0$ and consequently (see Section 7.4c) equation (9) has $n - r_i$ LIN solutions for u_i, where n is the order of A and r_i is the rank of $A - \lambda_i I$. When the characteristic roots are all different $r_i = n - 1$ for all i (this being a special case of the theorem in the next section). Thus (9) has just a single LIN solution for u_i, which can be obtained by giving an arbitrary value to one element of u_i in (9) and solving the resultant equations for the other $n - 1$ elements. This is done for each λ_i in turn, to obtain the respective u_i vectors.

Example. For our initial example

$$A = \begin{bmatrix} 1 & 4 \\ 9 & 1 \end{bmatrix},$$

the characteristic equation is

$$\begin{vmatrix} 1 - \lambda & 4 \\ 9 & 1 - \lambda \end{vmatrix} = 0 \qquad \text{or} \qquad (1 - \lambda)^2 - 36 = 0$$

and the characteristic roots of A are $\lambda_1 = -5$ and $\lambda_2 = 7$.

Denoting the elements of u_i by a_i and b_i, equation (9) is

$$\begin{bmatrix} 1 - \lambda_i & 4 \\ 9 & 1 - \lambda_i \end{bmatrix} \begin{bmatrix} a_i \\ b_i \end{bmatrix} = 0, \qquad \text{for} \qquad i = 1, 2.$$

For $\lambda_1 = -5$ this becomes

$$6a_1 + 4b_1 = 0$$

and

$$9a_1 + 6b_1 = 0.$$

Taking an arbitrary value of $a_1 = 2$ leads to the solution $b_1 = -3$ and consequently the corresponding characteristic vector is $u_1 = \begin{bmatrix} 2 \\ -3 \end{bmatrix}$.

Similarly, for the second characteristic root, $\lambda_2 = 7$, the equations are

$$-6a_2 + 4b_2 = 0,$$

and

$$9a_2 - 6b_2 = 0.$$

Taking an arbitrary value of $a_2 = 2$ leads to the solution $b_2 = 3$, so the characteristic vector is $u_2 = \begin{bmatrix} 2 \\ 3 \end{bmatrix}$.

The matrix of the characteristic vectors is

$$U = [u_1 \quad u_2] = \begin{bmatrix} 2 & 2 \\ -3 & 3 \end{bmatrix}$$

for which $\qquad |U| = 12 \qquad$ and $\qquad U^{-1} = \tfrac{1}{12}\begin{bmatrix} 3 & -2 \\ 3 & 2 \end{bmatrix}$

and, as in (7),

$$D = U^{-1}AU = \tfrac{1}{12}\begin{bmatrix} 3 & -2 \\ 3 & 2 \end{bmatrix}\begin{bmatrix} 1 & 4 \\ 9 & 1 \end{bmatrix}\begin{bmatrix} 2 & 2 \\ -3 & 3 \end{bmatrix} = \begin{bmatrix} -5 & 0 \\ 0 & 7 \end{bmatrix} = \begin{bmatrix} \lambda_1 & 0 \\ 0 & \lambda_2 \end{bmatrix}.$$

In solving the set of equations (9) for each of the two characteristic vectors u_1 and u_2 use was made of arbitrary values $a_1 = 2$ and $a_2 = 2$. We now demonstrate that this procedure of finding U and of having $D = U^{-1}AU$ as in (7) is unaffected by the choice of these arbitrary values. Suppose we had chosen $a_1 = 1$ and $a_2 = 3$. Then U would have been

$$U = \begin{bmatrix} 1 & 3 \\ -1.5 & 4.5 \end{bmatrix}$$

with $\qquad |U| = 9 \qquad$ and $\qquad U^{-1} = \tfrac{1}{9}\begin{bmatrix} 4.5 & -3 \\ 1.5 & 1 \end{bmatrix}$

and

$$D = U^{-1}AU = \tfrac{1}{9}\begin{bmatrix} 4.5 & -3 \\ 1.5 & 1 \end{bmatrix}\begin{bmatrix} 1 & 4 \\ 9 & 1 \end{bmatrix}\begin{bmatrix} 1 & 3 \\ -1.5 & 4.5 \end{bmatrix} = \begin{bmatrix} -5 & 0 \\ 0 & 7 \end{bmatrix} = \begin{bmatrix} \lambda_1 & 0 \\ 0 & \lambda_2 \end{bmatrix}$$

as before. This result is true in general; whenever A can be expressed as UDU^{-1}, solution of the equations (9), $(A - \lambda_i I)u_i = 0$, for u_i, given λ_i, can be obtained by use of an arbitrary value for an element of u_i.

Example. For

$$A = \begin{bmatrix} 1 & 4 & 1 \\ 2 & 1 & 0 \\ -1 & 3 & 1 \end{bmatrix}$$

the characteristic equation is

$$\begin{vmatrix} 1 - \lambda & 4 & 1 \\ 2 & 1 - \lambda & 0 \\ -1 & 3 & 1 - \lambda \end{vmatrix} = 0$$

which expands, by diagonal expansion (see Section 4.6), as

$$-\lambda^3 + \lambda^2(1 + 1 + 1) - \lambda \left(\begin{vmatrix} 1 & 4 \\ 2 & 1 \end{vmatrix} + \begin{vmatrix} 1 & 1 \\ -1 & 1 \end{vmatrix} + \begin{vmatrix} 1 & 0 \\ 3 & 1 \end{vmatrix} \right)$$

$$+ \begin{vmatrix} 1 & 4 & 1 \\ 2 & 1 & 0 \\ -1 & 3 & 1 \end{vmatrix} = 0,$$

and reduces to $\lambda(\lambda^2 - 3\lambda - 4) = 0$. Hence the characteristic roots of A are $\lambda = 0$, -1 and 4. Denoting the elements of u_i by a_i, b_i and c_i, equation (9) is

$$\begin{bmatrix} 1 - \lambda_i & 4 & 1 \\ 2 & 1 - \lambda_i & 0 \\ -1 & 3 & 1 - \lambda_i \end{bmatrix} \begin{bmatrix} a_i \\ b_i \\ c_i \end{bmatrix} = 0.$$

For $\lambda_1 = 0$ this becomes

$$a_1 + 4b_1 + c_1 = 0$$
$$2a_1 + b_1 \quad\quad = 0$$
$$-a_1 + 3b_1 + c_1 = 0.$$

Taking the arbitrary value $a_1 = 1$ leads to solutions $b_1 = -2$ and $c_1 = 7$, and the corresponding characteristic vector is $u_1 = \begin{bmatrix} 1 \\ -2 \\ 7 \end{bmatrix}$. Similarly for the second solution to the characteristic equation, $\lambda_2 = -1$, the equations are

$$2a_2 + 4b_2 + c_2 \quad = 0$$
$$2a_2 + 2b_2 \quad\quad = 0$$
$$-a_2 + 3b_2 + 2c_2 = 0,$$

and taking $a_2 = 2$ as the arbitrary value gives $b_2 = -2$ and $c_2 = 4$. Thus the second characteristic vector is $u_2 = \begin{bmatrix} 2 \\ -2 \\ 4 \end{bmatrix}$. Finally, for $\lambda_3 = 4$, the equations have solutions associated with the arbitrary value $a_3 = 3$ of $b_3 = 2$ and $c_3 = 1$. Hence $u_3 = \begin{bmatrix} 3 \\ 2 \\ 1 \end{bmatrix}$ and the matrix of the characteristic vectors is

$$U = [u_1 \; u_2 \; u_3] = \begin{bmatrix} 1 & 2 & 3 \\ -2 & -2 & 2 \\ 7 & 4 & 1 \end{bmatrix}.$$

It may be verified that $|U| = 40$, that

$$U^{-1} = \tfrac{1}{20} \begin{bmatrix} -5 & 5 & 5 \\ 8 & -10 & -4 \\ 3 & 5 & 1 \end{bmatrix}$$

and hence, as in (7),

$$D = U^{-1}AU = \begin{bmatrix} 0 & 0 & 0 \\ 0 & -1 & 0 \\ 0 & 0 & 4 \end{bmatrix} = \begin{bmatrix} \lambda_1 & 0 & 0 \\ 0 & \lambda_2 & 0 \\ 0 & 0 & \lambda_3 \end{bmatrix}.$$

Likewise $A = UDU^{-1}$ and

$$A^t = UD^tU^{-1} = \tfrac{1}{20} \begin{bmatrix} 1 & 2 & 3 \\ -2 & -2 & 2 \\ 7 & 4 & 1 \end{bmatrix} \begin{bmatrix} 0 & 0 & 0 \\ 0 & (-1)^t & 0 \\ 0 & 0 & 4^t \end{bmatrix} \begin{bmatrix} -5 & 5 & 5 \\ 8 & -10 & -4 \\ 3 & 5 & 1 \end{bmatrix}.$$

The reader will find it worthwhile to carry out these multiplications and satisfy himself that the results are as shown.

Illustration. Consider a machine (as described in Section 8.3) which can be in one of three states: (1) broken beyond repair, (2) in need of adjustment or (3) working properly. The probabilities of the machine changing from one state to another are arrayed in the transition probability matrix

$$P = \begin{bmatrix} 1 & 0 & 0 \\ \tfrac{1}{4} & \tfrac{1}{2} & \tfrac{1}{4} \\ \tfrac{1}{18} & \tfrac{8}{18} & \tfrac{9}{18} \end{bmatrix}.$$

Under these conditions we might ask ourselves: "What is the probability that the machine will be in State i after k time periods?" We may also wish to know the probability of being in State i when k becomes infinitely large (i.e. the steady-state probability; see Section 8.2). If x_0' is the initial state vector then the state vector k periods later is $x_k' = x_0'P^k$ (Section 2.5d), and so the problem involves deriving the kth power of P. As already indicated, equation (7) may be used for this problem. The calculations are summarized below.

The characteristic equation of P has roots 1, $\frac{1}{6}$ and $\frac{5}{6}$. The equations to find the characteristic vectors are

$$\begin{bmatrix} 1 - \lambda_i & 0 & 0 \\ \frac{1}{4} & \frac{1}{2} - \lambda_i & \frac{1}{4} \\ \frac{1}{18} & \frac{8}{18} & \frac{9}{18} - \lambda_i \end{bmatrix} \begin{bmatrix} a_i \\ b_i \\ c_i \end{bmatrix} = 0.$$

For $\lambda_1 = 1$, and taking $a_1 = 1$ as an arbitrary value, the characteristic

vector is $u_1 = \begin{bmatrix} 1 \\ 1 \\ 1 \end{bmatrix}$; with $\lambda_2 = \frac{1}{6}$ and $a_2 = 0$ the vector is $u_2 = \begin{bmatrix} 0 \\ 3 \\ -4 \end{bmatrix}$; and

for $\lambda_3 = \frac{5}{6}$ and $a_3 = 0$ the vector is $u_3 = \begin{bmatrix} 0 \\ 3 \\ 4 \end{bmatrix}$. Hence

$$U = \begin{bmatrix} 1 & 0 & 0 \\ 1 & 3 & 3 \\ 1 & -4 & 4 \end{bmatrix} \quad \text{with} \quad U^{-1} = \frac{1}{24} \begin{bmatrix} 24 & 0 & 0 \\ -1 & 4 & -3 \\ -7 & 4 & 3 \end{bmatrix}.$$

Thus

$$P^k = UD^kU^{-1} = \frac{1}{24} \begin{bmatrix} 1 & 0 & 0 \\ 1 & 3 & 3 \\ 1 & -4 & 4 \end{bmatrix} \begin{bmatrix} 1 & 0 & 0 \\ 0 & (\frac{1}{6})^k & 0 \\ 0 & 0 & (\frac{5}{6})^k \end{bmatrix} \begin{bmatrix} 24 & 0 & 0 \\ -1 & 4 & -3 \\ -7 & 4 & 3 \end{bmatrix},$$

which reduces to

$$P^k = \begin{bmatrix} 1 & 0 & 0 \\ 1 - r_k/2 - 3s_k/8 & r_k/2 & 3s_k/8 \\ 1 - r_k/2 - 2s_k/3 & 2s_k/3 & r_k/2 \end{bmatrix}$$

where $r_k = (\frac{5}{6})^k + (\frac{1}{6})^k$ and $s_k = (\frac{5}{6})^k - (\frac{1}{6})^k$.

From these expressions the elements of P^k are readily obtainable for any value of k. For example, if $k = 2$

$$r_2 = (\tfrac{5}{6})^2 + (\tfrac{1}{6})^2 = \tfrac{26}{36} \quad \text{and} \quad s_2 = (\tfrac{5}{6})^2 - (\tfrac{1}{6})^2 = \tfrac{24}{36} = \tfrac{2}{3},$$

and

$$P^2 = \begin{bmatrix} 1 & 0 & 0 \\ 1 - \frac{26}{72} - \frac{1}{4} & \frac{26}{72} & \frac{1}{4} \\ 1 - \frac{26}{72} - \frac{4}{9} & \frac{4}{9} & \frac{26}{72} \end{bmatrix} \doteq \begin{bmatrix} 1 & 0 & 0 \\ .39 & .36 & .25 \\ .20 & .44 & .36 \end{bmatrix}.$$

From the form of P^k given above, involving r_k and s_k which tend to zero when k is infinitely large, we see that the value of P^k becomes $\begin{bmatrix} 1 & 0 & 0 \\ 1 & 0 & 0 \\ 1 & 0 & 0 \end{bmatrix}$ as

k becomes infinitely large. This limiting form of P^k means that ultimately, irrespective of its initial state, the machine will be in State 1 (irreparably broken). Such a state, which ultimately collects a whole system, is called a trapping state or an absorbing state (see Section 8.3).

c. Characteristic roots not all different

As stated above, when the characteristic roots of A are all different, U is of full rank, U^{-1} exists and $A = UDU^{-1}$. When the characteristic roots are not all different, the matrix U^{-1} may or may not exist; when it does not, A cannot be expressed in the form $A = UDU^{-1}$. Conditions under which this form does exist are stated in the theorem that follows.

First, we define m_k to be the number of times λ_k is a root of the characteristic equation $|A - \lambda I| = 0$. We say that λ_k is a root having multiplicity m_k for $k = 1, \ldots, s$. This means that $(\lambda - \lambda_k)^{m_k}$ is a factor of $|A - \lambda I|$, and hence, because the characteristic equation is a polynomial of degree n in λ, $\sum\limits_{k=1}^{s} m_k = n$. The case when $m_k = 1$ for all values of k is included here, it being the situation already discussed, that of all characteristic roots being different.

Theorem.[1] For a square matrix A, of order n, whose characteristic roots are $\lambda_1, \lambda_2, \ldots, \lambda_s$ with multiplicities $m_1, m_2, \ldots, m_s$, where $\sum\limits_{k=1}^{s} m_k = n$, the necessary and sufficient condition under which A can be expressed in the form $A = UDU^{-1}$ where D is a diagonal matrix of all the n characteristic roots, is that the rank of $(A - \lambda_k I)$ be equal to $n - m_k$ for all $k = 1, 2, \ldots, s$.

The importance of the theorem is that for any root λ having multiplicity m the form $A = UDU^{-1}$ exists if and only if

$$r(A - \lambda I) = n - m;$$

and this includes the case of single roots, for which $m = 1$.

Example. For

$$A = \begin{bmatrix} -1 & -2 & -2 \\ 1 & 2 & 1 \\ -1 & -1 & 0 \end{bmatrix}$$

the characteristic equation reduces to $(\lambda - 1)^2(\lambda + 1) = 0$ giving roots 1, 1 and -1. Hence $\lambda = 1$ is a multiple root with multiplicity 2 for which the

---

[1] See Searle (1966, p. 175) for proof.

rank of $A - \lambda I$ must be investigated. Thus for $\lambda = 1$

$$A - \lambda I = \begin{bmatrix} -2 & -2 & -2 \\ 1 & 1 & 1 \\ -1 & -1 & -1 \end{bmatrix}$$

and its rank is $1 = 3 - 2 = n - m$. Therefore the form $A = UDU^{-1}$ exists. To find U, independent vectors u must be obtained such that for $\lambda = 1$,

$(A - \lambda I)u = 0$. Two such vectors are $u_1 = \begin{bmatrix} 1 \\ -1 \\ 0 \end{bmatrix}$ and $u_2 = \begin{bmatrix} 1 \\ 0 \\ -1 \end{bmatrix}$. For

$\lambda_3 = -1$ the equations $(A - \lambda_3 I)u_3 = 0$ have the solution $u_3 = \begin{bmatrix} 2 \\ -1 \\ 1 \end{bmatrix}$ and

hence $U = [u_1 \ u_2 \ u_3] = \begin{bmatrix} 1 & 1 & 2 \\ -1 & 0 & -1 \\ 0 & -1 & 1 \end{bmatrix}$. The reader should satisfy

himself that $D = U^{-1}AU$ is the appropriate diagonal matrix, and hence $A = UDU^{-1}$.

Example. For the matrix

$$A = \begin{bmatrix} 2 & -1 & 1 \\ 3 & 3 & -2 \\ 4 & 1 & 0 \end{bmatrix},$$

the characteristic equation has roots $\lambda = 1, 1$ and 3. For the multiple root of 1, with multiplicity $m = 2$, $A - \lambda I$ is

$$(A - I) = \begin{bmatrix} 1 & -1 & 1 \\ 3 & 2 & -2 \\ 4 & 1 & -1 \end{bmatrix},$$

which has rank 2 (row 3 equals row 1 plus row 2). Thus with $n = 3$ and $m = 2$, $r(A - I) = 2 \neq 3 - 2$, so the form $D = U^{-1}AU$ does not exist.

3. SOME PROPERTIES OF CHARACTERISTIC ROOTS

Properties of characteristic roots and vectors are utilized in many ways. A few are now discussed.

a. Powers of characteristic roots

If λ_i is a solution of the characteristic equation of A and if u_i is the corresponding characteristic vector, then by (4) $Au_i = \lambda_i u_i$. Pre-multiplication by A gives

$$A^2 u_i = A\lambda_i u_i = \lambda_i Au_i \tag{10}$$

because λ_i is a scalar, and using (4) again (10) becomes $A^2 u_i = \lambda_i^2 u_i$. This result is easily extended so that

$$A^k u_i = \lambda_i^k u_i, \tag{11}$$

where k is any positive integer. It is also true for negative integers where $A^{-k} = (A^{-1})^k$ if A^{-1} exists (see Exercise 14). By comparison it is apparent that if $Au = \lambda u$ because λ is a characteristic root and u the corresponding vector, then λ^k in $A^k u = \lambda^k u$ must be a characteristic root of A^k. Hence it is concluded that if λ is a characteristic root of A and if A is singular, λ^k is a characteristic root of A^k for positive integers k; and if A is nonsingular λ^k is a characteristic root of A^k for positive or negative integers k. In particular, when A is nonsingular and $k = -1$, if λ is a characteristic root of A, then $1/\lambda$ is a characteristic root of A^{-1}.

Example. If

$$A = \begin{bmatrix} 4 & 3 \\ 4 & 8 \end{bmatrix}, \quad A^2 = \begin{bmatrix} 28 & 36 \\ 48 & 76 \end{bmatrix} \quad \text{and} \quad A^{-1} = \tfrac{1}{20}\begin{bmatrix} 8 & -3 \\ -4 & 4 \end{bmatrix}.$$

The characteristic equations are as follows:

for A: $\lambda^2 - 12\lambda + 20 = 0$, with roots 2 and 10;

for A^2: $\lambda^2 - 104\lambda + 400 = 0$, with roots 4 and 100;

for A^{-1}: $20\lambda^2 - 12\lambda + 1 = 0$, with roots $\tfrac{1}{2}$ and $\tfrac{1}{10}$.

The characteristic roots of A^2 are the squares of those of A, and the characteristic roots of A^{-1} are the reciprocals of those of A.

b. Sum and product of characteristic roots

By diagonal expansion (Section 4.6) the equation

$$|A - \lambda I| = \begin{vmatrix} a_{11} - \lambda & a_{12} & a_{13} \\ a_{21} & a_{22} - \lambda & a_{23} \\ a_{31} & a_{32} & a_{33} - \lambda \end{vmatrix} = 0$$

reduces to $-\lambda^3 + (-\lambda)^2 \operatorname{tr}_1(A) + (-\lambda)\operatorname{tr}_2(A) + |A| = 0$

where $\operatorname{tr}_i(A)$ is the sum of the principal minors of order i in A (see Section 4.6). For A of order n this equation takes the form

$$(-\lambda)^n + (-\lambda)^{n-1} \operatorname{tr}_1(A) + (-\lambda)^{n-2} \operatorname{tr}_2(A) + \cdots + (-\lambda)\operatorname{tr}_{n-1}(A) + |A| = 0.$$

From this it can be shown (using the theory of equations) that

$$\sum_{i=1}^{n} \lambda_i = \text{tr}_1(A) \quad \text{and} \quad \prod_{i=1}^{n} \lambda_i = |A|.$$

Thus the sum of the characteristic roots of a matrix equals the trace of the matrix, and the product of the characteristic roots is the determinant of the matrix.

Example. The characteristic roots of $A = \begin{bmatrix} 1 & 4 & 1 \\ 2 & 1 & 0 \\ -1 & 3 & 1 \end{bmatrix}$ are 0, -1 and 4.

The sum is 3, which equals the trace of A, $1 + 1 + 1$, and their product is zero, equal to the determinant of A. (The first row equals the sum of the second and third rows.)

c. Non-zero characteristic roots

The conditions under which a matrix having multiple characteristic roots can be expressed in the form $U^{-1}AU = D$ have already been discussed. When it can be so expressed U is nonsingular and therefore

$$\text{rank}(D) = \text{rank}(A) = r.$$

But the rank of D is identical to the number of non-zero characteristic roots; hence the number of such roots is r, and we can write

$$U^{-1}AU = D = \begin{bmatrix} D_r^* & 0 \\ 0 & 0 \end{bmatrix}$$

where D_r^* is diagonal, of order r, its diagonal elements being the r non-zero characteristic roots of A.

Example. The characteristic roots of $\begin{bmatrix} 2 & 3 & 7 \\ 1 & 2 & 4 \\ 1 & 1 & 3 \end{bmatrix}$ are given by

$$\lambda(\lambda^2 - 7\lambda + 2) = 0.$$

Thus the number of non-zero roots is 2, the rank of the matrix.

d. Characteristic roots of a scalar product

If λ is a characteristic root of A and u is a corresponding characteristic vector, $Au = \lambda u$. Therefore, for a scalar c, $cAu = c\lambda u$, showing that $c\lambda$ is a characteristic root of cA; i.e. if λ is a characteristic root of A, $c\lambda$ is a characteristic root of cA. The equation $cAu = c\lambda u$ can also be written as $A(cu) = \lambda(cu)$, showing that if u is a characteristic vector of A so is cu.

Example. $A = \begin{bmatrix} 1 & 1 \\ 4 & 1 \end{bmatrix}$ has characteristic equation $(1 - \lambda)^2 = 4$ and hence

its characteristic roots are 3 and -1. But the characteristic equation of $5A$ is $(5 - \lambda)^2 = 100$, giving characteristic roots 15 and -5 which are 5 times those of A. The vector $\begin{bmatrix} 1 \\ 2 \end{bmatrix}$ is a characteristic vector of A corresponding to the characteristic root $\lambda = 3$; as is also the vector $\begin{bmatrix} 4 \\ 8 \end{bmatrix}$.

e. The Cayley-Hamilton theorem[1]

This theorem states that every square matrix satisfies its own characteristic equation. The theorem can be usefully applied when the characteristic equation is known, even if the roots are not. The nth and successive powers of A can be obtained as linear functions of A of power $n - 1$ and less. The inverse of A, if it exists, can also be obtained in a similar manner.

Example. For

$$A = \begin{bmatrix} 4 & 3 \\ 2 & 5 \end{bmatrix}, \qquad A^2 = \begin{bmatrix} 22 & 27 \\ 18 & 31 \end{bmatrix}, \qquad A^{-1} = \tfrac{1}{14}\begin{bmatrix} 5 & -3 \\ -2 & 4 \end{bmatrix}$$

and the characteristic equation of A is

$$\lambda^2 - 9\lambda + 14 = 0.$$

The Cayley-Hamilton theorem states that we may substitute A for λ:

$$A^2 - 9A + 14I = \begin{bmatrix} 22 & 27 \\ 18 & 31 \end{bmatrix} - 9\begin{bmatrix} 4 & 3 \\ 2 & 5 \end{bmatrix} + 14\begin{bmatrix} 1 & 0 \\ 0 & 1 \end{bmatrix} = \begin{bmatrix} 0 & 0 \\ 0 & 0 \end{bmatrix}.$$

Powers of A come from rewriting $A^2 - 9A + 14I = 0$ as $A^2 = 9A - 14I$.

Hence $\quad A^3 = 9A^2 - 14A \quad = 9(9A - 14I) - 14A \quad = 67A - 126I$

and $\quad A^4 = 67A^2 - 126A = 67(9A - 14I) - 126A = 477A - 938I$.

Similar expressions can be derived for the inverse A^{-1} if it exists. For $A^2 - 9A + 14I = 0$ is equivalent to

$$14I = 9A - A^2$$

so giving $\qquad A^{-1} = \tfrac{1}{14}(9I - A).$

Thus $\qquad (A^{-1})^2 = \tfrac{1}{196}(81I + A^2 - 18A)$

$$= \tfrac{1}{196}(67I - 9A),$$

and so on.

---

[1] See Searle (1966, p. 179) for proof.

Example. Consider the transition probability matrix in Section 2.5d:

$$P = \begin{bmatrix} .90 & .10 \\ .01 & .99 \end{bmatrix}.$$

Its characteristic equation is

$$\lambda^2 - 1.89\lambda + .89 = 0.$$

By the Cayley-Hamilton theorem, $P^2 - 1.89P + .89I = 0$. Rearranging,

$$P^2 = 1.89P - .89I;$$

hence $P^3 = 1.89P^2 - .89P$

$$= 1.89(1.89P - .89I) - .89P = 2.6821P - 1.6821I.$$

Any power of P may be derived by this means.

4. SYSTEMS OF LINEAR FIRST-ORDER DIFFERENCE EQUATIONS

Consider the system of first-order difference equations[1]

$$x_t = Ax_{t-1} + b \tag{12}$$

where x_t is a vector of variables measured at time t and its initial value is x_0. Our objectives are to write x_t as a function of x_0 and to ascertain if x_t approaches a limiting form x as t tends toward infinity.

Letting $t = 1, 2, \ldots$, we find

$$x_1 = Ax_0 + b$$
$$x_2 = Ax_1 + b = A(Ax_0 + b) + b = A^2x_0 + (A + I)b$$
$$x_3 = Ax_2 + b = A[A^2x_0 + (A + I)b] + b = A^3x_0 + (A^2 + A + I)b$$
$$\cdot$$
$$\cdot$$
$$\cdot$$
$$x_t = Ax_{t-1} + b = A^tx_0 + (A^{t-1} + A^{t-2} + \cdots + I)b.$$

From Section 5.6

$$I + A + A^2 + \cdots + A^{t-1} = (I - A^t)(I - A)^{-1}$$

provided $(I - A)^{-1}$ exists. Hence the general solution to (12) is

$$x_t = A^tx_0 + (I - A^t)(I - A)^{-1}b, \qquad t = 1, 2, \ldots. \tag{13}$$

---

[1] Goldberg (1958) has numerous examples of linear difference equations.

(The Cayley-Hamilton theorem is useful here in providing a method of obtaining A^t.) It follows from (13) that a sufficient condition for the vector x_t to approach a vector of constant values independent of x_0 as t tends toward infinity is that $A^t \to 0$ as $t \to \infty$. In that case the term $(I - A^t)$ in (13) approaches the identity matrix in the limit and the limiting form of x_t is $(I - A)^{-1}b$.

A Markov chain is a special case of (13). For then $A = P'$ is the transpose of a transition probability matrix and $b = 0$, so that

$$x_t = (P')^t x_0 = (P^t)' x_0.$$

Now, provided the steady-state probabilities exist (Section 8.3d), denote them by the vector x'. Then by definition

$$\lim_{t \to \infty} P^t = \begin{bmatrix} x' \\ x' \\ \cdot \\ \cdot \\ \cdot \\ x' \end{bmatrix}.$$

Therefore

$$\lim_{t \to \infty} x_t = \lim_{t \to \infty} (P^t)' x_0 = [x \quad x \quad \cdots \quad x] x_0.$$

But in the Markov chain situation x_0 is a state probability vector and so its elements sum to unity. Thus

$$\lim_{t \to \infty} x_t = [x \quad x \quad \cdots \quad x] x_0 = x;$$

i.e. the limiting form of x_t is the column vector x of steady-state probabilities, independent of the elements of x_0 (see Exercise 15).

The limiting form of A^t for the general case in (12) can be investigated by methods developed in Section 2. Assuming A is such that the theorem in Section 2c holds and the form $A = UDU^{-1}$ exists, then from (8)

$$A^t = UD^t U^{-1}$$

where D is the diagonal matrix whose diagonal elements are the characteristic roots, and the columns of U are the characteristic vectors corresponding to the roots. If $D^t \to 0$ as $t \to \infty$ then $A^t \to 0$ as $t \to \infty$. In order for this to occur each diagonal element of D (i.e., each characteristic root) must be less than 1 in absolute value. Thus a sufficient condition for a limiting form x to exist for the system of equations $x_t = Ax_{t-1} + b$ is that every characteristic root of A have an absolute value less than unity. If this condition holds then the limiting form of x_t as $t \to \infty$ is $x = (I - A)^{-1}b$.

Illustration. Metzler (1942, 1950) discusses the effect of foreign trade on national income. As a simple illustration we consider a system involving just two countries and hence define a series of 2×1 vectors: for example $y_t = \begin{bmatrix} y_{1t} \\ y_{2t} \end{bmatrix}$ is the vector representing national income in time period t for countries one and two. We make the following assumptions:

(i) National income y_t equals total consumption c_t plus net investment s_t plus exports x_t minus imports m_t:

$$y_t = c_t + s_t + x_t - m_t. \tag{14}$$

(ii) Consumption expenditures for domestically produced goods and services, d_t, equals total consumption minus imports:

$$d_t = c_t - m_t. \tag{15}$$

(iii) In each country, domestic consumption and imports are constant multiples of national income in the previous time period:

$$d_t = \begin{bmatrix} \alpha_1 & 0 \\ 0 & \alpha_2 \end{bmatrix} y_{t-1} \quad \text{and} \quad m_t = \begin{bmatrix} \beta_1 & 0 \\ 0 & \beta_2 \end{bmatrix} y_{t-1}. \tag{16}$$

(iv) Net investment is constant over time:

$$s_t = s_0 \quad \text{for all } t. \tag{17}$$

(v) Exports of each country equal imports of the other (because we have only two countries):

$$x_t = \begin{bmatrix} 0 & 1 \\ 1 & 0 \end{bmatrix} m_t. \tag{18}$$

Substituting from (15) and (17) into (14) gives

$$y_t = d_t + s_0 + x_t, \tag{19}$$

and on using (16) and (18) this becomes

$$y_t = \begin{bmatrix} \alpha_1 & 0 \\ 0 & \alpha_2 \end{bmatrix} y_{t-1} + s_0 + \begin{bmatrix} 0 & 1 \\ 1 & 0 \end{bmatrix} \begin{bmatrix} \beta_1 & 0 \\ 0 & \beta_2 \end{bmatrix} y_{t-1}$$

$$= By_{t-1} + s_0 \tag{20}$$

where $B = \begin{bmatrix} \alpha_1 & \beta_2 \\ \beta_1 & \alpha_2 \end{bmatrix}.$

Equation (20) is exactly the form $x_t = Ax_{t-1} + b$ considered in (12) and so therefore its solution follows (13):

$$y_t = B^t y_0 + (I - B^t)(I - B)^{-1} s_0 \qquad (21)$$

if $(I - B)^{-1}$ exists. Let us consider this for

$$B = \begin{bmatrix} 0.8 & 0.2 \\ 0.1 & 0.7 \end{bmatrix}.$$

Its characteristic roots are 0.6 and 0.9; and the corresponding characteristic vectors are $\begin{bmatrix} 1 \\ -1 \end{bmatrix}$ and $\begin{bmatrix} 2 \\ 1 \end{bmatrix}$. Thus $B = UDU^{-1}$ with $D = \begin{bmatrix} 0.6 & 0 \\ 0 & 0.9 \end{bmatrix}$, and since $D^t = \begin{bmatrix} 0.6^t & 0 \\ 0 & 0.9^t \end{bmatrix} \to \begin{bmatrix} 0 & 0 \\ 0 & 0 \end{bmatrix}$ as $t \to \infty$, we know $B^t \to \begin{bmatrix} 0 & 0 \\ 0 & 0 \end{bmatrix}$ as $t \to \infty$. Therefore, denoting the equilibrium income level vector as y,

$$y = (I - B)^{-1} s_0 = \frac{1}{0.04} \begin{bmatrix} 0.3 & 0.2 \\ 0.1 & 0.2 \end{bmatrix} s_0 = \begin{bmatrix} 7.5 & 5 \\ 2.5 & 5 \end{bmatrix} s_0.$$

If $s_0 = \begin{bmatrix} 100 \\ 200 \end{bmatrix}$, $y = \begin{bmatrix} 1,750 \\ 1,250 \end{bmatrix}$; i.e. $y_t = y_{t-1} = y$ satisfies $y_t = By_{t-1} + s_0$:

$$\begin{bmatrix} 1,750 \\ 1,250 \end{bmatrix} = \begin{bmatrix} 0.8 & 0.2 \\ 0.1 & 0.7 \end{bmatrix} \begin{bmatrix} 1,750 \\ 1,250 \end{bmatrix} + \begin{bmatrix} 100 \\ 200 \end{bmatrix} = \begin{bmatrix} 1,750 \\ 1,250 \end{bmatrix},$$

and $y_t \to y$ as $t \to \infty$. By substituting different import constants β_i in the matrix B [see equation (20)], the effect of various import policies on the equilibrium incomes of each country can be determined.

Linear difference equations of higher than first order can be converted into a system of first-order equations and then solved in that form. We demonstrate the procedure with a second-order equation.

Example. Consider the second-order difference equation

$$x_t - 3x_{t-1} + 2x_{t-2} = 0, \qquad \text{for } t = 2, 3, \ldots \qquad (22)$$

with initial conditions $x_0 = 0$ and $x_1 = 1$. The objective is to write (22) as a system of two first-order difference equations. This is done as follows: Let $y_t = x_{t-1}$ for $t = 1, 2, \ldots$. Then (22) may be written as

$$x_t = 3x_{t-1} - 2y_{t-1} \qquad \text{and} \qquad y_t = x_{t-1} \qquad \text{for } t = 2, 3, \ldots \qquad (23)$$

with initial conditions $x_1 = 1$ and $y_1 = 0$. Then, letting

$$v_t = \begin{bmatrix} x_t \\ y_t \end{bmatrix},$$

we may write (23) as a set of two first-order difference equations:

$$v_t = Av_{t-1}, \qquad \text{for } t = 2, 3, \ldots \tag{24}$$

where $\quad A = \begin{bmatrix} 3 & -2 \\ 1 & 0 \end{bmatrix}$, $\quad$ with initial conditions $v_1 = \begin{bmatrix} 1 \\ 0 \end{bmatrix}$.

From the general solution given in (13) with $b = 0$, we have

$$v_t = A^{t-1}v_1, \qquad t = 2, 3, \ldots.$$

Now

$$A^{t-1} = UD^{t-1}U^{-1} = \begin{bmatrix} 1 & 2 \\ 1 & 1 \end{bmatrix} \begin{bmatrix} 1 & 0 \\ 0 & 2 \end{bmatrix}^{t-1} \begin{bmatrix} -1 & 2 \\ 1 & -1 \end{bmatrix}$$

$$= \begin{bmatrix} 2^t - 1 & 2 - 2^t \\ 2^{t-1} - 1 & 2 - 2^{t-1} \end{bmatrix}$$

so that $\quad v_t = \begin{bmatrix} 2^t - 1 & 2 - 2^t \\ 2^{t-1} - 1 & 2 - 2^{t-1} \end{bmatrix} \begin{bmatrix} 1 \\ 0 \end{bmatrix} = \begin{bmatrix} 2^t - 1 \\ 2^{t-1} - 1 \end{bmatrix}.$

Since the first component of v_t is x_t, the solution to (22) is $x_t = 2^t - 1$.

5. DOMINANT CHARACTERISTIC ROOTS

Various theorems are to be found concerning the existence of a largest, or dominant, characteristic root of a matrix (see, for example, Part II of Chapter 4 in Frazer, Duncan and Collar, 1952). These theorems provide methods of finding the largest characteristic root of a matrix without having to solve the characteristic equation and dealing with matters such as multiple roots, multiple dominant roots and roots involving complex numbers as well as considering the problem of the actual existence of a dominant root. We shall not discuss these theorems here but will demonstrate a useful method for calculating the largest characteristic root *when it exists*. The existence of a real-valued dominant root is assumed throughout; on this basis a method for obtaining it is discussed.

The equation $A = UDU^{-1}$ has been established in (7). Writing

$$U = \{u_{ij}\}, \qquad U^{-1} = \{v_{ij}\} \qquad \text{and} \quad D = \{\lambda_i\}$$

for $i, j = 1, 2, \ldots, n$, where D is the diagonal matrix of the characteristic roots of A, the consequent result $A^k = UD^kU^{-1}$ can be written as

$$
A^k = \begin{bmatrix} u_{11} & \cdots & u_{1n} \\ \cdot & & \cdot \\ \cdot & & \cdot \\ \cdot & & \cdot \\ u_{n1} & \cdots & u_{nn} \end{bmatrix} \begin{bmatrix} \lambda_1^k & & \\ & \cdot & \\ & & \cdot \\ & & & \lambda_n^k \end{bmatrix} \begin{bmatrix} v_{11} & \cdots & v_{1n} \\ \cdot & & \cdot \\ \cdot & & \cdot \\ v_{n1} & \cdots & v_{nn} \end{bmatrix}
$$

$$
= \begin{bmatrix} \lambda_1^k u_{11} & \cdots & \lambda_n^k u_{1n} \\ \cdot & & \cdot \\ \cdot & & \cdot \\ \cdot & & \cdot \\ \lambda_1^k u_{n1} & \cdots & \lambda_n^k u_{nn} \end{bmatrix} \begin{bmatrix} v_{11} & \cdots & v_{1n} \\ \cdot & & \cdot \\ \cdot & & \cdot \\ v_{n1} & \cdots & v_{nn} \end{bmatrix}. \tag{25}
$$

Suppose that λ_1 is the dominant root, such that λ_1^k for some sufficiently large value of k is numerically so much greater than any of the values $\lambda_2^k, \ldots, \lambda_n^k$ that the latter may be taken as zero. Then (25) becomes

$$
A^k = \begin{bmatrix} \lambda_1^k u_{11} & 0 & \cdots & 0 \\ \cdot & & & \cdot \\ \cdot & & & \cdot \\ \lambda_1^k u_{n1} & 0 & \cdots & 0 \end{bmatrix} \begin{bmatrix} v_{11} & \cdots & v_{1n} \\ \cdot & & \cdot \\ \cdot & & \cdot \\ v_{n1} & \cdots & v_{nn} \end{bmatrix}
$$

$$
= \lambda_1^k \begin{bmatrix} u_{11} \\ \cdot \\ \cdot \\ \cdot \\ u_{n1} \end{bmatrix} [v_{11} \quad \cdots \quad v_{1n}]. \tag{26}
$$

Post-multiplying this by a non-null column vector x gives

$$
A^k x = \lambda_1^k \begin{bmatrix} u_{11} \\ \cdot \\ \cdot \\ \cdot \\ u_{n1} \end{bmatrix} [v_{11} \quad \cdots \quad v_{1n}] \begin{bmatrix} x_1 \\ \cdot \\ \cdot \\ \cdot \\ x_n \end{bmatrix} = \mu \lambda_1^k \begin{bmatrix} u_{11} \\ \cdot \\ \cdot \\ u_{n1} \end{bmatrix}
$$

where μ is the scalar $\mu = \sum\limits_{j=1}^{n} v_{1j} x_j$.

Defining

$$
w_k = \begin{bmatrix} w_{1k} \\ \cdot \\ \cdot \\ \cdot \\ w_{nk} \end{bmatrix} = \mu \lambda_1^k \begin{bmatrix} u_{11} \\ \cdot \\ \cdot \\ u_{n1} \end{bmatrix}
$$

then gives $A^k x = w_k$ and similarly $A^{k-1} x = w_{k-1}$. Now, from the definition of w_k, the ratio of the ith element of w_k to the ith element of w_{k-1} is

$$\frac{w_{i,k}}{w_{i,k-1}} = \frac{\mu \lambda_1^k \mu_{i1}}{\mu \lambda_1^{k-1} \mu_{i1}} = \lambda_1. \tag{27}$$

Hence, provided that λ_1 is the largest characteristic root and is such that for some sufficiently large value of k, λ_1^k predominates to a great enough extent over $\lambda_2^k, \ldots, \lambda_n^k$ that equation (26) is approximately true—provided, that these conditions are satisfied, we have shown that for some arbitrary non-null vector x, $A^k x$ and $A^{k-1} x$ are vectors w_k and w_{k-1} respectively, such that the ratio of their elements is λ_1. This means that repeated pre-multiplication of x by A will, after k multiplications, lead to a vector w_k such that the ratio of *each* of its elements to the corresponding element of w_{k-1} is the same for all elements, this ratio being the largest characteristic root of A. And furthermore, w_{k-1} is the corresponding characteristic vector, because

$$A w_{k-1} = A^k x = w_k = \lambda_1 w_{k-1}.$$

Although these results have been obtained by writing (26) as an exact equality, they are really only approximations. The degree of approximation depends upon k and upon the extent to which λ_1^k exceeds $\lambda_2^k, \ldots, \lambda_n^k$ that the latter can be assumed zero in (25) in the presence of λ_1^k. Increasing k increases the accuracy of the approximation, so that in carrying out this process it is possible to find the largest characteristic root to any required degree of accuracy by increasing k. The essential result, (27), is true for all values of i, that is, for all elements of w_k and w_{k-1}, namely of $A^k x$ and $A^{k-1} x$. In any numerical case it is unlikely that these ratios will be exactly equal for any particular value of k; they will differ, but the differences between them will decrease as the value of k is increased.

Example.

For the matrix $A = \begin{bmatrix} 1 & 1 \\ 4 & 1 \end{bmatrix}$ and the arbitrary vector $x = \begin{bmatrix} 1 \\ 1 \end{bmatrix}$ Table 1

shows the vector $w_k = A^k x$ for $k = 0, 1, \ldots, 6$, together with the ratios of corresponding elements, $w_{i,k}/w_{i,k-1}$. We see at once the decrease in the differences between the ratios of corresponding elements as k increases— and at $k = 6$ it is concluded that the largest root is 3. This is confirmed by solving the characteristic equation. Also, by studying the vector w_k as k increases, we see that its elements are approaching the ratio 1:2; thus a

characteristic vector associated with the dominant root of 3 is $\begin{bmatrix} 1 \\ 2 \end{bmatrix}$. This

TABLE 1.   EXAMPLE OF ITERATIVE VALUES OF A
DOMINANT LATENT ROOT

A				k			
	0	1	2	3	4	5	6
				$w_k = A^k x$			
$\begin{bmatrix} 1 & 1 \\ 4 & 1 \end{bmatrix}$	$\begin{bmatrix} 1 \\ 1 \end{bmatrix}$	$\begin{bmatrix} 2 \\ 5 \end{bmatrix}$	$\begin{bmatrix} 7 \\ 13 \end{bmatrix}$	$\begin{bmatrix} 20 \\ 41 \end{bmatrix}$	$\begin{bmatrix} 61 \\ 121 \end{bmatrix}$	$\begin{bmatrix} 182 \\ 365 \end{bmatrix}$	$\begin{bmatrix} 547 \\ 1{,}093 \end{bmatrix}$
				$w_{i,k}/w_{i,k-1}$			
$i = 1$	—	2	3.5	2.8	3.05	2.98	3.004
$i = 2$	—	5	2.6	3.2	2.95	3.02	2.995

may be verified by equation (4), $Au_i = \lambda u_i$:

$$\begin{bmatrix} 1 & 1 \\ 4 & 1 \end{bmatrix} \begin{bmatrix} 1 \\ 2 \end{bmatrix} = \begin{bmatrix} 3 \\ 6 \end{bmatrix} = 3 \begin{bmatrix} 1 \\ 2 \end{bmatrix}.$$

Demonstrating this method with a 2 × 2 matrix has kept the arithmetic to a minimum, but it can be applied to matrices of any order. The calculations become quite substantial for large-sized matrices, but with high-speed computers they are no longer a limiting problem. Although the true test for ending the iterative procedure is that $w_{i,k}/w_{i,k-1}$ should be the same for all i, within the limits of accuracy desired, two alternative procedures can be used that are not correct theoretically but are frequently satisfactory in practice. They demand continuing the iterative procedure until either

$$\sum_{i=1}^{n} \frac{w_{i,k}}{w_{i,k-1}} = \sum_{i=1}^{n} \frac{w_{i,k-1}}{w_{i,k-2}}$$

or

$$\frac{\sum_{i=1}^{n} w_{i,k}}{\sum_{i=1}^{n} w_{i,k-1}} = \frac{\sum_{i=1}^{n} w_{i,k-1}}{\sum_{i=1}^{n} w_{i,k-2}}.$$

Choice of which of these alternatives to use depends on the problem at hand and on matters of computing efficiency and numerical analysis which are beyond our scope here. Nevertheless, as alternative stopping procedures they are worth noting: one involves equating sums of ratios, the other involves ratios of sums.

Illustrations. Dominant characteristic roots have direct application in problems relating to age-distribution vectors and to transition probability matrices discussed in Sections 1 and 2. If x_0' is a vector describing the state of a machine at time 0 and P is the associated transition probability matrix, then the corresponding state vector at period n is $x_n' = x_0' P^n$. Now if the system tends toward a steady-state solution x (see Section 8.2) as n increases, the steady-state solution may be found by equating the state vector at periods n and $n + 1$ as follows:

$$x_{n+1}' = x_n' = x', \quad \text{say,}$$

so that in the steady state the relationship

$$x_{n+1}' = x_n' P$$

becomes
$$x' = x' P$$

equivalent to $(P' - I)x = 0$, or $P'x = x$. These equations emphasize that if a steady-state solution exists it corresponds to a characteristic root of unity, since the ratio w_k/w_{k-1} must be one for the relationship $P'x = x$ to hold. And since it can be shown that the transposes of transition probability matrices have characteristic roots whose absolute values are equal or less than unity, one that is unity is dominant.[1] Hence the characteristic vector corresponding to the dominant characteristic root of unity is the steady-state solution x. It is obtained by solving $(P' - I)x = 0$ together with the equation summing the elements of x to unity.

If there is no characteristic root that can be termed dominant in the sense used here the iterative procedure breaks down and tends to no limit. The matrix

$$A = \begin{bmatrix} 0 & 1 & 0 \\ 0 & 0 & 1 \\ 1 & 0 & 0 \end{bmatrix}$$

presents just such a case; it can be shown that $A^4 = A$, $A^5 = A^2$, $A^6 = I$, $A^7 = A$ and so on, with regular periodicity (see Section 8.3). Thus successive values of $w_k = A^k x$ for some arbitrary non-null vector x will exhibit similar periodicity, and ratios of corresponding elements $w_{i,k}/w_{i,k-1}$ will not approach a limiting value.

6. FACTORING THE CHARACTERISTIC EQUATION

The examples of obtaining characteristic roots that have been used so far have all involved solving either a cubic or a quadratic equation in λ. This is not difficult, but in finding characteristic roots of large matrices there arises

---

[1] See Exercise 3 for the 2×2 case, and Feller (1957, p. 384) for the general case.

the problem of having to solve polynomials of large degree, for example, a 20 × 20 matrix leads to a polynomial of degree 20. Should the dominant characteristic root be obtainable by the procedure just discussed, the second largest root can be found by a similar process after factoring out the dominant root by a method now to be described.

Operationally the procedure is as follows. Suppose λ_1 is the dominant characteristic root of A, and suppose u_1 is a characteristic vector corresponding to λ_1. Choose u_1 such that $u_1'u_1 = \lambda_1$. [This is achieved by obtaining any characteristic vector corresponding to λ_1, t_1 say, and deriving u_1 from t_1 as

$$u_1 = (\sqrt{\lambda_1/t_1't_1})t_1.$$

Then $u_1'u_1 = \lambda_1$.] The second largest characteristic root of A is then obtained as the largest characteristic root of $A - u_1u_1'$. The following theorem is the basis of this procedure.

Theorem.[1] If λ_1 is any characteristic root of A and u_1 is a corresponding characteristic vector with $u_1'u_1 = \lambda_1$, then the characteristic roots of $A - u_1u_1'$ except one root of zero are, with the exception of λ_1, those of A. That is, zero has replaced λ_1.

Example. The characteristic equation of

$$A = \begin{bmatrix} 1 & 1 & 0 \\ 3 & 1 & 2 \\ -10 & 9 & 1 \end{bmatrix}$$

reduces to

$$(\lambda - 2)(\lambda - 5)(\lambda + 4) = 0.$$

For the root $\lambda_1 = 5$ a characteristic vector is

$$t_1 = \begin{bmatrix} 2 \\ 8 \\ 13 \end{bmatrix}, \quad \text{for which} \quad u_1 = (\sqrt{\lambda_1/t_1't_1})t_1 = \sqrt{5/237}\begin{bmatrix} 2 \\ 8 \\ 13 \end{bmatrix}.$$

The characteristic equation for the matrix $A - u_1u_1'$ is then

$$|A - \lambda I - u_1u_1'| = 0$$

or

$$\left| \begin{bmatrix} 1 - \lambda & 1 & 0 \\ 3 & 1 - \lambda & 2 \\ -10 & 9 & 1 - \lambda \end{bmatrix} - \tfrac{5}{237}\begin{bmatrix} 4 & 16 & 26 \\ 16 & 64 & 104 \\ 26 & 104 & 169 \end{bmatrix} \right| = 0$$

----

[1] See Searle (1966, p. 189) for proof.

which simplifies to

$$\begin{vmatrix} 217 - 237\lambda & 157 & -130 \\ 631 & -83 - 237\lambda & -46 \\ -2,500 & 1,613 & -608 - 237\lambda \end{vmatrix} = 0.$$

On expansion this becomes

$$(237\lambda)^3 + (237\lambda)^2 474 - (237\lambda)449,352 = 0$$

reducing to

$$\lambda(\lambda - 2)(\lambda + 4) = 0$$

and the largest root of this equation is $\lambda = 2$. Hence $\lambda = 2$ is the second largest characteristic root of A.

In applying this technique to finding the second largest characteristic root of a matrix one would usually employ the methods of the previous section to find the dominant roots of A and $A - u_1 u_1'$.

7. APPENDIX: SYMMETRIC MATRICES

Just as reduction to equivalent canonical form (Sections 6.10 and 6.11) is a little different for symmetric matrices than it is for nonsymmetric ones, so also is this the case with reduction to canonical form under similarity. Thus if A is symmetric, $A = A'$, then the reduction $U^{-1}AU = D$ holds true for U being orthogonal ($U^{-1} = U'$, Section 5.5), so that $U'AU = D$. The matrix D is called the *canonical form under orthogonal similarity*. We now demonstrate this reduction, again differentiating between the two cases of characteristic roots all different and multiple characteristic roots.

a. Characteristic roots all different

If λ_i and λ_k are two different characteristic roots of a matrix A, with u_i and u_k being corresponding characteristic vectors,

$$Au_i = \lambda_i u_i \quad \text{and} \quad Au_k = \lambda_k u_k.$$

Using these equalities and noting that transposing a scalar leaves it unchanged, the following development can be made for a symmetric matrix $A = A'$. Consider the scalar

$$\lambda_i u_k' u_i = u_k'(\lambda_i u_i) = u_k' A u_i = u_i' A u_k$$
$$= u_i' \lambda_k u_k = \lambda_k u_i' u_k = \lambda_k u_k' u_i.$$

For $\lambda_i \neq \lambda_k$ this is true only if $u_k' u_i = 0$, and if all characteristic roots are different this will be the case for every pair of roots. Hence, for distinct characteristic roots, columns of $U = [u_1 \quad u_2 \quad \cdots \quad u_n]$ can be found such that each inner product of one with another is zero. By dividing each column by the square root of the sum of squares of each of its elements, a process known as *normalizing* a vector, the columns become such that the inner product of each with itself is 1. Hence $u_i'u_k = 0$ for $i \neq k$ and $u_i'u_i = 1$. Thus $U'U = I$, U is an orthogonal matrix and the reduction $U^{-1}AU = D$ becomes $U'AU = D$.

Example. The matrix

$$A = \begin{bmatrix} 3 & -6 & -4 \\ -6 & 4 & 2 \\ -4 & 2 & -1 \end{bmatrix}$$

has characteristic roots $\lambda = -1$, -4 and 11. Solving the equations $(A - \lambda_i I)u_i = 0$ for each characteristic root in turn, it will be found that the three characteristic vectors are $\begin{bmatrix} 1 \\ 2 \\ -2 \end{bmatrix}$, $\begin{bmatrix} 2 \\ 1 \\ 2 \end{bmatrix}$ and $\begin{bmatrix} 2 \\ -2 \\ -1 \end{bmatrix}$, so that the un-normalized form of U is

$$\begin{bmatrix} 1 & 2 & 2 \\ 2 & 1 & -2 \\ -2 & 2 & -1 \end{bmatrix}.$$

The process of normalizing each column requires dividing the first column by $\sqrt{1^2 + 2^2 + 2^2} = 3$. The second column is divided by $\sqrt{2^2 + 1^2 + 2^2} = 3$, and this is also the divisor for column 3. Hence the normalized form of U is

$$U = \tfrac{1}{3}\begin{bmatrix} 1 & 2 & 2 \\ 2 & 1 & -2 \\ -2 & 2 & -1 \end{bmatrix}, \quad U'U = I \text{ and } U'AU = \begin{bmatrix} -1 & 0 & 0 \\ 0 & -4 & 0 \\ 0 & 0 & 11 \end{bmatrix} = D.$$

b. Multiple characteristic roots

We have shown that for two distinct characteristic roots λ_i and λ_k of a symmetric matrix, characteristic vectors u_i and u_k exist such that $u_i'u_k = 0$. It remains to be shown that for a multiple characteristic root we can find a set of vectors that are mutually orthogonal. Suppose that λ_q is a multiple root with multiplicity m_q. Then for any A that can be expressed in the form $U^{-1}AU = D$, $r(A - \lambda_q I) = n - m_q$ (Section 2c), and $(A - \lambda_q I)u_q = 0$ has m_q LIN solutions u_q. Denote one solution by v_{q1}. Now consider solving

$$(A - \lambda_q I)u_q = 0$$

and

$$v_{q1}'u_q = 0$$

simultaneously for u_q. This set of equations has $m_q - 1$ LIN solutions for u_q and any one of them, v_{q2} say, is a characteristic vector of A and is orthogonal to the vector v_{q1}. It will also be orthogonal to the characteristic vector corresponding to any characteristic root different from λ_q, because of the orthogonality property of characteristic vectors corresponding to distinct characteristic roots of symmetric matrices. Similarly if $m_q > 2$ a third vector corresponding to λ_q can be obtained by solving

$$(A - \lambda_q I)u_q = 0$$
$$v_{q1}'u_q = 0$$

and
$$v_{q2}'u_q = 0$$

which have $m_q - 2$ LIN solutions for u_q. This process can be continued until m_q solutions are obtained, all orthogonal to each other and to the vectors corresponding to the other roots. Thus U is again orthogonal, and we have $U'AU = D$ with $U'U = I$.

Example.

$$A = \begin{bmatrix} -1 & -2 & 1 \\ -2 & 2 & -2 \\ 1 & -2 & -1 \end{bmatrix}$$

has characteristic roots $\lambda = 4, -2$ and -2. First we must consider the rank of $(A - \lambda_2 I)$ for the multiple root $\lambda_2 = -2$:

$$(A + 2I) = \begin{bmatrix} 1 & -2 & 1 \\ -2 & 4 & -2 \\ 1 & -2 & 1 \end{bmatrix}$$

has rank 1 which equals $3 - 2$, the order of A minus the multiplicity of the multiple root. Therefore characteristic roots of A exist such that $U'AU = D$. The characteristic vectors are found by solving $(A - \lambda_i I)u_i = 0$. For $\lambda_1 = 4$,

$(A - \lambda_1 I)u_1 = 0$ has the solution $u_1 = \begin{bmatrix} 1 \\ -2 \\ 1 \end{bmatrix}$, and for $\lambda_2 = -2$ the equations

$(A - \lambda_2 I)u_2 = 0$ reduce to

$$a_2 - 2b_2 + c_2 = 0 \tag{28}$$

where
$$u_2 = \begin{bmatrix} a_2 \\ b_2 \\ c_2 \end{bmatrix} = \begin{bmatrix} 1 \\ 1 \\ 1 \end{bmatrix} \tag{29}$$

is one solution. But $\lambda = -2$ is a multiple root and consequently a second solution to (28) is required, one whose inner product with (29) is zero. Thus a second characteristic vector corresponding to $\lambda = -2$ is derived by solving

$$a_3 - 2b_3 + c_3 = 0$$

and

$$a_3 + b_3 + c_3 = 0.$$

A solution is

$$u_3 = \begin{bmatrix} a_3 \\ b_3 \\ c_3 \end{bmatrix} = \begin{bmatrix} 1 \\ 0 \\ -1 \end{bmatrix},$$

and the un-normalized form of U is

$$[u_1 \quad u_2 \quad u_3] = \begin{bmatrix} 1 & 1 & 1 \\ -2 & 1 & 0 \\ 1 & 1 & -1 \end{bmatrix}.$$

The normalized form comes from dividing the columns by $\sqrt{6}$, $\sqrt{3}$ and $\sqrt{2}$, respectively, to give

$$U = \frac{1}{\sqrt{6}} \begin{bmatrix} 1 & \sqrt{2} & \sqrt{3} \\ -2 & \sqrt{2} & 0 \\ 1 & \sqrt{2} & -\sqrt{3} \end{bmatrix}.$$

It will be found that $U'U = I$ (i.e., U is orthogonal), and that $U'AU = D$ where D is diagonal with diagonal elements 4, -2 and -2.

c.   Real characteristic roots

Since the characteristic roots of a matrix of order n are the roots of the characteristic equation, a polynomial in λ of degree n, it is not true in general that all the roots are real numbers; some may be pairs of complex numbers such as $a + ib$ and $a - ib$ where $i = \sqrt{-1}$ and a and b are real numbers. But if A is symmetric and has no complex numbers as elements (i.e., all elements are real), there are no complex characteristic roots. [See Searle (1966, p. 192) for a proof.]

d.   Positive semi-definite symmetric matrices

Consider the general quadratic form discussed in Section 3.3, $z = x'Ax$, where A is symmetric of rank r. The characteristic roots of A are real, and provided that A can be expressed in canonical form under orthogonal similarity, this will be

$$U'AU = D = \begin{bmatrix} D_r^* & 0 \\ 0 & 0 \end{bmatrix}$$

where U is orthogonal and D_r^* is a diagonal matrix of the r non-zero characteristic roots of A. Suppose we make a linear transformation of variables (Section 3.1) from x to y, $y = U'x$, equivalent to $x = U'^{-1}y = Uy$. Then

$$z = x'Ax$$

$$= y'U'AUy$$

$$= y' \begin{bmatrix} D_r^* & 0 \\ 0 & 0 \end{bmatrix} y$$

$$= \sum_{i=1}^r \lambda_i y_i^2.$$

Now if z is positive semi-definite (Section 3.3), meaning that it is never negative, then $\sum_{i=1}^r \lambda_i y_i^2$ must be too, for all y_i, and hence all λ_i for $i = 1, \ldots, r$ must be positive; that is, the characteristic roots of a positive semi-definite symmetric matrix are positive real numbers.

Illustration. If the vector x represents a set of n normally and independently distributed random variables which have zero mean and variance-covariance matrix $\sigma^2 I$, so does the vector $y = U'x$ where U is orthogonal. Furthermore, for $z = x'Ax$ where A is positive semi-definite, symmetric and of rank r, z is distributed as $\sigma^2 \chi^2$ with r degrees of freedom. These results are shown as follows.

With $E(x) = 0$, $E(xx') = \sigma^2 I$, and $y = U'x$, the mean of y is

$$E(y) = E(U'x) = U'E(x) = 0$$

and its variance-covariance matrix is

$$E(yy') = E(U'xx'U) = U'E(xx')U = \sigma^2 U'IU = \sigma^2 I.$$

And for $z = x'Ax$, with A being positive semi-definite and symmetric of rank r, there exists a matrix U such that $U'U = I$ and $U'AU = D$, where D is diagonal with r real, positive non-zero elements in its diagonal, $\lambda_1, \ldots, \lambda_r$ say. Therefore for transformed variables $y = U'x$,

$$z = x'Ax = y'Dy = \sum_{i=1}^r \lambda_i y_i^2,$$

and the y's are a set of normally distributed random variables with zero mean and variance-covariance matrix $\sigma^2 I$. Thus each y_i^2 has the distribution $\sigma^2 \chi_1^2$, a chi-square distribution with 1 degree of freedom. And because the λ's are positive, z is the sum of r independent variables $\lambda_i y_i^2$ each distributed as $\lambda_i \sigma^2 \chi_1^2$. In cases where $\lambda_i \sigma^2 = 1$ (see Section 13.3), $x'Ax$ is thus distributed as χ_r^2.

8. EXERCISES

1. Find the characteristic roots and characteristic vectors of the following matrices. In each case combine the characteristic vectors into a matrix U and verify that $U^{-1}AU = D$ where D is the diagonal matrix of the characteristic roots.

$$B = \begin{bmatrix} 1 & 4 & 1 \\ 2 & 1 & 0 \\ -1 & 3 & 1 \end{bmatrix} \qquad C = \begin{bmatrix} 2 & -2 & 3 \\ 10 & -4 & 5 \\ 5 & -4 & 6 \end{bmatrix}$$

$$E = \begin{bmatrix} 7 & 4 & -1 \\ 4 & 7 & -1 \\ -4 & -4 & 4 \end{bmatrix} \qquad F = \begin{bmatrix} 9 & 15 & 3 \\ 6 & 10 & 2 \\ 3 & 5 & 1 \end{bmatrix}$$

$$G = \begin{bmatrix} 4 & -2 & 0 \\ -2 & 3 & -2 \\ 0 & -2 & 2 \end{bmatrix} \qquad H = \begin{bmatrix} 1 & 2 & 1 \\ 1 & 1 & 4 \\ 2 & -4 & 1 \end{bmatrix}$$

$$K = \begin{bmatrix} -1 & -2 & 1 \\ -2 & 2 & -2 \\ 1 & -2 & -1 \end{bmatrix} \qquad L = \begin{bmatrix} 1 & 4 & 2 \\ 4 & 1 & 2 \\ 8 & 8 & 1 \end{bmatrix}$$

$$M = \begin{bmatrix} -9 & 2 & 6 \\ 2 & -9 & 6 \\ 6 & 6 & 7 \end{bmatrix} \qquad N = \begin{bmatrix} 2 & 7 & 1 \\ 2 & 3 & 8 \\ 9 & 4 & 4 \end{bmatrix}$$

2. Of the matrices in Exercise 1, show that $F = B^2 + B$, and that if λ is a characteristic root of B then $\lambda^2 + \lambda$ is a characteristic root of F. Show, generally, that if F is a polynomial of a matrix A, each characteristic root of F is that same polynomial of a characteristic root of A.

3. For the transpose of the general 2×2 transition probability matrix

$$P' = \begin{bmatrix} p_{11} & 1 - p_{22} \\ 1 - p_{11} & p_{22} \end{bmatrix},$$

factor the characteristic equation to obtain the two characteristic roots of P' and verify that neither root can ever exceed 1 in absolute value.

4. Find the characteristic roots of

$$T = \begin{bmatrix} \frac{1}{2} & \frac{1}{2} & 0 \\ \frac{1}{4} & \frac{1}{2} & \frac{1}{4} \\ 0 & \frac{1}{2} & \frac{1}{2} \end{bmatrix}$$

and show that

$$T^k = \begin{bmatrix} \frac{1}{4} + (\frac{1}{2})^{k+1} & \frac{1}{2} & \frac{1}{4} - (\frac{1}{2})^{k+1} \\ \frac{1}{4} & \frac{1}{2} & \frac{1}{4} \\ \frac{1}{4} - (\frac{1}{2})^{k+1} & \frac{1}{2} & \frac{1}{4} + (\frac{1}{2})^{k+1} \end{bmatrix}.$$

5. Lucas (1967) considers a theoretical model for demand for capital goods which uses the stock adjustment pattern known as the "flexible accelerator." In a simplified version of his model, let x_1^* represent the desired stock of capital and x_2^* represent the desired stock of labor. Then if the vector x^* reflects the desired levels of capital and labor while the vector x_n represents the actual levels of capital and labor at period n, the following adjustment equations may be considered:

$$x_{n+1} - x_n = B(x^* - x_n) \qquad \text{or} \qquad x_{n+1} = (I - B)x_n + Bx^*$$

where B is a 2 × 2 "adjustment coefficient matrix." As a specific example consider

$$B = \begin{bmatrix} 0.1 & -0.5 \\ 0.3 & 0.9 \end{bmatrix}.$$

Compute the relevant characteristic roots and verify that their absolute values are less than unity, thus ensuring that the above adjustment process converges to x^* as $n \to \infty$.

6. A retailer has the following inventory policy for two products: order the amount ordered last month plus an adjustment factor times the difference $(x_i^* - x_{n,i})$, where x_i^* is a target inventory for product i and $x_{n,i}$ is last month's order size for product i. The following data are given:

Product	x_i^*	Last Month's Order	Adjustment Factor
1	20	15	0.5
2	50	40	0.1

(a) Write, in vector form, a difference equation for the amount ordered (of the two products) in month n as a function of the previous month's orders.

(b) Given the above data, find the characteristic roots of the relevant matrix in (a), verifying that their absolute values are less than unity, thus ensuring convergence.

(c) Letting the adjustment factor for product 1 be 1.5 instead of 0.5, what are the characteristic roots? Are they both less than unity in absolute value? Describe what happens to the first element of x_n as $n \to \infty$.

7. Ezekiel (1938), in developing his "cobweb theorem," presents a simple market equilibrium model that can be represented by a system of difference equations. The model, where all symbols are scalars, is

$$D_t = -m_a p_t + b_a \quad \text{with} \quad m_a > 0 \quad \text{and} \quad b_a > 0;$$
$$S_{t+1} = m_s p_t + b_s \quad \text{with} \quad m_s > 0 \quad \text{and} \quad b_s > 0;$$

and $\qquad D_{t+1} = S_{t+1},$

where D_t is the quantity demanded, S_t is the quantity supplied and p_t is the market price, all in period t, and m_a, m_s, b_a and b_s are parameters. The fact that S_{t+1} is a function of p_t leads the dynamic market mechanism to changing prices over time.

(a) Derive the first-order difference equation governing price over time.

(b) Write p_t as a function of p_0, the initial market price.

(c) What condition must hold in order for p_t to approach a finite limit as $t \to \infty$? If that condition holds, what is that limit?

(d) Will the price oscillate over time?

(e) Suppose $m_a = 2$, $m_s = 1$, $b_a = 4$, $b_s = 1$ and $p_0 = 2$. Write a general expression for p_t as a function of time.

8. Samuelson's multiplier-accelerator model (see Exercise 6.13) yielded the following second-order difference equation governing the behavior of national income over time:

$$y_t = (\alpha + \alpha\beta)y_{t-1} - \alpha\beta y_{t-2} + g_t$$

where α is the marginal propensity to consume, β is the accelerator coefficient, and g_t represents government expenditures in period t.

Suppose $\alpha = 0.92$ and $\beta = \frac{12}{23}$. If $g_0 = 160$ and $y_{-2} = y_{-1} = y_0 = 2,000$, then national income will be at equilibrium. Now suppose $g_1 = 176$.

(a) Derive a system of first-order difference equations from the single second-order difference equation given above after substituting for α, β and g_t, $t = 0, 1$. [*Hint:* $\alpha\beta = 0.48$.]

(b) Will y_t approach some finite limit as $t \to \infty$? If so, what is that limit?

(c) Solve this system of equations to find y_t as a function of t.

9. Show that when $a = d$ the characteristic vectors of $\begin{bmatrix} a & b \\ c & d \end{bmatrix}$ are of the form $\begin{bmatrix} -b \\ a - \lambda_1 \end{bmatrix}$ and $\begin{bmatrix} -b \\ a - \lambda_2 \end{bmatrix}$ where λ_1 and λ_2 are the characteristic roots, and verify that $2a - \lambda_1 = \lambda_2$.

10. Use the iterative procedure to establish, correct to 2 decimal places, the dominant characteristic root for two of the matrices in Exercise 1, and by using the factorization procedure of Section 6 find the other two characteristic roots.

11. Factor the characteristic equation of

$$A = \begin{bmatrix} a & b & b \\ b & a & b \\ b & b & a \end{bmatrix}$$

using the characteristic root $\lambda = a + 2b$. Carry out the same process using the root $\lambda = a - b$.

12. Express $A = \begin{bmatrix} 1 & 0 & 0 \\ u & v & 0 \\ x & y & z \end{bmatrix}$ in the form $A = UDU^{-1}$ where $D = \begin{bmatrix} 1 & 0 & 0 \\ 0 & v & 0 \\ 0 & 0 & z \end{bmatrix}$. In

doing so, notice that $\begin{bmatrix} 1 & 0 & 0 \\ a & 1 & 0 \\ b & c & 1 \end{bmatrix}^{-1} = \begin{bmatrix} 1 & 0 & 0 \\ -a & 1 & 0 \\ ac - b & -c & 1 \end{bmatrix}$.

13. Use the result of Exercise 12 to show that the kth power of the transition probability matrix

$$P = \begin{bmatrix} 1 & 0 & 0 \\ \tfrac{1}{2} & \tfrac{1}{2} & 0 \\ \tfrac{1}{2} - c(1-c) & 2c(1-c) & \tfrac{1}{2} - c(1-c) \end{bmatrix}$$

is, for $z = \tfrac{1}{2} + c(c-1)$,

$$P^k = \begin{bmatrix} 1 & 0 & 0 \\ 1 - (\tfrac{1}{2})^k & (\tfrac{1}{2})^k & 0 \\ 1 + z^k - (\tfrac{1}{2})^{k-1} & (\tfrac{1}{2})^{k-1} - 2z^k & z^k \end{bmatrix}.$$

14. Prove that the relationship $A^k u_i = \lambda_i^k u_i$ holds for any negative integer k if A^{-1} exists.

15. (a) Using the results of Exercise 3 for the transpose of the general 2×2 transition probability matrix

$$P' = \begin{bmatrix} p_{11} & 1 - p_{22} \\ 1 - p_{11} & p_{22} \end{bmatrix}$$

find the diagonal matrix of characteristic roots D and the corresponding matrix U such that $P' = UDU^{-1}$.

(b) Assuming that all elements of P are positive, use (a) to show that as $n \to \infty$ P^n tends toward a limiting form. What property does the limiting matrix have?

REFERENCES

Chow, G. C. (1957). *Demand for Automobiles in the United States: A Study in Consumer Durables*. North-Holland Publishing Company, Amsterdam.

Ezekiel, M. (1938). The cobweb theorem. *Quarterly Journal of Economics*, **52**, 255–280.

Feller, W. (1957). *An Introduction to Probability Theory and Its Applications*. Second Edition, Wiley, New York.

Frazer, R. A., W. J. Duncan, and A. R. Collar (1952). *Elementary Matrices*. Cambridge University Press.

Goldberg, S. (1958). *Introduction to Difference Equations*. Wiley, New York.

Lucas, R. E., Jr. (1967). Optimal investment policy and the flexible accelerator. *International Economic Review*, **8**, 78–85.

Metzler, L. A. (1942). Underemployment equilibrium in international trade. *Econometrica*, **10**, 97–112.

Metzler, L. A. (1950). A multiple-region theory of income and trade. *Econometrica*, **18**, 329–354.

Searle, S. R. (1966). *Matrix Algebra for the Biological Sciences*. Wiley, New York.

CHAPTER 13

SPECIAL TOPICS

This chapter makes a variety of special topics available for reference use. Their treatment in terms of the notation used in this book should prove helpful to readers wishing to progress further with matrix algebra, for the topics discussed will be found useful in extending many aspects of the work described earlier in the book. The advanced reader may therefore wish to spend more time on this chapter than the beginner.

1. NORMALITY AND ORTHOGONALITY

An orthogonal matrix has been defined in Section 5.5: it is a square matrix whose product with its transpose is the identity matrix. Such a matrix has several properties of interest; before discussing them we give some definitions that lead up to orthogonality.

The *norm* of a vector is the square root of the sum of squares of its elements. Thus if $x' = \{x_i\}$ for $i = 1, 2, \ldots, n$, then the norm of the vector x is $\sqrt{x'x} = \sqrt{\sum_{i=1}^{n} x_i^2}$. If the norm of a vector is unity, i.e. if $\sqrt{x'x} = 1$, the vector is said to be a *normal vector*.

Two vectors of the same order, x and y say, are said to be *orthogonal* if their inner product is zero; i.e. if $x'y = y'x = 0$, x and y are orthogonal. If, in addition, both vectors are normal, they are described as being *orthonormal*. Hence if $x'x = 1, y'y = 1$ and $x'y = 0$, then x and y are orthonormal. A natural extension of this is to a set of vectors $z_1, z_2, z_3, \ldots$. If they are all normal, $z_i'z_i = 1$ for all i, and if they are orthogonal in pairs, $z_i'z_j = 0$ for all $i \neq j$, then they form an *orthonormal set of vectors*.

[*333*]

When the rows of a square matrix are an orthonormal set of vectors the matrix is orthogonal. For

$$A = \begin{bmatrix} x_1' \\ x_2' \\ \cdot \\ \cdot \\ \cdot \\ x_n' \end{bmatrix} \qquad \text{and} \qquad A' = [x_1 \quad \cdots \quad x_n],$$

$AA' = I$ if, and only if, $x_i'x_i = 1$ for all i and $x_i'x_j = 0$ for $i \neq j$. Thus the matrix A is orthogonal. Furthermore, when the rows of a square matrix are orthonormal the columns are also orthonormal. This is so because with $AA' = I$, $|AA'| = |A||A'| = 1$ (see Section 4.5) and so $|A| \neq 0$. Therefore A^{-1} exists and hence from $AA' = I$ we have $A' = A^{-1}$. Thus we may write $A'A = I$, showing that the columns of A are also orthonormal. From this we note that the property of orthogonality implies more in a matrix than it does in a pair of vectors. With a matrix orthogonality implies not only pair-wise orthogonality of rows (and columns) but also normality of the rows (and columns). This and other properties of orthogonal matrices are stated below.

i. The inner product of any row (column) of an orthogonal matrix with itself is 1, and its inner product with any other row (column) is zero.

ii. A product of orthogonal matrices is itself orthogonal: if A and B are orthogonal $AA' = BB' = I$, and the product $(AB)(AB)'$ readily simplifies to I.

iii. The determinant of every orthogonal matrix is either $+1$ or -1. Taking the determinant of both sides of the equation $AA' = I$ yields this result.

iv. If λ is a characteristic root of an orthogonal matrix then so is $1/\lambda$. This is so because the characteristic equation $|A - \lambda I| = 0$ can be rearranged to obtain $|I - \lambda A'| = 0$ for $AA' = I$ by pre-multiplying the matrix $(A - \lambda I)$ by A'. Hence $|(1/\lambda)I - A'| = 0$, and transposing gives $|A - (1/\lambda)I| = 0$; i.e. $1/\lambda$ is a characteristic root of A. A consequence of this result is that all orthogonal matrices of odd order have at least one characteristic root equal to $+1$ or -1, the remaining roots occurring in pairs, λ and $1/\lambda$.

v. If S is a matrix such that $S' = -S$, then $P = (I - S)(I + S)^{-1}$ is orthogonal. The reader should satisfy himself that $P'P = I$.

Matrices of the form $S' = -S$ are called *skew-symmetric*, or sometimes just *skew*. They are square and have all elements on one side of the diagonal equal to minus their counterparts on the other side, the diagonal elements being zero. If a skew-symmetric matrix is of odd order its determinant is zero and it has no inverse; if it is of even order the determinant is a perfect square and the inverse is skew. Any square matrix M can be expressed uniquely as the sum of a symmetric and a skew-symmetric matrix: $S = \frac{1}{2}(M + M')$ and $T = \frac{1}{2}(M - M')$. S is symmetric and T is skew.

Demonstration and proof of these properties are left as an exercise for the reader.

2. MATRICES HAVING ALL ELEMENTS EQUAL

On many occasions it happens that matrices can be partitioned into sub-matrices each of which has all its elements the same. On other occasions matrices can be expressed as linear functions of such matrices. In either case it is advantageous to make use of certain properties of these equi-element matrices.

First consider the vector whose every element is unity. When of order r it is often denoted by $\mathbf{1}_r$, with the subscript omitted whenever context permits. For example

$$\mathbf{1}_4' = [1 \quad 1 \quad 1 \quad 1].$$

(Bold-face type is used here to distinguish the vector $\mathbf{1}$ from the scalar 1.) Products involving $\mathbf{1}$ lead to summations of the elements of the other vectors and matrices involved. Thus if $x = \{x_i\}$ for $i = 1, \ldots, r$, then

$$\mathbf{1}_r'x = x'\mathbf{1}_r = \sum_{i=1}^{r} x_i.$$

Application of this result to the matrix $A_{r \times c}$ shows that $\mathbf{1}_r'A_{r \times c}$ is a row vector of the column totals of A and $A_{r \times c}\mathbf{1}_c$ is a column vector of the row totals of A. Similarly, $\mathbf{1}_r'A_{r \times c}\mathbf{1}_c$ is a scalar, the sum of all elements of A, and $\mathbf{1}_r'\mathbf{1}_r = r$.

The product $\mathbf{1}_r\mathbf{1}_c'$ is a matrix of order $r \times c$ with every element equal to 1. We denote this by $J_{r \times c}$; i.e.

$$\mathbf{1}_r\mathbf{1}_c' = J_{r \times c}.$$

Then λJ is a matrix having every element equal to λ. For example,

$$J_{2 \times 3} = \begin{bmatrix} 1 & 1 & 1 \\ 1 & 1 & 1 \end{bmatrix} \quad \text{and} \quad 5J_{2 \times 3} = \begin{bmatrix} 5 & 5 & 5 \\ 5 & 5 & 5 \end{bmatrix}.$$

The following properties are easily demonstrated:

$$J_{r \times c}' = J_{c \times r},$$
$$J_{r \times c}J_{c \times k} = cJ_{r \times k},$$
and
$$J_{r \times c}J_{c \times r} = cJ_r,$$

where J_r is square, of order r, with all elements unity. Properties pertaining to J_r are:

$$J_r \text{ is symmetric,} \quad \text{i.e.,} \quad J_r' = J_r;$$
$$J_r^2 = rJ_r, \quad \text{and} \quad J_r^k = r^{k-1}J_r;$$
$$|J_r| = 0, \quad \text{and so } J_r^{-1} \text{ does not exist;}$$
$$\text{rank}(J_r) = 1, \quad \text{and the only non-zero characteristic root is } r.$$

Pre-multiplication of any matrix A by a J matrix leads to a matrix having every row the same, the elements being the column totals of A; and post-multiplication by J yields a matrix having every column the same, the elements being the row totals of A. These two results lead to

$$J_{r \times s} A_{s \times t} J_{t \times w} = \left(\sum_{i=1}^{s} \sum_{j=1}^{t} a_{ij} \right) J_{r \times w}.$$

Example. For

$$A = \begin{bmatrix} 2 & 5 & -4 \\ 3 & 4 & 6 \end{bmatrix},$$

$$J_{4 \times 2} A = \begin{bmatrix} 1 & 1 \\ 1 & 1 \\ 1 & 1 \\ 1 & 1 \end{bmatrix} \begin{bmatrix} 2 & 5 & -4 \\ 3 & 4 & 6 \end{bmatrix} = \begin{bmatrix} 5 & 9 & 2 \\ 5 & 9 & 2 \\ 5 & 9 & 2 \\ 5 & 9 & 2 \end{bmatrix},$$

$$A J_{3 \times 2} = \begin{bmatrix} 2 & 5 & -4 \\ 3 & 4 & 6 \end{bmatrix} \begin{bmatrix} 1 & 1 \\ 1 & 1 \\ 1 & 1 \end{bmatrix} = \begin{bmatrix} 3 & 3 \\ 13 & 13 \end{bmatrix},$$

and
$$J_{4 \times 2} A J_{3 \times 2} = 16 J_{4 \times 2}.$$

A useful extension of these J matrices is a matrix which is a linear function of I_r and J_r, namely,

$$V_r = a I_r + b J_r,$$

where a and b are non-zero scalars. For example

$$V_3 = \begin{bmatrix} a+b & b & b \\ b & a+b & b \\ b & b & a+b \end{bmatrix}.$$

The reader might satisfy himself that by writing V_r symbolically as

$$V_r = V_r(a, b),$$

the following results are true:

$$V_r' = V_r,$$

$$[V_r(a, b)]^{-1} = V_r\left(\frac{1}{a}, \frac{-b}{a(a + rb)} \right),$$

$$|V_r(a, b)| = a^{r-1}(a + rb),$$

$$V_r(a, b) V_r(x, y) = V_r(ax, ay + bx + rby),$$

and from this last

$$[V_r(a, b)]^2 = V_r(a^2, 2ab + rb^2).$$

By replacing a by $a - \lambda$ in the expression for $|V_r(a, b)|$ it is seen that $V_r(a, b)$ has one characteristic root equal to $(a + rb)$ and $(r - 1)$ roots equal to a.

Illustration. It is shown in Section 3.3 that the sum of squares

$$SS = \sum_{i=1}^{n} (x_i - \bar{x})^2$$

can be expressed as

$$SS = x'(I - U_n)x$$

where x is a vector of n scalar observations $x_1, x_2, \ldots, x_n$, and U_n is of order n with every element equal to $1/n$. Hence

$$I - U_n = I_n - \frac{1}{n} J_n$$

$$= V_n(1, -1/n).$$

Therefore, from the results just derived,

$$|I - U_n| = 1^{n-1}(1 - n/n) = 0;$$
$$(I - U_n)^2 = V_n[1^2, 2(1)(-1/n) + n(-1/n)^2]$$
$$= V_n(1, -1/n)$$
$$= (I - U_n);$$

and $(I - U_n)$ has one characteristic root equal to zero and $(n - 1)$ roots equal to 1. Hence its rank is $n - 1$.

Should x be a vector of independent random variables each having mean μ and variance σ^2, x_i can be written as $x_i = \mu + e_i$ and so

$$SS = \Sigma(x_i - \bar{x})^2 = x'(I - U_n)x$$
$$= \Sigma(e_i - \bar{e})^2 = e'(I - U_n)e$$

where e is the vector of random terms e_i. Now $I - U_n$ has the characteristic root $\lambda = 1$ with multiplicity $n - 1$; but for $\lambda = 1$, $(I - U_n - \lambda I)$ reduces to $-U_n$, which has rank $1 = n - (n - 1)$. Hence (Section 12.2c) $I - U_n$ can be reduced to canonical form. Furthermore, if the distribution of the x_i's is normal, e represents a vector of normally distributed variables having zero mean and variance-covariance matrix $\sigma^2 I$. The illustration of Section 12.7d therefore applies to $SS = e'(I - U_n)e$, and consequently SS is distributed as $\sigma^2 \chi_{n-1}^2$, a chi-square distribution with $n - 1$ degrees of freedom.

3. IDEMPOTENT MATRICES

In Chapters 7 and 11 use is made of a matrix H for which $H^2 = H$; such a matrix is said to be *idempotent*. Thus if A is idempotent $A^2 = A$, $A^3 = A$, $A^4 = A$, and so on. Since A^2 exists only when A is square, all idempotent matrices are square.

Example.

If
$$A = \begin{bmatrix} 2 & 4 & 6 \\ 4 & 8 & 12 \\ -3 & -6 & -9 \end{bmatrix}$$

$A^2 = A$ and for any positive integer k, $A^k = A$.

Graybill (1961) gives an extensive discussion of the properties of idempotent matrices, some of which we now summarize. It is assumed in all cases that A is idempotent.

i. The only nonsingular idempotent matrix is the identity matrix. (If A^{-1} exists, multiplying both sides of the equation $A^2 = A$ by A^{-1} leads to $A = I$.)

ii. $I - A$ is idempotent.

$$[(I - A)^2 = I + A^2 - 2A = I + A - 2A = I - A.]$$

iii. If A and B are idempotent AB is also, provided $AB = BA$.

$$[(AB)^2 = ABAB = A^2B^2 \text{ if } BA = AB, \text{ and } A^2B^2 = AB.]$$

iv. If P is orthogonal and A idempotent, $P'AP$ is idempotent.

$$[(P'AP)^2 = P'APP'AP = P'A^2P = P'AP.]$$

v. The characteristic roots of an idempotent matrix are either 0 or 1. [If λ and u are a characteristic root and corresponding vector of A, $Au = \lambda u$, and $A^2u = \lambda^2 u$; but because $A^2 = A$, $A^2u = Au$ and hence $\lambda^2 u = \lambda u$, i.e., $(\lambda^2 - \lambda)u = 0$. But u is non-null, and so $\lambda^2 - \lambda = 0$, the only solutions to which are $\lambda = 0$ or 1.]

vi. The number of characteristic roots of A equal to 1 is the same as the rank of A. [If $r(A) = r$ and D is the diagonal matrix of the characteristic roots of A, $r(D) = r(A) = r$. Therefore, since the only non-zero elements in D are 1's there must be r of them.]

vii. The trace of an idempotent matrix equals its rank. [Trace (A) equals the sum of the characteristic roots, which by (vi) above must be r.]

viii. A general form for an idempotent matrix is $A = X(YX)^{-1}Y$ provided that $(YX)^{-1}$ exists.

$$[A^2 = X(YX)^{-1}YX(YX)^{-1}Y = X(YX)^{-1}Y = A.]$$

A general symmetric form is $A = X(X'X)^{-1}X'$ provided $X'X$ is nonsingular.

The properties just listed apply to any idempotent matrix. Applications of symmetric idempotent matrices to problems in mathematical statistics are discussed in some detail in Graybill (1961). In this context idempotency plays an important part in establishing the distributional properties of quadratic forms. We cite two theorems of Graybill's as illustration; they are but two from a lengthy array of theorems that he considers.

Illustrations.

Theorem 1. (Theorem 4.19 of Graybill, 1961.) If x is a vector of random variables having zero means and variance-covariance matrix $\sigma^2 I$, the expected value of $x'Ax$, where A is idempotent of rank r, is $r\sigma^2$.

Proof. From the general expression for a quadratic form given in Section 3.3,

$$E(x'Ax) = E\left(\sum_i x_i^2 a_{ii} + \sum\sum_{i \neq j} x_i x_j a_{ij}\right)$$

$$= \sum_i a_{ii}(Ex_i^2) + \sum\sum_{i \neq j} a_{ij}E(x_i x_j).$$

And since the x's have zero means, are independent and have variance σ^2 this leads to

$$E(x'Ax) = \sum_i a_{ii}\sigma^2 + \sum\sum_{i \neq j} a_{ij}(0)$$

$$= \sigma^2 \operatorname{tr}(A)$$

$$= \sigma^2 \sum (\text{characteristic roots of } A)$$

and with A being idempotent of rank r this gives

$$E(x'Ax) = r\sigma^2.$$

Notice that no knowledge of the form of the distribution of the x's is required, only that their means be zero, that they be independent and have uniform variance.

A simple use of Theorem 1 is in the sum of squares considered at the end of the previous section:

$$SS = \sum_{i=1}^{n}(x_i - \bar{x})^2 = x'(I - U_n)x = e'(I - U_n)e.$$

The vector e is now the x vector of Theorem 1 and therefore, because $I - U_n$ is idempotent of rank $n - 1$,

$$
\begin{aligned}
E(\text{SS}) &= E[e'(I - U_n)e] \\
&= \sigma^2 \operatorname{rank}(I - U_n) \\
&= (n - 1)\sigma^2.
\end{aligned}
$$

Theorem 2. (Theorem 4.6 of Graybill, 1961.) If x is a vector of random variables independently and normally distributed with zero means and unit variance, the quadratic form $x'Ax$ is distributed as χ^2 with r degrees of freedom, provided that A is an idempotent symmetric matrix of rank r.

Proof. It is shown in Section 12.7d that $x'Ax$, where A is symmetric, can be expressed as a sum $\sum_{i=1}^{r} \lambda_i y_i^2$ where the $\lambda_i y_i^2$ are independent random variables distributed as $\lambda_i \sigma^2 \chi_1^2$, every λ_i being positive. Under the conditions of the theorem the x's have unit variance so therefore $\sigma^2 = 1$, and A is idempotent so therefore $\lambda_i = 1$ [property (v) just given]. Hence $\Sigma \lambda_i y_i^2$ is a sum of independent variables each distributed as χ_1^2 and so their sum has the χ_r^2 distribution.

The sum of squares SS provides a simple illustration of Theorem 2. If, in addition to having zero mean and variance matrix $\sigma^2 I$, the vector e represents a vector of normally distributed variables, then Theorem 2 applies and we conclude that SS is distributed as $\sigma^2 \chi^2$ with $n - 1$ degrees of freedom.

At first sight these procedures may appear somewhat clumsy for deriving results that are well known. But the methods and notation involved apply universally to any such problem that can be expressed in matrix format. It is for this reason that matrix algebra is being used on an increasingly wider scale for investigating distributional properties of both linear functions and quadratic forms that arise in analysis of variance and other statistical techniques.

4. NILPOTENT MATRICES

A matrix A is said to be nilpotent of index p if $A^p = 0$ but $A^{p-1} \neq 0$.

Example.

$$
A = \begin{bmatrix} 1 & 2 & 5 \\ 2 & 4 & 10 \\ -1 & -2 & 5 \end{bmatrix}, \quad A^2 = \begin{bmatrix} 0 & 0 & 0 \\ 0 & 0 & 0 \\ 0 & 0 & 0 \end{bmatrix}.
$$

The matrix A is nilpotent of index 2.

Matrices for which $A^p = I$ also exist. For example, if

$$A = \begin{bmatrix} 0 & 0 & 6 \\ \frac{1}{2} & 0 & 0 \\ 0 & \frac{1}{3} & 0 \end{bmatrix}$$

then $A^3 = I$. Perhaps the name "uni-potent" might be appropriate for such matrices. The partitioned matrix

$$A = \begin{bmatrix} I & B \\ 0 & -I \end{bmatrix}$$

is always of this form. No matter what matrix is used for B, $A^2 = I$ and $A = A^{-1}$.

5. A VECTOR OF DIFFERENTIAL OPERATORS

This section and the next make simple use of differential calculus, and should be omitted by readers not familiar with this branch of mathematics.

Sometimes a function of several variables has to be differentiated with respect to each of the variables concerned. In such cases it is often convenient to write the derivatives as a column vector. Since the most common usage of this situation is likely to involve only linear and quadratic functions of variables we confine ourselves to these. For example, if

$$\lambda = 3x_1 + 4x_2 + 9x_3$$

where λ, x_1, x_2 and x_3 are scalar variables, we write the derivatives of λ with respect to x_1, x_2 and x_3 as

$$\frac{\partial \lambda}{\partial x} = \begin{bmatrix} \dfrac{\partial \lambda}{\partial x_1} \\[2mm] \dfrac{\partial \lambda}{\partial x_2} \\[2mm] \dfrac{\partial \lambda}{\partial x_3} \end{bmatrix} = \begin{bmatrix} \dfrac{\partial}{\partial x_1} \\[2mm] \dfrac{\partial}{\partial x_2} \\[2mm] \dfrac{\partial}{\partial x_3} \end{bmatrix} \lambda = \begin{bmatrix} 3 \\ 4 \\ 9 \end{bmatrix}.$$

The second vector shows how the symbol $\dfrac{\partial}{\partial x}$ can represent a whole *vector of differential operators*. Noting that

$$\lambda = 3x_1 + 4x_2 + 9x_3$$

$$= \begin{bmatrix} 3 & 4 & 9 \end{bmatrix} \begin{bmatrix} x_1 \\ x_2 \\ x_3 \end{bmatrix}$$

$$= a'x,$$

so defining vectors a and x, we therefore see that with $\dfrac{\partial}{\partial x}$ representing a column of differential operators, $\partial\lambda/\partial x$ is

$$\frac{\partial}{\partial x}(a'x) = \frac{\partial}{\partial x}(x'a) = a. \tag{1}$$

Suppose this principle is applied in turn to every element of a vector $y = Ax$ expressed as

$$\begin{bmatrix} y_1 \\ y_2 \\ \cdot \\ \cdot \\ \cdot \\ y_n \end{bmatrix} = \begin{bmatrix} a_1'x \\ a_2'x \\ \cdot \\ \cdot \\ \cdot \\ a_n'x \end{bmatrix}, \tag{2}$$

where the y_i's are the elements of y and $a_1', a_2', \ldots, a_n'$ are the rows of the matrix A. The vector of operators $\dfrac{\partial}{\partial x}$ is now to be used on the elements of y. For each element it yields a column vector of partial derivatives, $\partial y_i/\partial x$, for $i = 1, 2, \ldots, n$. Suppose we write the n values of this vector alongside one another, forming a matrix; then the expression $\partial y/\partial x$ of the vector $y = Ax$ is found to be the matrix A'. For,

$$\begin{aligned} \frac{\partial y}{\partial x} &= \begin{bmatrix} \dfrac{\partial y_1}{\partial x} & \dfrac{\partial y_2}{\partial x} & \cdots & \dfrac{\partial y_n}{\partial x} \end{bmatrix} \\ &= \begin{bmatrix} \dfrac{\partial a_1'x}{\partial x} & \dfrac{\partial a_2'x}{\partial x} & \cdots & \dfrac{\partial a_n'x}{\partial x} \end{bmatrix}, \quad \text{from (2)} \\ &= [a_1 \quad a_2 \quad \cdots \quad a_n], \quad \text{from (1)} \\ &= A' \tag{3} \end{aligned}$$

from the definition of the a_i' as rows of A. Hence

$$\frac{\partial}{\partial x}(Ax) = A'.$$

Similarly, if $\alpha_1, \alpha_2, \ldots, \alpha_n$ are the columns of A

$$x'A = [x'\alpha_1 \quad x'\alpha_2 \quad \cdots \quad x'\alpha_n]$$

and

$$\frac{\partial}{\partial x}(x'A) = A.$$

Note from (3) that the complete expression for $\partial y/\partial x$ in the case of A being 3×3, for example, is

$$\frac{\partial y}{\partial x} = \frac{\partial}{\partial x}(Ax) = \begin{vmatrix} \dfrac{\partial y_1}{\partial x_1} & \dfrac{\partial y_2}{\partial x_1} & \dfrac{\partial y_3}{\partial x_1} \\[2mm] \dfrac{\partial y_1}{\partial x_2} & \dfrac{\partial y_2}{\partial x_2} & \dfrac{\partial y_3}{\partial x_2} \\[2mm] \dfrac{\partial y_1}{\partial x_3} & \dfrac{\partial y_2}{\partial x_3} & \dfrac{\partial y_3}{\partial x_3} \end{vmatrix} = A'. \tag{4}$$

Example. For

$$y = Ax = \begin{bmatrix} 2 & 6 & -1 \\ 3 & -2 & 4 \\ 3 & 4 & 7 \end{bmatrix} \begin{bmatrix} x_1 \\ x_2 \\ x_3 \end{bmatrix},$$

i.e.,

$$\begin{bmatrix} y_1 \\ y_2 \\ y_3 \end{bmatrix} = \begin{bmatrix} 2x_1 + 6x_2 - x_3 \\ 3x_1 - 2x_2 + 4x_3 \\ 3x_1 + 4x_2 + 7x_3 \end{bmatrix},$$

the value of $\partial y/\partial x$ given by (4) is

$$\frac{\partial y}{\partial x} = \begin{bmatrix} 2 & 3 & 3 \\ 6 & -2 & 4 \\ -1 & 4 & 7 \end{bmatrix} = A'.$$

Consider the quadratic form of order n:

$$y = x'Ax$$
$$= \sum_i \sum_j a_{ij}x_ix_j,$$

the summations being for values of i and j from 1 through n. Differentiating with respect to the elements of x gives $\partial y/\partial x$ as

$$\frac{\partial y}{\partial x} = \begin{vmatrix} \dfrac{\partial}{\partial x_1} \\[2mm] \dfrac{\partial}{\partial x_2} \\[2mm] \dfrac{\partial}{\partial x_3} \\[1mm] \cdot \\ \cdot \\ \cdot \\ \dfrac{\partial}{\partial x_n} \end{vmatrix} \sum_i \sum_j a_{ij}x_ix_j = \begin{bmatrix} \sum_i a_{1j}x_j + \sum_i a_{i1}x_i \\[2mm] \sum_j a_{2j}x_j + \sum_i a_{i2}x_i \\[1mm] \cdot \\ \cdot \\ \cdot \\ \sum_j a_{nj}x_j + \sum_i a_{in}x_i \end{bmatrix} = \begin{bmatrix} a_1'x \\ a_2'x \\ \cdot \\ \cdot \\ \cdot \\ a_n'x \end{bmatrix} + \begin{bmatrix} \alpha_1'x \\ \alpha_2'x \\ \cdot \\ \cdot \\ \cdot \\ \alpha_n'x \end{bmatrix}$$

where a_i' and α_i are the ith row and column respectively of A. Hence the two vectors are Ax and $A'x$ and so

$$\frac{\partial}{\partial x}(x'Ax) = Ax + A'x.$$

For symmetric A, $A = A'$ and thus

$$\frac{\partial}{\partial x}(x'Ax) = 2Ax. \tag{5}$$

Example. For

$$y = x'Ax$$

$$= \begin{bmatrix} x_1 & x_2 & x_3 \end{bmatrix} \begin{bmatrix} 1 & 3 & 5 \\ 3 & 4 & 7 \\ 5 & 7 & 9 \end{bmatrix} \begin{bmatrix} x_1 \\ x_2 \\ x_3 \end{bmatrix}$$

$$= x_1^2 + 6x_1x_2 + 10x_1x_3 + 4x_2^2 + 14x_2x_3 + 9x_3^2,$$

$$\frac{\partial y}{\partial x} = \begin{bmatrix} \dfrac{\partial y}{\partial x_1} \\[2mm] \dfrac{\partial y}{\partial x_2} \\[2mm] \dfrac{\partial y}{\partial x_3} \end{bmatrix} = \begin{bmatrix} 2x_1 + 6x_2 + 10x_3 \\ 6x_1 + 8x_2 + 14x_3 \\ 10x_1 + 14x_2 + 18x_3 \end{bmatrix} = 2Ax.$$

Illustration. The method of least squares in statistics involves minimizing the sum of squares of the elements of the vector e in the equation

$$y = Xb + e. \tag{6}$$

If $e = \{e_i\}$ for $i = 1, 2, \ldots, n$, the sum of squares is

$$S = \sum_{i=1}^{n} e_i^2$$

$$= e'e$$

$$= (y - Xb)'(y - Xb)$$

$$= y'y - b'X'y - y'Xb + b'X'Xb$$

and because $y'Xb$ is a scalar it equals its transpose, $b'Xy$, and hence

$$S = y'y - 2b'X'y + b'X'Xb.$$

Minimizing this with respect to elements of the vector b involves equating to zero the expression $\partial S/\partial b$ where $\partial/\partial b$ is a vector of differential operators of the nature just discussed. Using results (1) and (5) gives

$$\frac{\partial S}{\partial b} = -2X'y + 2X'Xb \tag{7}$$

and equating this to a null vector leads to

$$b = (X'X)^{-1}X'y, \tag{8}$$

provided that $(X'X)^{-1}$ exists.

6. JACOBIANS

Matrices of first- and second-order partial derivatives are often used in solving maximization and minimization problems in business and economic models. An example of the matrix of first-order partial derivatives

$$\left\{\frac{\partial y_i}{\partial x_j}\right\} \qquad \text{for} \quad i, j = 1, 2, \ldots, n$$

is shown in equation (4). The determinant of this matrix, call it J, is known as a *Jacobian*. It applies to any situation where variables denoted by y's are functions of other variables x, and as such is sometimes referred to as the Jacobian of the functions y_i with respect to the variables x_j. When the relationship between the x's and y's is linear and can be expressed as $y = Tx$ then equation (4) shows that $J = |T'|$; i.e., the Jacobian is the determinant of the transpose of T, or equivalently the determinant of T.

The Jacobian finds widespread use in the transformation of variables in integral calculus. For an integral involving the differentials $dy_1, dy_2, \ldots,$ dy_n, a change of variables is made from $y_1, y_2, \ldots, y_n$ to $x_1, x_2, \ldots, x_n$ by substituting for the y's in terms of the x's and by replacing the differentials by $|J| dx_1, dx_2, \ldots, dx_n$ where J is the Jacobian of the y's with respect to the x's, and $|J|$ is its absolute value.

Another matrix of partial derivatives is the Hessian. If y is a function of n variables $x_1, x_2, \ldots, x_n$, the matrix of second-order partial derivatives of y with respect to the x's is called a Hessian:

$$H = \left\{\frac{\partial^2 y}{\partial x_i \, \partial x_j}\right\}, \qquad \text{for} \quad i, j = 1, 2, \ldots, n.$$

Illustration. In considering the problem of a consumer maximizing utility subject to a budget constraint, Henderson and Quandt (1958) derive conditions that must be met in order to insure a maximum by analyzing the principal minors of an augmented Hessian. Samuelson (1947, Chapter 6) employs the Jacobian in variate transformations concerning a wide range of problems in the theory of consumer choice and profit maximization.

7. MATRIX FUNCTIONS

It is sometimes useful to define matrix functions analogous to scalar algebraic functions. Consider the matrix function e^K defined as

$$e^K = I + K + K^2/2 + K^3/(3!) + \cdots$$
$$= I + \sum_{r=1}^{\infty} K^r/(r!),$$

analogous to a scalar power series expansion. The matrix K must be square in order for this definition to hold and then e^K is a matrix of the same order as K, a power series in K, in fact.

Another matrix function is given by the relationship

$$R = \log_e P \tag{9}$$

where R and P are matrices. To obtain R directly a power series might again be invoked from scalar algebra,

$$\log_e x = \log [1 - (1 - x)] = - \sum_{i=1}^{\infty} (1 - x)^{2i-1}/(2i - 1),$$

but as this holds true only for limited values of x it is preferable to take (9) as being equivalent to

$$P = e^R$$
$$= I + \sum_{k=1}^{\infty} R^k/(k!),$$

as already discussed. This can be used to establish a relationship between the elements of P and R:

$$p_{ij} = \delta_{ij} + \sum_{k=1}^{\infty} r_{ij}^{(k)}/(k!)$$

where $P = \{p_{ij}\}$, $\delta_{ij} = 1$ if $i = j$ and 0 otherwise, and $r_{ij}^{(k)}$ is the ijth element of R^k. In some cases this result reduces to a relatively simple form (see Exercise 10).

8. DIRECT SUMS

The direct sum of two matrices A and B is defined as

$$A \oplus B = \begin{bmatrix} A & 0 \\ 0 & B \end{bmatrix}$$

where the zeros are null matrices of appropriate order. The definition applies whether or not A and B are of the same order; for example

$$[1 \quad 2 \quad 3] \oplus \begin{bmatrix} 6 & 7 \\ 8 & 9 \end{bmatrix} = \begin{bmatrix} 1 & 2 & 3 & 0 & 0 \\ 0 & 0 & 0 & 6 & 7 \\ 0 & 0 & 0 & 8 & 9 \end{bmatrix}.$$

The transpose of a direct sum is the direct sum of the transposes. The rank of a direct sum is the sum of the ranks, as is evident from the definition of rank (Section 6.4). It is clear from the definition of a direct sum that $A \oplus (-A) \neq 0$ unless A is null. Furthermore,

$$(A \oplus B) + (C \oplus D) = (A + C) \oplus (B + D)$$

only if the necessary conditions of conformability for addition are met. Similarly

$$(A \oplus B)(C \oplus D) = AC \oplus BD$$

provided that conformability for multiplication is satisfied. Should A and B be nonsingular, the preceding result leads to

$$(A \oplus B)^{-1} = A^{-1} \oplus B^{-1}.$$

The direct sum $A \oplus B$ is square either if both A and B are square or if the number of rows in the two matrices equals the number of columns. The determinant of $A \oplus B$ equals $|A|\,|B|$ if both A and B are square, but otherwise it is zero (as may be seen from considering the Laplace expansion of $|A \oplus B|$). When both A and B are square the characteristic roots of their direct sum are the characteristic roots of A and B, as is evident from the characteristic equation of $A \oplus B$. If u and v are characteristic vectors of A and B respectively, then $\begin{bmatrix} u \\ 0 \end{bmatrix}$ and $\begin{bmatrix} 0 \\ v \end{bmatrix}$ are characteristic vectors of $A \oplus B$. However, when A and B are not square but $A \oplus B$ is, it appears that the only conclusion to be drawn about the characteristic roots of $A \oplus B$ is that at least d of them will be zero where d is the difference between the number of rows and columns in A (or B). The characteristic vectors will be found as for any matrix.

9. DIRECT PRODUCTS

The direct product of two matrices $A_{p \times q}$ and $B_{m \times n}$ is defined as

$$A_{p \times q} * B_{m \times n} = \begin{bmatrix} a_{11}B & \cdots & a_{1q}B \\ & & \\ \cdot & & \cdot \\ \cdot & & \cdot \\ \cdot & & \cdot \\ a_{p1}B & \cdots & a_{pq}B \end{bmatrix}. \tag{10}$$

It is clear from this that the direct product can be partitioned into as many sub-matrices as there are elements of A, each sub-matrix being B multiplied by an element of A. The elements of the direct product consist of all possible products of an element of A multiplied by an element of B, and its order is $pm \times qn$. For example,

$$[1 \quad 2 \quad 3] * \begin{bmatrix} 6 & 7 \\ 8 & 9 \end{bmatrix} = \begin{bmatrix} 6 & 7 & 12 & 14 & 18 & 21 \\ 8 & 9 & 16 & 18 & 24 & 27 \end{bmatrix}.$$

The transpose of a direct product is the direct product of the transposes—as is evident from transposing equation (10). Note, however, that $(A * B)' = A' * B'$, in contrast to $(AB)' = B'A'$. As expressed in (10), $A * B$ consists of p sets of m rows each. If the rank of B is b there will be b independent rows in each of these sets. But if the rank of A is a, all rows of the last $p - a$ sets will be linear combinations of all rows of the first a sets. Therefore only the b rows in each of the first a sets are independent of each other, and so the number of independent rows in $A * B$ is ab. Hence the rank of $A * B$ is ab, the product of the rank of A and the rank of B.

Direct products are frequently called Kronecker products [Cornish (1957) and Vartak (1955), for example], although MacDuffee (1933), in an extensive discussion of direct products, does not mention the name. Some writers call $A * B$ the *right direct product* to distinguish it from $B * A$ which is then called the *left direct product*. On some occasions $B * A$ will be found defined as the right-hand side of equation (10), but the definition given there, of $A * B$, is the one usually employed, and is the definition used in this book for direct product. The relationship between $A * B$ and $B * A$ is detailed in Section 8.10 of Searle (1966).

The following results are of interest.

i. For x and y being vectors:

$$x' * y = yx' = y * x'.$$

ii. For D being a diagonal matrix of order k:

$$D * A = d_1 A \oplus d_2 A \oplus \cdots \oplus d_k A,$$

but $A_{p \times q} * D_k = (A * I_k)[D \oplus D \oplus \cdots \oplus D$, for q terms].

iii. For λ being a scalar:

$$\lambda * A = \lambda A = A * \lambda = A\lambda.$$

iv. When using partitioned matrices,

$$[A_1 \quad A_2] * B = [A_1 * B \quad A_2 * B],$$

but $\qquad A * [B_1 \quad B_2] \neq [A * B_1 \quad A * B_2].$

v. Provided that conformability conditions for regular matrix multiplication are satisfied

$$(A * B)(X * Y) = AX * BY.$$

vi. For A and B both square and nonsingular

$$(A * B)^{-1} = A^{-1} * B^{-1}.$$

These results are readily verified with simple numerical examples. It is recommended that the reader satisfy himself as to their validity.

In contemplating the determinant of $A * B$ we need be concerned only with the situations in which both A and B are square, because when they are not, consideration of rank reveals that $A * B$ will not be of full rank, and its determinant is therefore zero. If A and B are square, of order p and m respectively, then from result (v) just given

$$A_p * B_m = (A_p * I_m)(I_p * B_m)$$

and hence $$|A_p * B_m| = |A_p * I_m|\, |I_p * B_m|. \qquad (11)$$

Now it can be shown[1] that either

$$|A * B| = |B * A| \text{ or } |A * B| = -|B * A|.$$

The latter cannot be true because, if it were, putting $B = A$ would give $|A * A| = -|A * A|$. Therefore

$$|A * B| = |B * A|,$$

and applying this to (11) gives

$$|A_p * B_m| = |I_m * A_p|\, |I_p * B_m|.$$

But from result (ii) $$|I_m * A_p| = |A|^m,$$

so that $$|A_p * B_m| = |A|^m\, |B|^p.$$

The characteristic equation of $A * B$ exists only when $A * B$ is square. If A and B are both square with characteristic vectors u and v corresponding to characteristic roots λ and μ respectively, then $Au = \lambda u$ and $Bv = \mu v$. Therefore

$$Au * Bv = \lambda u * \mu v$$

and so $$(A * B)(u * v) = \lambda \mu (u * v).$$

Hence the characteristic roots of $A * B$ are the products of the characteristic roots of A with those of B, and the corresponding characteristic vectors are the direct products of the characteristic vectors of A and B. This result means

---

[1] See Searle (1966, pp. 216–217).

that if $U^{-1}AU = D_a$ and $V^{-1}BV = D_b$ are the canonical forms under similarity, then the canonical reduction of $A * B$ is

$$(U^{-1} * V^{-1})(A * B)(U * V) = D_a * D_b.$$

When $A * B$ is square but A and B are not, the only conclusion to be drawn about characteristic roots of $A * B$ is that $r(A) \cdot r(B)$ of them will be non-zero.

Estimating contrasts in factorial experiments is a situation where direct products can be of great use. In particular, the interactions between linear, quadratic, cubic and higher-order effects of the factors can be expressed in terms of direct products involving these effects of each factor separately. Indications of this procedure are shown in Section 8.10 of Searle (1966).

10. TRACE OF A MATRIX PRODUCT

The trace of a matrix is defined in Section 1.6. Thus if $A = \{a_{ij}\}$ for $i, j = 1, 2, \ldots, n$, the trace of A is $\text{tr}(A) = \sum_{i=1}^{n} a_{ii}$, the sum of the diagonal elements. We now show that for a product matrix $\text{tr}(AB) = \text{tr}(BA)$ and hence $\text{tr}(ABC) = \text{tr}(BCA) = \text{tr}(CAB)$. Note that $\text{tr}(AB)$ exists only if AB is square, which occurs only when A is $r \times c$ and B is $c \times r$. Then if $AB = \{(ab)_{ij}\}$,

$$\text{tr}(AB) = \sum_{i=1}^{r}(ab)_{ii} = \sum_{i=1}^{r} \left(\sum_{j=1}^{c} a_{ij}b_{ji} \right)$$

$$= \sum_{j=1}^{c} \left(\sum_{i=1}^{r} b_{ji}a_{ij} \right) = \sum_{j=1}^{r} (ba)_{jj}$$

$$= \text{tr}(BA),$$

Extension to products of three or more matrices is obvious.

11. EXERCISES

1. Evaluate the following determinants using results of this chapter:

$$\begin{vmatrix} 0 & -1 & 7 \\ 1 & 0 & 3 \\ -7 & -3 & 0 \end{vmatrix}, \quad \begin{vmatrix} 0 & -2 & 4 & 6 \\ 2 & 0 & 3 & -7 \\ -4 & -3 & 0 & 9 \\ -6 & 7 & -9 & 0 \end{vmatrix} \quad \text{and} \quad \begin{vmatrix} 0 & 16 & -1 & 2 \\ -16 & 0 & 7 & 0 \\ 1 & -7 & 0 & 6 \\ -2 & 0 & -6 & 0 \end{vmatrix}.$$

2. (a) For

$$A = \begin{bmatrix} 0 & -a & b & -c \\ a & 0 & -d & e \\ -b & d & 0 & -f \\ c & -e & f & 0 \end{bmatrix}$$

explain why

$$|I + A| = 1 + (a^2 + b^2 + c^2 + d^2 + e^2 + f^2) + |A|.$$

(b) Evaluate $|A|$.

3. Derive orthogonal matrices using the following matrices and results from Section 1:

$$\begin{bmatrix} 0 & -1 & 1 \\ 1 & 0 & 2 \\ -1 & -2 & 0 \end{bmatrix} \quad \text{and} \quad \begin{bmatrix} 0 & 1 & -2 \\ -1 & 0 & 5 \\ 2 & -5 & 0 \end{bmatrix}.$$

Evaluate their determinants, show that their product is orthogonal, find their characteristic roots and reduce each to canonical form.

4. Derive a symmetric idempotent matrix using $X = \begin{bmatrix} 1 & 2 \\ 0 & -1 \\ -1 & 0 \end{bmatrix}$; find its rank

and characteristic roots. Denoting it by M, show that $P'MP$ is idempotent where P is one of the orthogonal matrices derived in Exercise 3.

5. For the J matrices defined in Section 2 show that
   (a) $J_{1 \times m} J_{m \times 1} \quad = m,$
   (b) $J_{m \times 1} J_{1 \times n} \quad = J_{m \times n},$
   (c) $J_{1 \times m} J_m J_{m \times 1} = m^2,$
   (d) $J_m(aI_m + bJ_m) = (a + mb)J_m.$

6. Show that, with respect to x_1 and x_2, the matrix of first partial derivatives of

$$y_1 = 6x_1^2 x_2 + 2x_1 x_2 + x_2^2 \quad \text{and} \quad y_2 = 2x_1^3 + x_1^2 + 2x_1 x_2$$

is the same as the Hessian of

$$y = 2x_1^3 x_2 + x_1^2 x_2 + x_1 x_2^2.$$

7. Express the quadratic form

$$7x_1^2 + 4x_1 x_2 - 5x_2^2 - 6x_2 x_3 + 3x_3^2 + 6x_1 x_3$$

in the form $x'Ax$ where $A = A'$, and show that

$$\frac{\partial}{\partial x}(x'Ax) = 2Ax.$$

8. (a) Show that $X(YX)^{-1}Y$ is idempotent and is the identity matrix when X and Y are diagonal.
   (b) Show that a symmetric orthogonal matrix equals its inverse. Construct such a matrix, of order 2.
   (c) If $W = R^{-1} - R^{-1}Z(Z'R^{-1}Z + D^{-1})^{-1}Z'R^{-1}$ show that $(R + ZDZ')$ is the inverse of W.

(d) If $D = X'(XV^{-1}X')^{-1}XV^{-1}$ show that where V is symmetric

   i. $D' = V^{-1}DV$,

   ii. D is idempotent,

   iii. $(I - D')V^{-1}(I - D) = (I - D')V^{-1} = V^{-1}(I - D)$.

(e) Prove that the trace of $X(YX)^{-1}Y$ equals the number of rows in Y.

9. If K is a square matrix of order n, what values of K make e^K equal to the following?

   (a) eI (b) $I - K + eK$

   (c) I (d) $e^\lambda I$

   (e) $I + K$ (f) $I + K(e^n - 1)/n$

10. With

$$R = \begin{bmatrix} -a & a \\ b & -b \end{bmatrix}$$

show that

$$R^n = [-(a + b)]^{n-1}R;$$

and for

$$P = \begin{bmatrix} p_1 & 1 - p_1 \\ p_2 & 1 - p_2 \end{bmatrix}$$

show that the functional relationship

$$R = \log_e P$$

implies

$$a = \frac{-(1 - p_1)\log_e(p_1 - p_2)}{1 - p_1 + p_2}$$

and

$$a + b = -\log_e(p_1 - p_2).$$

11. Find the characteristic roots of $A = \begin{bmatrix} 4 & -2 \\ 5 & -3 \end{bmatrix}$ and of $B = \begin{bmatrix} 2 & -1 \\ 2 & 5 \end{bmatrix}$ and

show that their products are the characteristic roots of $A * B$.

12. For vectors x and y, matrices A and B, and scalars μ and λ, show that

   (a) $(x' * B)(A * y) = Byx'A$,

   (b) $(\lambda * y)(x' * \mu) = \mu\lambda yx'$,

   (c) $(B * B)(x * x) = (B * Bx)x \neq B(x * Bx)$,

   (d) $(A \oplus B)^{-1}$ exists only if A^{-1} and B^{-1} exist.

13. For matrices A, B and C show that

$$A * (B \oplus C) \neq (A * B) \oplus (A * C)$$

but

$$(A \oplus B) * C = (A * C) \oplus (B * C).$$

REFERENCES

Cornish, E. A. (1957). An application of the Kronecker product of matrices in multiple regression. *Biometrics*, **13**, 19–27.

Graybill, F. A. (1961). *An Introduction to Linear Statistical Models*. Vol. I, McGraw-Hill, New York.

Henderson, J. M., and R. E. Quandt (1958). *Microeconomic Theory: A Mathematical Approach*. McGraw-Hill, New York.

MacDuffee, C. C. (1933). *The Theory of Matrices*. Springer, Berlin. (Also, Chelsea Publishing Company, New York, 1946, 1956.)

Samuelson, P. A. (1947). *Foundations of Economic Analysis*. Harvard University Press, Cambridge, Mass.

Searle, S. R. (1966). *Matrix Algebra for the Biological Sciences*. Wiley, New York.

Vartak, M. N. (1955). On an application of Kronecker product of matrices to statistical designs. *Annals of Mathematical Statistics*, **26**, 420–438.

INDEX OF APPLICATIONS

GENERAL INDEX

357